T0371688

Computational Problems for Physics

Computational Problems
for Physics
With Guided Solutions Using Python

Rubin H. Landau, Manuel José Páez

CRC Press
Taylor & Francis Group
Boca Raton London New York

CRC Press is an imprint of the
Taylor & Francis Group, an **informa** business

CRC Press
Taylor & Francis Group
6000 Broken Sound Parkway NW, Suite 300
Boca Raton, FL 33487-2742

© 2020 by Taylor & Francis Group, LLC
CRC Press is an imprint of Taylor & Francis Group, an Informa business

No claim to original U.S. Government works

International Standard Book Number-13: 978-1-1387-0591-3 (Hardback)

This book contains information obtained from authentic and highly regarded sources. Reasonable efforts have been made to publish reliable data and information, but the author and publisher cannot assume responsibility for the validity of all materials or the consequences of their use. The authors and publishers have attempted to trace the copyright holders of all material reproduced in this publication and apologize to copyright holders if permission to publish in this form has not been obtained. If any copyright material has not been acknowledged please write and let us know so we may rectify in any future reprint.

Except as permitted under U.S. Copyright Law, no part of this book may be reprinted, reproduced, transmitted, or utilized in any form by any electronic, mechanical, or other means, now known or hereafter invented, including photocopying, microfilming, and recording, or in any information storage or retrieval system, without written permission from the publishers.

For permission to photocopy or use material electronically from this work, please access www.copyright.com (http://www.copyright.com/) or contact the Copyright Clearance Center, Inc. (CCC), 222 Rosewood Drive, Danvers, MA 01923, 978-750-8400. CCC is a not-for-profit organization that provides licenses and registration for a variety of users. For organizations that have been granted a photocopy license by the CCC, a separate system of payment has been arranged.

Trademark Notice: Product or corporate names may be trademarks or registered trademarks, and are used only for identification and explanation without intent to infringe.

Visit the Taylor & Francis Web site at
http://www.taylorandfrancis.com

and the CRC Press Web site at
http://www.crcpress.com

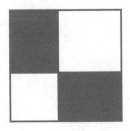

Table of contents

Acknowledgments

Thank you our former students for being experimental subjects as we developed computational physics courses and problems over these last couple of decades. And those of you who have moved on with your lives but found the time to get back to tell us how much you have benefitted from our courses and books; bless you!

While we have tried to give credit to the books that have provided motivation and materials for many of this book's problems, please forgive us if we have forgotten to mention you. After all, 20 years is a long time. In particular, we have probably made as our own materials from the pioneering works of Gould & Tobochnik, Koonin, and Press et al..

Our lives have been enriched by the invaluable friendship, encouragement, helpful discussions, and experiences we have had with many colleagues and students over the years. With uncountable sadness we particularly remember our deceased colleagues and friends Paul Fink, Cristian Bordeianu, and Jon Maestri, some of the most good-natured and good people we could have ever hoped to know. We are particularly indebted to Hans Kowallik, Guillermo Avendaño-Franco, Saturo S. Kano, Shashi Phatak, Oscar A. Restrepo, Jaime Zuluaga, Viktor Podolskiy, and Bruce Sherwood for their technical help and most of all their friendship.

The authors wish to express their gratitude to Lou Han, for the encouragement and support throughout the realization of this book, and to Titus Beu and Veronica Rodriguez who have helped in the final production.

Finally, we extend our gratitude to our families, whose reliable support and encouragement are lovingly accepted, as always.

Series Preface

There can be little argument that computation has become an essential element in all areas of physics, be it via simulation, symbolic manipulations, data manipulations, equipment interfacing, or something with which we are not yet familiar. Nevertheless, even though the style of teaching and organization of subjects being taught by physics departments have changed in recent times, the actual content of the courses has been slow to incorporate the new-found importance of computation. Yes, there are now speciality courses and many textbooks in *Computational Physics*, but that is not the same thing as incorporating computation into the very heart of a modern physics curriculum so that the physics being taught today more closely resembles the physics being done today. Not only will such integration provide valuable professional skills to students, but it will also help keep physics alive by permitting new areas to be studied and old problems to be solved.

This series is intended to provide undergraduate and graduate level textbooks for a modern physics curriculum in which computation is incorporated within the traditional subjects of physics, or in which there are new, multidisciplinary subjects in which physics and computation are combined as a "computational science." The level of presentation will allow for their use as primary or secondary textbooks for courses that wish to emphasize the importance of numerical methods and computational tools in science. They will offer essential foundational materials for students and instructors in the physical sciences as well as academic and industry professionals in physics, engineering, computer science, applied math, and biology.

Titles in the series are targeted to specific disciplines that currently lack a textbook with a computational physics approach. Among these subject areas are condensed matter physics, materials science, particle physics, astrophysics, mathematical methods of computational physics, quantum mechanics, plasma physics, fluid dynamics, statistical physics, optics, biophysics, electricity and magnetism, gravity, cosmology, and high-performance computing in physics. We aim for a presentation that is concise and practical, often including solved problems and examples. The books are meant for teaching, although researchers may find them useful as well. In select cases, we have allowed more advanced, edited works to be included when they share the spirit of the series — to contribute to wider application of computational tools in the classroom as well as research settings.

Although the series editors had been all too willing to express the need for change in the physics curriculum, the actual idea for this series came from the series manager, Lou Han, of Taylor & Francis Publishers. We wish to thank him sincerely for that, as well as for encouragement and direction throughout the project.

Steve Gottlieb, *Bloomington*
Rubin H. Landau, *Corvallis*
Series Editors

Preface

As seems true in many areas, practicing scientists now incorporate powerful computational techniques as key elements in their work. In contrast, physics courses often include computational tools only to illustrate the physics, with little discussion of the method behind the tools, and of the limits to a simulation's reliability and precision. Yet, just as a good researcher would not believe a physics results if the mathematics behind it were not solid, so we should not believe a physics results if the computation behind it is not understood and reliable. While specialty courses and textbooks in Computational Physics are an important step in the right direction, we see an additional need to incorporate modern computational techniques throughout the Physics curriculum. In addition to enhancing the learning process, computational tools are valuable tools in their own right, especially considering the broad areas in which physics graduates end up working.

The authors have spent over two decades trying to think up computational problems and demonstrations for physics courses, both as part of the development of our Computational Physics texts, and as material to present as tutorials at various professional meetings and institutions. This book is our effort at collecting those problems and demos, adding to them, and categorizing them so that they may extend what has traditionally been used for homework and demonstrations throughout the physics curriculum.

Our assumed premise is that learning to compute scientifically requires you to get your hands dirty and to open up that black box of a program. Our preference is that the reader use a compiled language since this keeps her closer to the basic algorithms, and more likely to be able to estimate the numerical error in the answer (essential for science). Nevertheless, programming from scratch can be time consuming and frustrating, and so we provide many sample codes as models for the problems at hand. However, readers or instructors may prefer to approach our problems with a problem solving environment such as Sage, Maple, or Mathematica, in which case our codes can serve as templates.

We often present simple pseudocodes in the text, with full Python code listings at the end of each chapter (most numeric, but some symbolic).[1] The Python language

---

[1] Please note that copying and pasting a code from a pdf listing is not advisable because the formatting, to which Python is sensitive, is not preserved in the process. One needs to open the .py

plus its family of packages comprise a veritable ecosystem for computing [CiSE(07,11)]. Python is free, robust, portable, universal, and provides excellent visualization via the *MatPlotLib* and *VPython* packages (the latter also called by its original name *Visual*). We find Python the easiest compiled language for education, with excellent applications in research and development. Further details are provided in Chapter 1.

Each chapter in the text contains a Chapter Overview with highlights as to what is to follow. Chapters 1 and 2 review background materials used throughout the rest of the book. Chapter 1 covers basic computational methods, including floating point numbers and their errors, integration, differentiation, random numbers generation, and the solution to ordinary and partial differential equations. Chapter 2 covers fundamental numerical analysis tools, including Fourier analysis, noise reduction, wavelet analysis, principal components analysis, root searching, least-squares fitting, and fractal dimension determination. Although some of the problems and demos in Chapters 1 and 2 may find use in Mathematical Methods of Physics courses, those chapters are meant as review or reference.

This book cover multiple areas and levels of physics with the chapters organized by subject. Most of the problems are at an upper-division undergraduate level, which should by fine for many graduate courses as well, but with a separate chapter aimed at entry-level courses. We leave it to instructors to decide which problems and demos may work best in their courses, and to modify the problems for their own purposes. In all cases, the introductory two chapters are important to cover initially.

We hope that you find this book useful in changing some of what is studied in physics. If you have some favorite problems or demos that you think would enhance the collection, or some suggestions for changes, please let us know.

RHL, rubin@science.oregonstate.edu   Tucson, November 2017
MJP, mpaezenator@gmail.com   Medellín, November 2017

---

version of the code with an appropriate code editor.

About the Authors

Rubin Landau is a Distinguished Professor Emeritus in the Department of Physics at Oregon State University in Corvallis and a Fellow of the American Physical Society (Division of Computational Physics). His research specialty is computational studies of the scattering of elementary particles from subatomic systems and momentum space quantum mechanics. Landau has taught courses throughout the undergraduate and graduate curricula, and, for over 20 years, in computational physics. He was the founder of the OSU Computational Physics degree program, an Executive Committee member of the APS Division of Computational Physics, and the AAPT Technology Committee. At present Landau is the Education co-editor for AIP/IEEE *Computing in Science & Engineering* and co-editor of this Taylor & Francis book series on computational physics. He has been a member of the XSEDE advisory committee and has been part of the Education Program at the SuperComputing (SC) conferences for over a decade.

Manuel José Páez-Mejia has been a Professor of Physics at Universidad de Antioquia in Medellín, Colombia since January 1969. He has been teaching courses in Modern Physics, Nuclear Physics, Computational Physics, Numerical Methods, Mathematical Physics, and Programming in Fortran, Pascal, and C languages. He has authored scientific papers in nuclear physics and computational physics, as well as texts on the C Language, General Physics, and Computational Physics (coauthored with Rubin Landau and Cristian Bordeianu). In the past, he and Landau conducted pioneering computational investigations of the interactions of mesons and nucleons with few-body nuclei. Professor Paez has led workshops in Computational Physics throughout Latin America, and has been Director of Graduate Studies in Physics at the Universidad de Antioquia.

Web Materials

The Python codes listed in the text are available on the CRC Press website at
`http://www.crcpress.com/product/isbn/??`
We have also created solutions to many problems in a variety of computer languages, and they, as well as these same Python codes, are available on the Web at
`http://science.oregonstate.edu/~landaur/Books/CPbook/eBook/Codes/`.
Updates of the programs will be posted on the websites.

Background material for this book of problems is probably best obtained from Computational Physics text books (particularly those by the authors!). In addition, Python notebook versions of every chapter in our CP text [LPB(15)] are available at
`http://physics.oregonstate.edu/~landaur/Books/CPbook/eBook/Notebooks/`.
As discussed in the documentation there, the notebook environment permits the reader to run codes and follow links while reading the book electronically.

Furthermore, most topics from our CP text are covered in video lecture modules at
`http://science.oregonstate.edu/~landaur/Books/CPbook/eBook/Lectures/`.

General System Requirements

The first chapter of the text provides details about the Python ecosystem for computing and the packages that we use in the text. Basically, a modern version of Python and its packages are needed.

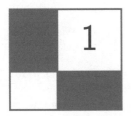

1

Computational Basics for Physics

1.1   Chapter Overview

There is no way that a single chapter or two can provide the background necessary for the proper use of computation in physics. So let's hope that this chapter is just a review. (If not, you may want to look at some of the related video lecture modules at http://physics.oregonstate.edu/~landaur/Books/CPbook/eBook/Lectures/.) In this chapter we cover computing basics, starting with some of the tools available as part of the Python ecosystem, and particularly for visualization and matrix manipulations. There follows a discussion of number representations and the limits and consequences of using floating-point numbers (often absent in Computer Science classes). We then review some basic numerical methods for differentiation, integration, and random number generation. We end with a discussion of the algorithms for solving ordinary and partial differential equations, techniques used frequently in the text. Problems are presented for many of these topics, and doing them would be a good way to get started!

Most every problem in this book requires some visualization. Our sample programs tend to do this with either the Matplotlib or VPython (formerly Visual) package, or both (§1.2.1). Including visualization in the programs does make them longer, though having embedded visualization speeds up the debugging and learning processes, and is more fun. In any case, the user always has the option of outputting the results to a data file and then visualizing them separately with a program such as gnuplot or Grace.

1.2   The Python Ecosystem

This book gives solution codes in Python, with similar codes in other languages available on the Web. Python is free, robust, portable, and universal, and we find it the easiest compiled language for education. It contains high-level, built-in data types, which make matrix manipulations and graphics easy, and there are a myriad of free packages and powerful libraries which make it all around excellent for scientific work, even symbolic manipulation. For learning Python, we recommend the online tutorials [Ptut(14), Pguide(14), Plearn(14)], the books by Langtangen [Lang-tangen(08), Langtangen(09)], and the *Python Essential Reference* [Beazley(09)].

The Python language plus its family of packages comprise a veritable ecosystem for computing [CiSE(07,11)]. To include package `PackageName` in your program, you use either an `import PackageName` statement, which loads the entire package, or to load a specific method include a `from PackageName` statement at the beginning of your program; for example,

```
from vpython import *
y1 = gcurve(color = blue)
```

In our work we use the **packages**

Matplotlib (Mathematics Plotting Library)	`http://matplotlib.org`
NumPy (Numerical Python)	`http://www.numpy.org/`
SciPy (Scientific Python)	`http://scipy.org`
SymPy (Symbolic Python)	`http://sympy.org`
VPython (Python with Visual package)	`http://vpython.org/`

Rather than search the Web for packages, we recommend the use of *Python Package Collections*, which are collections of Python packages that have been engineered and tuned to work well together, and that can be installed in one fell swoop. We tend to use

Anaconda	`https://store.continuum.io/cshop/anaconda/`
Enthought Canopy	`https://www.enthought.com/products/canopy/`
Spyder (in Anaconda)	`https://pythonhosted.org/spyder/`

1.2.1   Python Visualization Tools

VPython, the nickname for Python plus the Visual package, is particularly useful for creating 3-D solids, 2-D plots, and animations. In the past, and with some of our programs, we have used the "classic" version of VPython, which is accessed via importing the module `visual`. That version will no longer be supported in the future and so we have (mostly) converted over to VPython 7, but have included the classic (VPython 6) versions in the `Codes` folder as well[1]. The Visual package is accessed in VPython 7 by importing the module `vpython`. (Follow VPython's online instructions to load VPython 7 into Spyder or notebooks.)

In Figure 1.1 we present two plots produced by the program `EasyVisualVP.py` given in Listing 1.1.[2] Notice that the plotting technique with VPython is to create first a

---

[1]For some programs we provide several versions in the Codes folder: those with a "Vis" suffix use the classic Visual package, those with a "VP" suffix use the newer VPython package, and those with "Mat", or no suffix, use Matplotlib.

[2]We remind the reader that copying and pasting a program from a pdf listing is not advisable because the formatting, to which Python is sensitive, is not preserved in the process. One needs to open the `.py` version of the program with an appropriate code editor.

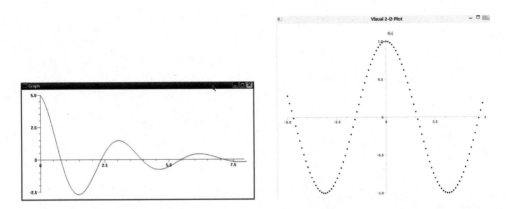

Figure 1.1. Screen dumps of two x-y plots produced by our program `EasyVisualVP.py` using the VPython package. The *left* plot uses default parameters while the *right* plot uses user-supplied options.

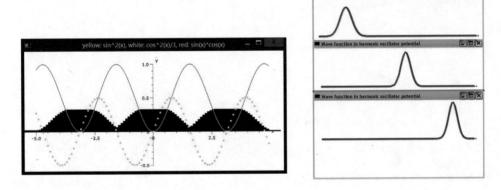

Figure 1.2. *Left:* Output from the program 3GraphVP.py that places three different types of 2-D plots on one graph using VPython. *Right* Three frames from a VPython animation of a quantum mechanical wave packet produced with HOmov.py.

plot object, and then to add the points one at a time to the object. In contrast, Matplotlib creates a vector of points and plots the entire vector in one fell swoop.

The program `3GraphVP.py` in Listing 1.2 places several plots in the same figure and produces the graph on the left of Figure 1.2. There are vertical bars created with `gvbars`, dots created with `gdots`, and a curve created with `gcurve` (colors appear only as shades of gray in the paper text). Creating animations with VPython is essentially just making the same 2-D plot over and over again, with each one at a slightly differing time. Three frames produced by `HOmov.py` are shown on the right of Figure 1.2. The part which makes the animation is simple:

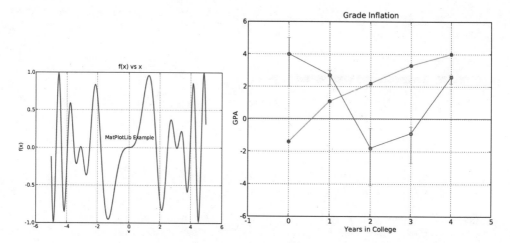

Figure 1.3. Matplotlib plots. *Left:* Output of EasyMatPlot.py showing a simple, x-y plot. *Right:* Output from GradesMatPlot.py that places two sets of data points, two curves, and unequal upper and lower error bars, all on one plot.

```
PlotObj= curve(x=xs, color=color.yellow, radius=0.1)
...
while True:
    rate(500)
    RePsi[1:-1] =...
    ImPsi[1:-1] =..
    PlotObj.y = 4*(RePsi**2 + ImPsi**2)
```

The package Matplotlib is a powerful plotting package for 2-D and 3-D graphs and data plots of various sorts. It uses the sophisticated numerics of NumPy and LAPACK [Anderson et al.(113)] and commands similar to MATLAB™. It assumes that you have placed the x and y values you wish to plot into 1-D arrays (vectors), and then plots these vectors with a single call. In `EasyMatPlot.py`, given in Listing 1.3, we import Matplotlib as the `pylab` library:

```
from pylab import *
```

Then we calculate and input arrays of the x and y values

```
x = arange(Xmin, Xmax, DelX)    # x array in range + increment
y = -sin(x)*cos(x)     # y array as function of x array
```

where the # indicates the beginning of a comment. As you can see, NumPy's `arange` method constructs an array covering "a range" between `Xmax` and `Xmin` in steps of `DelX`. Because the limits are floating-point numbers, so too will be the individual x_i's. And because x is an array, y = -sin(x)*cos(x) is automatically one too! The actual plotting is performed with a dash '-' used to indicate a line, and `lw=2` to set its width.

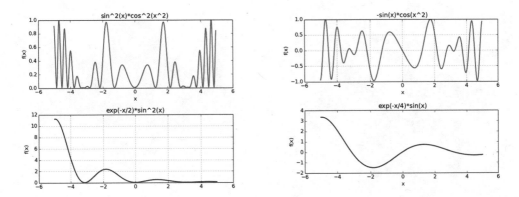

Figure 1.4. *Left* and *Right* columns show two separate outputs, each of two figures, produced by `MatPlot2figs.py`. (We used the slider button to add some space between the red and blue plots.)

The result is shown on the left of Figure 1.3, with the desired labels and title. The `show()` command produces the graph on your desktop.

In Listing 1.4 we give the code `GradesMatplot.py`, and on the right of Figure 1.3 we show its output. Here we repeat the `plot` command several times in order to plot several data sets on the same graph and to plot both the data points and the lines connecting them. We import Matplotlib (pylab), and then import NumPy, which we need for the `array` command. Because we have imported two packages, we add the `pylab` prefix to the `plot` commands so that Python knows which package to use. A horizontal line is created by plotting an array with all y values equal to zero, unequal lower and upper error bars are included as well as grid lines.

Often the science is clearer if there are several curves in one plot, and, several plots in one figures. Matplotlib lets you do this with the `plot` and the `subplot` commands. For example, in `MatPlot2figs.py` in Listing 1.5 and Figure 1.4, we have placed two curves in one plot, and then output two different figures, each containing two plots. The key here is repetition of the `subplot` command:

```
figure(1)                                            #  1st figure
subplot(2,1,1)          # 1st subplot, 2 rows, 1 column
subplot(2,1,2)                                    # 2nd subplot
```

If you want to visualize a function like the dipole potential

$$V(x,y) = [B + C/(x^2 + y^2)^{3/2}]x, \qquad (1.1)$$

you need a 3-D visualization in which the mountain height $z = V(x,y)$, and the x and y axes define the plane below the mountain. The impression of three dimensions is obtained by shading, parallax, and rotations with the mouse, and other tricks. In

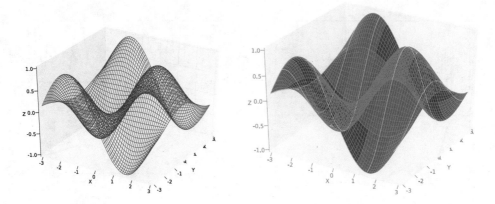

Figure 1.5. *Left:* A 3-D wire frame. *Right:* a surface plot with wire frame. Both are produced by the program `Simple3Dplot.py` using Matplotlib.

Figure 1.5 left we show a wire-frame plot and in Figure 1.5 right a surface-plus-wire-frame plot. These are obtained from the program `Simple3Dplot.py` in Listing 1.6. The `meshgrid` method sets up grid matrix from the x and y coordinate vectors, and then constructs the $Z(x, y)$ surface with another vector operation.

A *scatter plot* is a useful way to visualize individual points (x_i, y_j, z_k) in 3-D. In Figure 1.6 left we show two such plots created with `PondMatPlot.py` in Listing 1.7 and `Scatter3dPlot.py` in Listing 1.8. Here the 111 indicates a $1 \times 1 \times 1$ grid.

1. As shown in Figure 1.7, a beam of length $L = 10$ m and weight $W = 400$ N rests on two supports a distance $d = 2$ m apart. A box of weight $W_b = 800$ N, initially above the left support, slides frictionlessly to the right with a velocity $v = 7$ m/s.

   a. Write a program that calculates the forces exerted on the beam by the right and left supports as the box slides along the beam.
   b. Extend your program so that it creates an animation showing the forces and the position of the block as the box slides along the beam. In Listing 1.10 we present our code `SlidingBox.py` that uses the Python Visual package, and in Figure 1.7 left we present a screen shot captured from this code's animation. Modify it for the problem at hand.
   c. Extend the two-support problem to a box sliding to the right on a beam with a third support under the right edge of the beam.

2. As shown on the left of Figure 1.8, a two kilogram mass is attached to two 5-m strings that pass over frictionless rollers. There is a student holding the end of each string. Initially the strings are vertical, but then move apart as the students move at a constant 4 m/s, one to the right and one to the left.

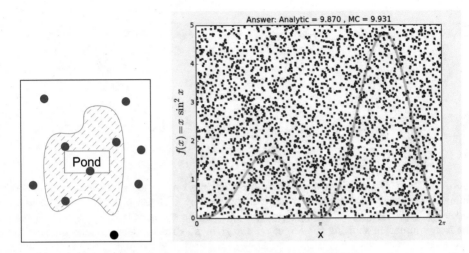

Figure 1.6. *Left:* Throwing stones into a pond as a technique for measuring its area. The ratio of "hits" to total number of stones thrown equals the ratio of the area of the pond to that of the box. *Right:* The evaluation of an integral via a Monte Carlo (stone throwing) technique of the ratio of areas.

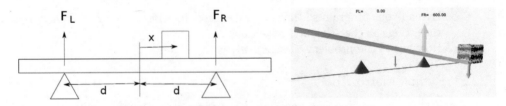

Figure 1.7. *Left:* A beam and a box supported at two points. *Right:* A screen shot from the animation showing the forces on the beam as the weight moves.

(a) What is the initial tension of each cord?

(b) Compute the tension in each cord as the students move apart.

(c) Can the students ever get both strings to be horizontal?

(d) Extend your program so that it creates an animation showing the tensions in the string as the students move apart. In Listing 1.9 we present our code using the Python Visual package, and in Figure 1.8 right we present a screen shot captured from this code's animation. Modify the code as appropriate for your problem.

3. As shown on the right of Figure 1.8, masses $m_1 = m_2 = 1$kg are suspended by two strings of length $L = 2.5m$ and connected by a string of length $s = 1m$.

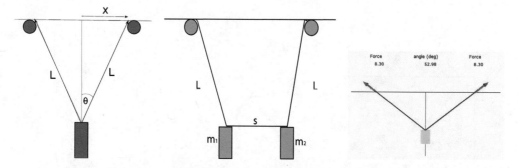

Figure 1.8. *Left:* A mass is suspended from two strings on frictionless rollers, with students pulling horizontally at the end of each string. *Center:* A mass m_1 and a second m_2 are suspended by two strings of length L and connected by a string of length s. *Right:* A screen shot from animation showing the forces on two strings as the students pulling on each mass move apart.

As before, the strings of length L are being pulled horizontally over frictionless pulleys.

a. Write a program that computes the tension of each cord as the students move apart.

b. Extend your program so that it creates an animation showing the tensions in the strings as the students move apart.

c. How would the problem change if $m_1 \neq m_2$?

1.2.2 Python Matrix Tools

Dealing with many numbers at one time is a prime strength of computation, and computer languages have abstract data types for just that purpose. A *list* is Python's built-in data type for a sequence of numbers or objects kept in a definite order. An *array* is a higher-level data type available with the *NumPy* package, and can be manipulated like a vector. Square brackets with comma separators [1, 2, 3] are used for lists, with square brackets also used to indicate the index for a list item:

```
>>> L = [1, 2, 3]                          # Create list
>>> L[0]                               # Print element 0
1
>>> L                              # Print entire list
 [1, 2, 3]
>>> L[0]= 5                          # Change element 0
>>> len(L)                              # Length of list
3
```

NumPy arrays convert Python lists into arrays, which can be manipulated like vectors:

```
>>> vector1 = array([1, 2, 3, 4, 5])           # Fill array wi list
>>> print('vector1 =', vector1)
```

```
vector1 =  [1 2 3 4 5]
>>> vector2 = vector1 + vector1                         # Add 2 vectors
>>> print('vector2=', vector2)
vector2= [ 2  4  6  8 10]
>>> matrix1 = array(([0,1],[1,3]))                   # An array of arrays
>>> print(matrix1)
[[0 1]
 [1 3]]
>>> print  (matrix1 * matrix1)                      # Matrix multiply
 [[0 1]
  [1 9]]
```

When describing NumPy arrays, the number of "dimensions", `ndim`, means the number of indices, which can be as high as 32. What might be called the "size" or "dimensions" of a matrix is called the *shape* of a NumPy array:

```
>>> import numpy as np
>>> np.arange(12)                                      # List 12 ints
array([ 0,  1,  2,  3,  4,  5,  6,  7,  8,  9, 10, 11])
>>> np.arange(12).reshape((3,4))          # Reshape to 3x4
array([[ 0,  1,  2,  3],
       [ 4,  5,  6,  7],
       [ 8,  9, 10, 11]])
>>> a = np.arange(12).reshape((3,4))
>>> a.shape
(3L, 4L)
>>> a.ndim                                          # Dimension?
2
>>> a.size                                       # Number of elements?
12
```

If you want to form the familiar matrix product from two arrays, you use the `dot` function, whereas the asterisk ∗ is used for an element-by-element product:

```
>>> matrix1= array( [[0,1], [1,3]])
>>> matrix1
array([[0, 1],
       [1, 3]])
>>> print ( dot(matrix1,matrix1) )              # Matrix or dot product
[[ 1  3]
 [ 3 10]]
>>> print (matrix1 * matrix1)               # Element-by-element product
[[0 1]
 [1 9]]
```

Rather than writing your own matrix routines, for the sake of speed and reliability we recommend the use of well established libraries. Although the array objects of NumPy are not the same as mathematical matrices, there is the `LinearAlgebra` package that treats 2-D arrays as mathematical matrices. Consider the standard solution of linear equations

$$A\mathbf{x} = \mathbf{b}, \tag{1.2}$$

where we have used a bold character to represent a vector. For example,

```
>>> from numpy import *
>>> from numpy.linalg import*
>>> A = array( [ [1,2,3],  [22,32,42],  [55,66,100] ] ) # Array of arrays
>>> print ('A =', A)
A = [[  1   2   3]
     [ 22  32  42]
     [ 55  66 100]]
```

We solve (1.2) with NumPy's `solve` command, and then test the solution:

```
>>> b = array([1,2,3])
>>> from numpy.linalg import solve
>>> x = solve(A, b) # Finds solution
>>> print ('x =', x)
x = [ -1.4057971  -0.1884058   0.92753623] # The solution
>>> print ('Residual =',  dot(A, x) - b)              # LHS-RHS
Residual = [4.44089210e-16   0.00000000e+00  -3.55271368e-15]
```

A direct, if not most efficient, way to solve (1.2) is to calculate the inverse A^{-1}, and then multiply through by the inverse, $\mathbf{x} = A^{-1}\mathbf{b}$:

```
>>> from numpy.linalg import inv
>>>  dot(inv(A), A)                                   # Test inverse
array([[  1.00000000e+00,  -1.33226763e-15,  -1.77635684e-15],
       [  8.88178420e-16,   1.00000000e+00,   0.00000000e+00],
       [ -4.44089210e-16,   4.44089210e-16,   1.00000000e+00]])
>>> print ('x =', multiply(inv(A), b))
x = [-1.4057971  -0.1884058   0.92753623]             # Solution
>>> print ('Residual =',  dot(A, x) - b)
Residual = [  4.44089210e-16   0.00000000e+00  -3.55271368e-15]
```

To solve the eigenvalue problem,

$$I\omega = \lambda\omega, \tag{1.3}$$

we call the `eig` method (as in `Eigen.py`):

```
>>> from numpy import*
>>> from numpy.linalg import eig
>>> I = array( [[2./3,-1./4], [-1./4,2./3]] )
>>> print('I =\n', I)
    I =  [[ 0.66666667 -0.25      ]
          [-0.25        0.66666667]]
>>> Es, evectors = eig(A)                      # Solve eigenvalue problem
>>> print('Eigenvalues =', Es, '\n Eigenvector Matrix =\n', evectors)
    Eigenvalues =   [ 0.91666667  0.41666667]
    Eigenvector Matrix =   [[ 0.70710678  0.70710678]
                            [-0.70710678  0.70710678]]
>>> Vec = array([ evectors[0, 0], evectors[1, 0] ] )
>>> LHS = dot(I, Vec)
>>> RHS = Es[0]*Vec
>>> print('LHS - RHS =', LHS-RHS) # Test for 0
       LHS - RHS = [  1.11022302e-16  -1.11022302e-16]
```

1. Find the numerical inverse of

$$A = \begin{bmatrix} +4 & -2 & +1 \\ +3 & +6 & -4 \\ +2 & +1 & +8 \end{bmatrix}. \tag{1.4}$$

   a. Check your inverse in both directions; that is, check that $AA^{-1} = A^{-1}A = I$.
   b. Note the number of decimal places to which this is true as this gives you some idea of the precision of your calculation.
   c. Determine the number of decimal places of agreement there is between your numerical inverse and the analytic result:

$$A^{-1} = \frac{1}{263} \begin{bmatrix} +52 & +17 & +2 \\ -32 & +30 & +19 \\ -9 & -8 & +30 \end{bmatrix}. \tag{1.5}$$

2. Consider the matrix A again, here being used to describe three simultaneous linear equations, $A\mathbf{x} = \mathbf{b}$. Solve for three different $\mathbf{x}$ vectors appropriate to the three different $\mathbf{b}$'s:

$$\mathbf{b}_1 = \begin{bmatrix} +12 \\ -25 \\ +32 \end{bmatrix}, \quad \mathbf{b}_2 = \begin{bmatrix} +4 \\ -10 \\ +22 \end{bmatrix}, \quad \mathbf{b}_3 = \begin{bmatrix} +20 \\ -30 \\ +40 \end{bmatrix}.$$

3. Consider the matrix $A = \begin{bmatrix} \alpha & \beta \\ -\beta & \alpha \end{bmatrix}$, where you are free to use any values you want for α and β. Show numerically that the eigenvalues and eigenvectors are the complex conjugates

$$\mathbf{x}_{1,2} = \begin{bmatrix} +1 \\ \mp i \end{bmatrix}, \quad \lambda_{1,2} = \alpha \mp i\beta. \tag{1.6}$$

1.2.3   Python Algebraic Tools

Symbolic manipulation software represents a supplementary, yet powerful, approach to computation in physics [Napolitano(18)]. Python distributions often contain the symbolic manipulation packages *Sage* and *SymPy*, which are quite different from each other. *Sage* is in the same class as Maple and MATHEMATICA and is beyond what we care to cover in this book. In contrast, the *SymPy* package runs very much like any other Python package from within a Python shell. For example, here we use Python's interactive shell to import methods from SymPy and then take some analytic derivatives:

```
>>> from sympy import *
>>> x, y = symbols('x y')
>>> y = diff(tan(x),x);  y
```

```
      tan^2(x) + 1
>>> y = diff(5*x**4 + 7*x**2, x, 1);   y         #  dy/dx  1 optional
    20 x^3 + 14 x
>>> y = diff(5*x**4+7*x**2, x, 2);   y           #  d^2y/dx^2
    2  (30 x^2 + 7)
```

The `symbols` command declares the variables x and y as algebraic, and the `diff` command takes the derivative with respect to second argument (the third argument is order of derivative). Here are some expansions:

```
>>> from sympy import *
>>> x, y = symbols('x y')
>>> z = (x + y)**8; z
    (x + y)^8
>>> expand(z)
    x^8 + 8 x^7 y + 28 x^6 y^2 + 56 x^5 y^3 + 70 x^4 y^4
                    + 56 x^3 y^5 + 28 x^2 y^6 + 8 x y^7 + y^8
```

SymPy knows about infinite series, and about different expansion points:

```
>>> sin(x).series(x, 0)                          # Sin x series
    x - x^3/6 + x^5/120 + \mathcal{O}(x^6)$
>>> sin(x).series(x,10)                          #  sin x about x=10
 sin(10) + x cos(10) - x^2  sin(10)/2 - x^3  cos(10)/6
        + x^4  sin(10)/24 + x^5  cos(10)/120 +O(x^6)
>>> z = 1/cos(x); z                 # Division, not an inverse
    $1/\cos(x)$
>>> z.series(x, 0)                     # Expand 1/cos x about 0
    1 + x^2/2 + 5 x^4/24 + O(x^6)
```

A classic difficulty with computer algebra systems is that the produced answers may be correct though not in a simple enough form to be useful. SymPy has functions such as `simplify`, tellfactor, `collect,` `cancel`, and `apart` which often help:

```
>>> factor(x**2 -1)
    (x - 1) (x + 1)                                   # Well done
>>> factor(x**3 - x**2 + x - 1)
    (x - 1) (x^2 + 1)
>>> simplify((x**3 + x**2 - x - 1)/(x**2 + 2*x + 1))
    x - 1
>>> factor(x**3+3*x**2*y+3*x*y**2+y**3)
    (x + y)^3                                         # Much better!
>>> simplify(1 + tan(x)**2)
    cos(x)^{(-2)}
```

1.3   Dealing with Floating Point Numbers

Scientific computations must account for the limited amount of computer memory used to represent numbers. Standard computations employ integers represented in *fixed-point* notation and other numbers in *floating-point* or scientific notation. In Python, we usually deal with 32 bit integers and 64 bit floating point numbers (called double precision in other languages). Doubles have approximately 16 decimal places of precision and magnitudes in the range

$$4.9 \times 10^{-324} \leq \text{double precision} \leq 1.8 \times 10^{308}. \tag{1.7}$$

If a double becomes larger than 1.8×10^{308}, a fault condition known as an *overflow* occurs. If the double becomes smaller than 4.9×10^{-324}, an underflow occurs. For overflows, the resulting number may end up being a machine-dependent pattern, not a number (NAN), or unpredictable. For underflows, the resulting number is usually set to zero.

Because a 64-bit floating point number stores the equivalent of only 15–16 decimal places, floating-point computations are usually approximate. For example, on a computer

$$3 + 1.0 \times 10^{-16} = 3. \tag{1.8}$$

This loss of precision is measured by defining the *machine precision* ϵ_m as the maximum positive number that can be added to a stored 1.0 without changing that stored 1.0:

$$1.0_c + \epsilon_m \overset{\text{def}}{=} 1.0_c, \tag{1.9}$$

where the subscript c is a reminder that this is a computer representation of 1. So, except for powers of 2, which are represented exactly, we should assume that all floating-point numbers have an error in the fifteenth place.

1.3.1 Uncertainties in Computed Numbers

Errors and uncertainties are integral parts of computation. Some errors are computer errors arising from the limited precision with which computers store numbers, or because of the approximate nature of algorithm. An algorithmic error may arise from the replacement of infinitesimal intervals by finite ones or of infinite series by finite sums, such as,

$$\sin(x) = \sum_{n=1}^{\infty} \frac{(-1)^{n-1} x^{2n-1}}{(2n-1)!} \simeq \sum_{n=1}^{N} \frac{(-1)^{n-1} x^{2n-1}}{(2n-1)!} + \mathcal{E}(x, N), \tag{1.10}$$

where $\mathcal{E}(x, N)$ is the approximation error. A reasonable algorithm should have $\mathcal{E}$ decreasing as N increases.

A common type of uncertainty in computations that involve many steps is *round-off* errors. These are accumulated imprecisions arising from the finite number of digits in floating-point numbers. For the sake of brevity, imagine a computer that kept just four decimal places. It would then store 1/3 as 0.3333 and 2/3 as 0.6667, where the computer has "rounded off" the last digit in 2/3. Accordingly, even a simple subtraction can be wrong:

$$2 \left(\frac{1}{3} \right) - \frac{2}{3} = 0.6666 - 0.6667 = -0.0001 \neq 0. \tag{1.11}$$

So although the result is small, it is not 0. Even with full 64 bit precision, if a calculation gets repeated millions or billions of times, the accumulated error answer may become large.

Actual calculations are often a balance. If we include more steps then the approximation error generally follows a power-law decrease. Nevertheless the relative round-off error after N steps tends to accumulate randomly, approximately like $\sqrt{N}\epsilon_m$. Because the total error is the sum of both these errors, eventually the ever-increasing round-off error will dominate. As rule of thumb, as you increase the number of steps in a calculation you should watch for the answer to converge or stabilize, decimal place by decimal place. Once you see what looks like random noise occurring in the last digits, you know round-off error is beginning to dominate, and you should probably step back a few steps and quit. An example is given in Figure 1.9.

1. Write a program that determines your computer's underflow and overflow limits (within a factor of 2). Here's a sample pseudocode

```
under = 1.
over = 1.
begin do N times
     under = under/2.
     over = over * 2.
     write out: loop number, under, over
end do
```

   a. Increase N if your initial choice does not lead to underflow and overflow.
   b. Check where under- and overflow occur for floating-point numbers.
   c. Check what are the largest and the most negative integers. You accomplish this by continually adding and subtracting 1.

2. Write a program to determine the machine precision ϵ_m of your computer system within a factor of 2. A sample pseudocode is

```
eps = 1.
begin do N times
  eps = eps/2.
  one = 1. + eps
end do
```

   a. Determine experimentally the machine precision of floats.
   b. Determine experimentally the machine precision of complex numbers.

1.4   Numerical Derivatives

Although the mathematical definition of the derivative is simple,

$$\frac{dy(t)}{dt} \stackrel{\text{def}}{=} \lim_{h \to 0} \frac{y(t+h) - y(t)}{h}, \tag{1.12}$$

it is not a good algorithm. As h gets smaller, the numerator to fluctuate between 0 and machine precision ϵ_m, and the denominator approaches zero. Instead, we use the Taylor series expansion of a $f(x+h)$ about x with h kept small but finite. In the *forward-difference* algorithm we take

$$\left. \frac{dy(t)}{dt} \right|_{FD} \simeq \frac{y(t+h) - y(t)}{h} + \mathcal{O}(h). \tag{1.13}$$

This $\mathcal{O}(h)$ error can be cancelled off by evaluating the function at a half step less than and a half step greater than t. This yields the *central-difference derivative*:

$$\left. \frac{dy(t)}{dt} \right|_{CD} \frac{y(t+h/2) - y(t-h/2)}{h} + \mathcal{O}(h^2). \tag{1.14}$$

The central-difference algorithm for the second derivative is obtained by using the central-difference algorithm on the corresponding expression for the first derivative:

$$\left. \frac{d^2y(t)}{dt^2} \right|_{CD} \simeq \frac{y'(t+h/2) - y'(t-h/2)}{h} \simeq= \frac{y(t+h) + y(t-h) - 2y(t)}{h^2}. \tag{1.15}$$

1. Use forward- and central-difference algorithms to differentiate the functions $\cos t$ and e^t at $t = 0.1, 1.,$ and 100.

    a. Print out the derivative and its relative error $\mathcal{E}$ as functions of h. Reduce the step size h until it equals machine precision $h \simeq \epsilon_m$.
    b. Plot $\log_{10} |\mathcal{E}|$ versus $\log_{10} h$ and check whether the number of decimal places obtained agrees with the estimates in the text.

2. Calculate the second derivative of $\cos t$ using the central-difference algorithms.

    a. Test it over four cycles, starting with $h \simeq \pi/10$ and keep reducing h until you reach machine precision

1.5 Numerical Integration

Mathematically, the Riemann definition of an integral is the limit

$$\int_a^b f(x) \, dx = \lim_{h \to 0} \sum_{i=1}^{(b-a)/h} f(x_i) h. \tag{1.16}$$

Numerical integration is similar, but approximates the integral as the a finite sum over rectangles of height $f(x)$ and widths (or weights) w_i:

$$\int_a^b f(x) \, dx \simeq \sum_{i=1}^{N} f(x_i) w_i. \tag{1.17}$$

Equation (1.17) is the standard form for all integration algorithms: the function $f(x)$ is evaluated at N points in the interval $[a, b]$, and the function values $f_i \equiv f(x_i)$ are summed with each term in the sum weighted by w_i. The different integration algorithms amount to different ways of choosing the points x_i and weights w_i. If you are free to pick the integration points, then our suggested algorithm is Gaussian quadrature. If the points are evenly spaced, then Simpson's rule makes good sense.

The trapezoid and Simpson integration rules both employ $N - 1$ boxes of width h evenly-spaced throughout the integration region $[a, b]$:

$$x_i = a + ih, \qquad h = \frac{b - a}{N - 1}, \qquad i = 0, N - 1. \tag{1.18}$$

For each interval, the trapezoid rule assumes a trapezoid of width h and height $(f_i + f_{i+1})/2$, and, accordingly, approximates the area of each trapezoid as $\frac{1}{2}hf_i + \frac{1}{2}hf_{i+1}$. To apply the trapezoid rule to the entire region $[a, b]$, we add the contributions from all subintervals:

$$\int_a^b f(x)\, dx \simeq \frac{h}{2}f_1 + hf_2 + hf_3 + \cdots + hf_{N-1} + \frac{h}{2}f_N, \tag{1.19}$$

where the endpoints get counted just once, but the interior points twice. In terms of our standard integration rule (1.17), we have

$$w_i = \left\{ \frac{h}{2}, h, \ldots, h, \frac{h}{2} \right\} \qquad \text{(Trapezoid Rule)}. \tag{1.20}$$

In `TrapMethods.py` in Listing 1.15 we provide a simple implementation.

Simpson's rule is also for evenly spaced points of width h, though with the heights given by parabolas fit to successive sets of three adjacent integrand values. This leads to the approximation:

$$\int_a^b f(x)dx \simeq \frac{h}{3}f_1 + \frac{4h}{3}f_2 + \frac{2h}{3}f_3 + \frac{4h}{3}f_4 + \cdots + \frac{4h}{3}f_{N-1} + \frac{h}{3}f_N. \tag{1.21}$$

In terms of our standard integration rule (1.17), this is

$$w_i = \left\{ \frac{h}{3}, \frac{4h}{3}, \frac{2h}{3}, \frac{4h}{3}, \ldots, \frac{4h}{3}, \frac{h}{3} \right\} \qquad \text{(Simpson's Rule)}. \tag{1.22}$$

Because the fitting is done with sets of three points, *the number of points N must be odd for Simpson's rule.*

In general, you should choose an integration rule that gives an accurate answer using the least number of integration points. For the trapezoid and Simpson rules the errors vary as

$$\mathcal{E}_t = O\left(\frac{[b - a]^3}{N^2} \right) \frac{d^2 f}{dx^2}, \qquad \mathcal{E}_s = O\left(\frac{[b - a]^5}{N^4} \right) \frac{d^4 f}{dx^4}, \tag{1.23}$$

where the derivatives are evaluated someplace within the integration region. So unless the integrand has behavior problems with its derivatives, Simpson's rule should converge more rapidly than the trapezoid rule and with less error. While it seems like one might need only to keep increasing the number of integration points to obtain better accuracy, relative round-off error tends to accumulate, and, after N integration points, grows like

$$\epsilon_{ro} \simeq \sqrt{N}\epsilon_m, \tag{1.24}$$

where $\epsilon_m \simeq 10^{15}$ is the machine precision (discussed in §1.3). So even though the error in the algorithm can be made arbitrary small, the total error, that is, the error due to algorithm plus the error due to round-off, eventually will increase like $\sqrt{N}$.

1.5.1   Gaussian Quadrature

Gauss figured out a way of picking the N points and weights in (1.17) so as to make an integration over [-1,1] exact if $g(x)$ is a polynomial of degree $2N-1$ or less. To accomplish this miraculous feat, the x_i's must be the N zeros of the Legendre polynomial of degree N, and the weights related to the derivatives of the polynomials [LPB(15)]:

$$P_N(x_i) = 0, \qquad w_i = \frac{2}{(1-x_i^2)[P_N'(x_i)]^2}. \tag{1.25}$$

Not to worry, we supply a program that determines the points and weights. If your integration range is [a,b] and not [-1,+1], they will be scaled as

$$x_i' = \frac{b+a}{2} + \frac{b-a}{2}x_i, \qquad w_i' = \frac{b-a}{2}w_i. \tag{1.26}$$

In general, Gaussian quadrature will produce higher accuracy than the trapezoid and Simpson rules for the same number of points, and is our recommended integration method.

Our Gaussian quadrature code `IntegGaussCall.py` in Listing 1.16 requires the value for precision `eps` of the points and weights to be provided by the user. Overall precision is usually increased by increasing the number of points used. The points and weights are generated by the method `GaussPoints.py`, which will be included automatically in your program via the `from GaussPoints import GaussPoints` statement.

1.5.2   Monte Carlo (Mean Value) Integration

Monte Carlo integration is usually simple, but not particularly efficient. It is just a direct application of the *mean value theorem*:

$$I = \int_a^b dx\, f(x) = (b-a)\langle f \rangle. \tag{1.27}$$

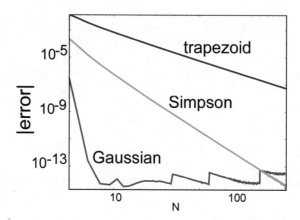

Figure 1.9. A log-log plots of the relative error in an integration using the trapezoid rule, Simpson's rule, and Gaussian quadrature *versus* the number of integration points N.

The mean is determined by *sampling* the function $f(x)$ at random points within the integration interval:

$$\langle f \rangle \simeq \frac{1}{N} \sum_{i=1}^{N} f(x_i) \quad \Rightarrow \quad \int_a^b dx \, f(x) \simeq (b-a) \frac{1}{N} \sum_{i=1}^{N} f(x_i). \tag{1.28}$$

The uncertainty in the value obtained for the integral I after N samples of $f(x)$ is measured by the standard deviation σ_I. If σ_f is the standard deviation of the integrand f in the sampling, then for a normal distribution of random number we would have

$$\sigma_I \simeq \frac{1}{\sqrt{N}} \sigma_f. \tag{1.29}$$

So, for large N the error decreases as $1/\sqrt{N}$. In Figure 1.6 left we show a scatter plot of the points used in a Monte Carlo integration by the code `PondMapPlot.py` in Listing 1.7.

Before you actually use random numbers to evaluate integrals, we recommend that you work through §1.6.2 to be sure your random number generator is working properly.

On the left of Figure 1.6 we show a pond whose area we wish to determine. We can determine the area of such an irregular figure by throwing stones in the air (generating random (x, y) values), and counting the number of splashes N_{pond} as well as the number of stones lying on the ground N_{box}. The area of the pond is then given by the simple ratio:

$$A_{pond} = \frac{N_{pond}}{N_{pond} + N_{box}} A_{box}. \tag{1.30}$$

1. Write a program to integrate a function for which you know the analytic answer so that you can determine the error. Use

      a. the trapezoid rule,

      b. the Simpson rule,

      c. Gaussian quadrature, and

      d. Monte Carlo integration.

2. Compute the relative error $\epsilon = |(\text{numerical-exact})/\text{exact}|$ for each case, and make a log-log plot of relative error versus N as we do in Figure 1.9. You should observe a steep initial power-law drop-off of the error, which is characteristic of an algorithmic error decreasing with increasing N. Note that the ordinate in the plot is the negative of the number of decimal places of precision in your integral.

3. The algorithms should stop converging when round-off error starts to dominate, as noted by random fluctuations in the error. Estimate the number of decimal places of precision obtained for each of the three rules.

4. Use sampling to compute π (number hits in unit circle/total number hits = π/Area of box).

5. Evaluate the following integrals:

      a. $\int_0^1 e^{\sqrt{x^3 + 5\,x}}\, dx$

      b. $\int \frac{1}{x^2 + 2\,x + 4}\, dx$

      c. $\int_0^\infty e^{(-x^2)}\, dx$

1.6   Random Number Generation

Randomness or chance occurs in different areas of physics. For example, quantum mechanics and statistical mechanics are statistical by nature, and so randomness enters as one of the key assumptions of statistics. Or, looked at the other way, random processes such as the motion of molecules were observed early on, and this led to theory of statistical mechanics. In addition to the randomness in nature, many computer calculations employ *Monte Carlo* methods that include elements of chance to either simulate random physical processes, such as thermal motion or radioactive decay, or in mathematical evaluations, such as integration.

    Randomness describes a lack of predicability or regular pattern. Mathematically, we define a sequence $r_1, r_2, \ldots$ as *random* if there are no short- or long-range correlations among the numbers. This does not necessarily mean that all the numbers in the sequence are equally likely to occur; that is called *uniformity*. As a case in point, 0, 2, 4, 6, ... is uniform though probably not random. If $P(r)\, dr$ is the probability of finding r in the interval $[r, r + dr]$, a uniform distribution has $P(r) = $ a constant.

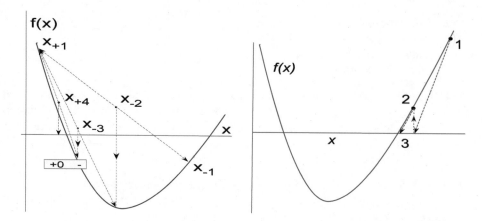

Figure 1.10. *Left:* A graphical representation of three steps involved in solving for a zero of $f(x)$ using the bisection algorithm. *Right:* Two steps shown for the Newton-Raphson method of root finding in which the function is approximated as a straight line, and then the intercept of that line is determined.

Computers, being deterministic, cannot generate truly random numbers. However, they can generate *pseudorandom numbers*, and the built-in generators are often very good at this. The `random` module in Python produces a sequence of random numbers, and can be used after an `import random` statement. The module permits many options, with the simple command `random.random()` returning the next random floating point number, distributed uniformly in the range [0.0, 1.0). But if you look hard enough, you are sure to find correlations among the numbers.

The *linear congruential* or *power residue* method is the common way of generating a pseudorandom sequence of numbers:

$$r_{i+1} \overset{\text{def}}{=} (a\,r_i + c)\,\text{mod}\,M = \text{remainder}\left(\frac{a\,r_i + c}{M}\right). \tag{1.31}$$

Wikipedia has a table of common choices, for instance, $m = 2^{48}$, $a = 25214903917$, $c = 11$. Here *mod* is a function (% sign in Python) for modulus or *remaindering*, which is essentially a bit-shift operation that results in the least significant part of the input number and hence counts on the randomness of round-off errors.

Your computer probably has random-number generators that should be better than one computed by a simple application of the power residue method. In Python we use `random.random()`, the Mersenne Twister generator. To initialize a random sequence, you need to plant a seed (r_0), or in Python say `random.seed(None)`, which seeds the generator with the system time, which would differ for repeated executions. If random numbers in the range $[A, B]$ are needed, you only need to scale, for example,

$$x_i = A + (B - A)r_i, \quad 0 \le r_i \le 1, \quad \Rightarrow \quad A \le x_i \le B. \tag{1.32}$$

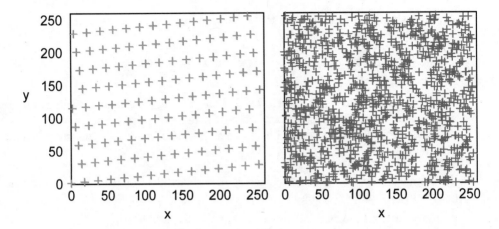

Figure 1.11. *Left:* A plot of successive random numbers $(x, y) = (r_i, r_{i+1})$ generated with a deliberately "bad" generator. *Right:* A plot generated with the built-in random number generator. While the plot on the right is not proof that the distribution is random, the plot on the left is proof enough that the distribution is not random.

1.6.1 Tests of Random Generators

A good general rule, before starting a full calculation, is to check your random number generator by itself. Here are some ways:

- Look at a print out of the numbers and check that they fall within the desired range and that they look different from each other.

- A simple plot of r_i versus i (Figure 1.12) may not prove randomness, though it may disprove it as well as showing the range of numbers.

- Make an x-y plot of $(x_i, y_i) = (r_{2i}, r_{2i+1})$. If your points have noticeable regularity (Figure 1.11 left), the sequence is not random. Random points (Figure 1.11 right) should uniformly fill a square with no discernible pattern (a cloud).

- A simple test of uniformity, though not randomness, evaluates the kth moment of a distribution

$$\langle x^k \rangle = \frac{1}{N} \sum_{i=1}^{N} x_i^k \simeq \int_0^1 dx \, x^k P(x) \simeq \frac{1}{k+1} + O\left(\frac{1}{\sqrt{N}}\right), \qquad (1.33)$$

where the approximate value is good for a continuous uniform distribution. If the deviation from (1.33) varies as $1/\sqrt{N}$, then you *also* know that the distribution is random since this assumes randomness.

- Another simple test determines the kth order correlation of a sequence:

$$C(k) = \frac{1}{N}\sum_{i=1}^{N} x_i \, x_{i+k}, \simeq \int_0^1 dx \int_0^1 dy \, xy = \frac{1}{4}, \quad (k = 1, 2, \ldots), \qquad (1.34)$$

where the approximate value will hold if the random numbers are distributed with a constant joint probability, which we take to be 1. Here again, if the deviation from (1.34) varies as $1/\sqrt{N}$, then you *also* know that the distribution is random.

1. To help see why we recommend using an industrial strength random-number generator, try instead using the linear congruential method (1.31).
   a. Start with an unwise choice: $(a, c, M, r_1) = (57, 1, 256, 10)$.
   b. Determine the *period*, that is, how many numbers are generated before the sequence repeats.
   c. Look for correlations in this unwise choice by observing clustering on a plot of successive pairs $(x_i, y_i) = (r_{2i-1}, r_{2i})$, $i = 1, 2, \ldots$. (Do *not* connect the points with lines.)
   d. Plot r_i versus i and comment on its comparison to Figure 1.12.
   e. Now get serious and test the linear congruential method using one of the set of constants given by Wikipedia.

2. Test the built-in random-number generator on your computer by using a variety of the tests discussed above.

3. Compare the scatter plot you obtain using the built-in random-number generator with that of your "good" linear congruential method

1.6.2   Central Limit Theorem

Before you start using random numbers in computations, it is a good idea to verify that your generated randomness agrees with the statistical meaning of randomness. A way to do that is to test if your generated numbers obey the *central limit theorem*. One of the things that the theorem tells us is that when a large number N of independent, random variables are added together, their sum tends toward a normal (Gaussian) distribution:

$$\rho(x) = \frac{1}{\sigma\sqrt{2\pi}}e^{-(x-\langle x \rangle)^2/2\sigma^2}, \qquad (1.35)$$

where $\langle x \rangle$ is the mean value of x and σ is the standard deviation:

$$\langle x \rangle = \frac{1}{N}\sum_{i=1}^{N} x_i, \qquad \sigma = \frac{1}{N}\sum_{i=1}^{N}(x - \langle x \rangle)^2. \qquad (1.36)$$

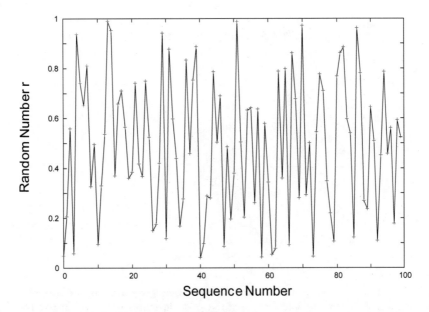

Figure 1.12. A plot of a uniform pseudorandom sequence r_i versus i. The points are connected to make it easier to follow the order. While this does not prove that a distribution is random, it at least shows the range of values and that there is fluctuation.

Thus, in addition to checking your random number generator, the theorem also provides a simple algorithm for producing random numbers distributed with a Gaussian weighting about the mean.

1. Generate and sum 1000 uniformly distributed random numbers in range [0,1].

2. Create a list, and place your sum as the first element in a list. After creating many sums and filling the list with them, check that their average value is 1/2, and that there is a normal distribution of sum values around the average.

3. Repeat steps 1 and 2 for $N = 9999$ (a 10,000 element list).

4. Compute the mean value of all of the sums in your list.

5. Compute σ via (1.36).

6. Make a histogram plot of all the sums with 100 bins covering all of the values obtained. Choose plotting options such that the distribution is normalized, that is, has unit value when integrated over. (In Matplotlib this is done with the option `normed=True`.)

7. Plot on the same graph the normal distribution (1.35) using the value for the standard deviation σ that you computed in a previous step.

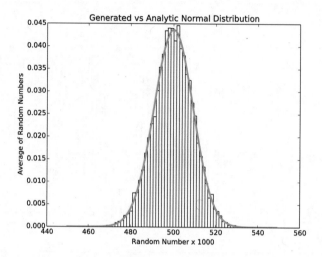

Figure 1.13. A normal distribution of random numbers generated as the sum of uniformly distributed random numbers, and a curve of the analytic expression for a normal distribution.

8. Run your program 100 times and note how the histograms for each run tend to fluctuate around the normal distribution curve.

9. Take and plot the average of your 100 runs and compare that to (1.35).

Our program `CentralValue.py` in Listing 1.11 produced the results shown in Figure 1.13.

1.7 Ordinary Differential Equations Algorithms

Many of the problems in this book require the solution of an ordinary differential equation (ODE). It is easy to solve almost every ODE with one or two simple algorithms, and because the same algorithms work for nonlinear ODEs, the physics will not be restricted to linear systems with small displacements. As also used in classical dynamics, a standard form for a single ODE, or a set of simultaneous ODEs, of *any order* expresses them as N simultaneous first-order ODEs in N unknowns $y^{(0)} - y^{(N-1)}$. The unknowns are combined into a single N-dimensional vector $\mathbf{y}$, with the

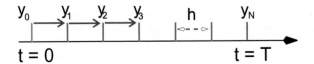

Figure 1.14. A sequence of uniform steps of length h taken in solving a differential equation. The solution starts at time $t = 0$ and is integrated in steps of h until $t = T$.

RHS's combined into the vector $\mathbf{f}$:

$$\frac{d\mathbf{y}(t)}{dt} = \mathbf{f}(t, \mathbf{y}), \tag{1.37}$$

$$\mathbf{y} = \begin{bmatrix} y^{(0)}(t) \\ y^{(1)}(t) \\ \ddots \\ y^{(N-1)}(t) \end{bmatrix}, \qquad \mathbf{f} = \begin{bmatrix} f^{(0)}(t, \mathbf{y}) \\ f^{(1)}(t, \mathbf{y}) \\ \ddots \\ f^{(N-1)}(t, \mathbf{y}) \end{bmatrix}. \tag{1.38}$$

The rule indicated by (1.37) is that the RHS function $\mathbf{f}(t, \mathbf{y})$ may *not* contain any explicit derivatives, although individual components of $y^{(i)}$ may be equal to derivatives.

To see how this works, we start with Newton's law with an arbitrary force, and define each level of derivative as a new variable:

$$\frac{d^2 x}{dt^2} = \frac{1}{m} F\left(t, x, \frac{dx}{dt}\right), \tag{1.39}$$

$$y^{(0)}(t) \stackrel{\text{def}}{=} x(t), \qquad y^{(1)}(t) \stackrel{\text{def}}{=} \frac{dx}{dt} = \frac{dy^{(0)}(t)}{dt}, \tag{1.40}$$

$$\Rightarrow \quad \frac{dy^{(0)}}{dt} = y^{(1)}(t) = f^{(0)}, \qquad \frac{dy^{(1)}}{dt} = \frac{1}{m} F(t, y^{(0)}, y^{(1)}) = f^{(1)}. \tag{1.41}$$

1.7.1   Euler & Runge-Kutta Rules

As illustrated in Figure 1.14, an ODE is solved numerically by starting with an initial value of the dependent variable $y(t = 0) \equiv y_0$, and using the derivative function $f(t, y)$ to advance y_0 one small step h forward in time to $y(t = h) \equiv y_1$. The algorithm then just keeps repeating itself, treating the new values for y as new initial conditions for the next step. *Euler's rule* does this via a straightforward application of the forward-difference algorithm for the derivative:

$$\mathbf{y}_{n+1} \simeq \mathbf{y}_n + h\mathbf{f}(t_n, \mathbf{y}_n) \qquad \text{(Euler's Rule)}, \tag{1.42}$$

where $y_n \equiv y(t_n)$ is the value of y at time t_n. Aside from its use in initiating other algorithms, Euler's method is not accurate enough for scientific work. As with the forward-difference derivative, the error in Euler's rule is $\mathcal{O}(h^2)$, which is equivalent to ignoring the effect of acceleration on the position of a projectile, but including it on the projectile's velocity.

In contrast to Euler's rule, the fourth-order *Runge-Kutta algorithm*, rk4, has proven to be robust and capable of industrial strength work. It is our recommended approach for solving ODEs. The algorithm is based upon the formal integral of the differential equation:

$$\frac{dy}{dt} = f(t, y) \quad \Rightarrow \quad y(t) = \int f(t, y) \, dt \quad \Rightarrow \quad y_{n+1} = y_n + \int_{t_n}^{t_{n+1}} f(t, y) \, dt. \quad (1.43)$$

The critical idea here is to expand the integrand around the midpoint of the integration interval and thereby obtain $\mathcal{O}(h^4)$ precision via the cancellation of the h and h^3 terms. The price paid for the improved precision is having to approximate three derivatives and the unknown y at the middle of the interval [Press et al.(94)]:

$$\mathbf{y}_{n+1} = \simeq \mathbf{y}_n + \frac{1}{6}(\mathbf{k}_1 + 2\mathbf{k}_2 + 2\mathbf{k}_3 + \mathbf{k}_4), \quad (1.44)$$

$$\mathbf{k}_1 = h\mathbf{f}(t_n, \mathbf{y}_n), \qquad \mathbf{k}_2 = h\mathbf{f}\left(t_n + \frac{h}{2}, \mathbf{y}_n + \frac{\mathbf{k}_1}{2}\right),$$

$$\mathbf{k}_3 = h\mathbf{f}\left(t_n + \frac{h}{2}, \mathbf{y}_n + \frac{\mathbf{k}_2}{2}\right), \quad \mathbf{k}_4 = h\mathbf{f}(t_n + h, \mathbf{y}_n + \mathbf{k}_3).$$

The program `rk4Call.py` in Listing 1.12 uses rk4 to solve $\ddot{y} = -100y - 2\dot{y} + 100\sin(3t)$, and plots up the results using *VPython*. As an alternative, the program `rk4Duffing.py` in Listing 1.13 solves $\ddot{y} = -2\gamma\dot{y} - \alpha y - \beta y^3 + F\cos\omega t$, and plots up the results using *Mathplotlib*. The details of the rk4 algorithm remain unchanged regardless of the ODE and are contained in the function `rk4Algor.py`. Do *not* modify it! Instead, *modify only* the force function `f()` in `rk4Call.py` that contains the RHS (1.38) of the ODE. Note, after the first call to `rk4Algor.py`, the compiled version of it, `rk4Algor.pyc`, will be left in the working directory. The statement `from rk4Algor import rk4Algor` at the beginning of `rk4Call.py` includes the precompiled method. Here is a pseudocode version of `rk4Call.py`:

```
#  Pseudocode for rk4Call.py  for ODE y" = -100y-2y'+ 100 sin(3t)
import packages
Initialize variables, y(0) = position, y(1) = velocity
h = (Tend-Tstart)/N                             # Time step
Define f(t, y)                        # Function with RHS's
while (t < Tend)
    call rk4Algor                     # y_new = y_old + Delta
    t = t + h
    plot each new point
```

Here is a pseudocode version of `rk4Duffing.py`:

```
# Pseudocode for rk4Duffing.py
import packages
Declare  yy = all positions, vy = all velocities, tt = all t's
Define f(t, y)                  # RHS Force Function
i = 0, initialize y[0], y[1]
for  0 < t < 100
    store tt[i], yy[i] = y[0], vy[i] = y[1]
    call rk4Algor              # rk4:  y_new = y_old + Delta y
    i = i + 1
plot yy, vy  vectors
```

1.8   Partial Differential Equations Algorithms

There is no one algorithm that can be applied to all of the different types of partial differential equations. Although all PDE solutions we discuss apply finite difference approximations for derivatives, the details depend upon the equation and boundary conditions. Accordingly, we present the different techniques within the context of their physics use: *relaxation* for Laplace and Poisson's equations, §5.2.1, and *time stepping* for the heat and wave equations, §7.2, §4.2. As an example, which we cover more fully in §7.2, the heat equation,

$$\frac{\partial T(x,t)}{\partial t} = \kappa \frac{\partial^2 T(x,t)}{\partial x^2}, \tag{1.45}$$

is a PDE in space and time. When the time and space derivatives are approximated in terms of finite differences, the PDE becomes the finite difference equation

$$\frac{T(x,t+\Delta t) - T(x,t)}{\Delta t} = \alpha \frac{T(x+\Delta x,t) + T(x-\Delta x,t) - 2T(x,t)}{\Delta x^2}. \tag{1.46}$$

We form an algorithm from this equation by reordering it such that the temperature at an earlier time j can be stepped forward to yield the temperature at a later time $j+1$:

$$T_{i,j+1} = T_{i,j} + \eta \left[T_{i+1,j} + T_{i-1,j} - 2T_{i,j} \right]. \tag{1.47}$$

1.9   Code Listings

```
# EasyVisualVP.py  :  VPthon,  simple  graph  object

from vpython import *                    # Import Vpython

graph1=graph(align='left',width=400, height=400,
    background=color.white,foreground=color.black)
Plot1=gcurve(color=color.red)                    # gcurve method
for x in arange(0,8.1,0.1):                       # x range
    Plot1.plot(pos=(x,5*cos(2*x)*exp(-0.4*x)))
graph2=graph(align='right',width=400, height=400,
    background=color.white,foreground=color.black,
```

```
    title='2-D Plot', xtitle='x', ytitle='f(x)')
Plot2=gdots(color=color.black)                        # Dots
for x in arange(-5,5,0.1):
    Plot2.plot(pos=(x,cos(x)))                        # plot dots
```

Listing 1.1. **EasyVisualVP.py** uses VPython to produce the two plots in Figure 1.1.

```
# 3GraphVP.py: Vpython package, 3 plots, with bars, dots & curve

from vpython import *

string="blue: sin^2(x), black= cos^2(x), cyan: sin(x)*cos(x)"
graph1=graph(title=string, xtitle='x', ytitle='y',
  background=color.white, foreground=color.black)
y1 = gcurve(color=color.blue)          # curve
y2 = gvbars(color=color.black)         # vertical bars
y3 = gdots(color=color.cyan)           # dots
for x in arange(-5,5,0.1):             # arange for  plots
    y1.plot(pos=(x,sin(x)**2))
    y2.plot(pos=(x,cos(x)*cos(x)/3.))
    y3.plot(pos=(x,sin(x)*cos(x)))
```

Listing 1.2. **3GraphVP.py** produces a 2-D x-y plot with the VPython package.

```
# EasyMatPlot.py: Simple use of matplotlib's plot command

from pylab import *                        # Load Matplotlib

Xmin = -5.;    Xmax = +5.;   Npoints= 500
DelX = (Xmax - Xmin) / Npoints
x = arange(Xmin, Xmax, DelX)
y =  sin(x) * sin(x*x)                       # Function of array
print ('arange => x[0], x[1],x[499]=%8.2f %8.2f %8.2f'
  %(x[0],x[1],x[499]))
print ('arange => y[0], y[1],y[499]=%8.2f %8.2f %8.2f'
  %(y[0],y[1],y[499]))
print ("\n Doing plotting, look for Figure 1" )
xlabel('x');          ylabel('f(x)');          title(' f(x) vs x')
text(-1.75,  0.75, 'MatPlotLib \n Example')  # Text on plot
plot(x, y, '-', lw=2)
grid(True)                                    # Form grid
show()
```

Listing 1.3. **EasyMatPlot.py** produces a 2-D x-y plot using the Matplotlib package (which includes the NumPy package).

```
# GradeMatPlot.py: Matplotlib plot multi-data sets

import pylab as p                                    # Matplotlib
from numpy import*

p.title('Grade Inflation')                           # Title and labels
p.xlabel('Years in College')
p.ylabel('GPA')
xa = array([-1, 5])                                  # For horizontal line
ya = array([0, 0])                                   #   "          "
p.plot(xa, ya)                                       # Draw horizontal line
x0 = array([0, 1, 2, 3, 4])                          # Data set 0 points
y0 = array([-1.4, +1.1, 2.2, 3.3, 4.0])
p.plot(x0, y0, 'bo')                                 # Data set 0 = blue circles
p.plot(x0, y0, 'g')                                  # Data set 0 = line
x1 = arange(0, 5, 1)                                 # Data set 1 points
y1 = array([4.0, 2.7, -1.8, -0.9, 2.6])
p.plot(x1, y1, 'r')
errTop = array([1.0, 0.3, 1.2, 0.4, 0.1])            # Asymmetric error bars
errBot = array([2.0, 0.6, 2.3, 1.8, 0.4])
p.errorbar(x1, y1, [errBot, errTop], fmt = 'o')      # Plot error bars
p.grid(True)                                         # Grid line
p.show()
```

Listing 1.4. **GradesMatPlot.py** produces an *x-y* plot using the Matplotlib package.

```
# MatPlot2figs.py: plot of 2 subplots on 1 fig, 2 separate figs

from pylab import *                                  # Load matplotlib

Xmin = -5.0;        Xmax = 5.0;        Npoints= 500
DelX= (Xmax-Xmin)/Npoints                            # Delta x
x1 = arange(Xmin, Xmax, DelX)                        # x1 range
x2 = arange(Xmin, Xmax, DelX/20)                     # Different x2 range
y1 = -sin(x1)*cos(x1*x1)                             # Function 1
y2 =  exp(-x2/4.)*sin(x2)                            # Function 2
print("\n Now plotting, look for Figures 1 & 2 on desktop")
figure(1)        # Fig 1
subplot(2,1,1)                                       # 1st subplot in first figure
plot(x1, y1, 'r', lw=2)
xlabel('x'); ylabel('f(x)'); title('$-\sin(x)*\cos(x^2)$')
grid(True)                                           # Form grid
subplot(2,1,2)                                       # 2nd subplot in first figure
plot(x2, y2, '-', lw=2)
xlabel('x')                                          # Axes labels
ylabel('f(x)')
title('exp(-x/4)*sin(x)')
figure(2)  # Fig 2
subplot(2,1,1)                                       # 1st subplot in 2nd figure
plot(x1, y1*y1, 'r', lw=2)
xlabel('x'); ylabel('f(x)'); title('$\sin^2(x)*\cos^2(x^2)$')  ↵

     # form grid
subplot(2,1,2)                                       # 2nd subplot in 2nd figure
plot(x2, y2*y2, '-', lw=2)
xlabel('x'); ylabel('f(x)'); title('$\exp(-x/2)*\sin^2(x)$')
grid(True)
show()
```

Listing 1.5. **MatPlot2figs.py** produces the two figures shown in Figure 1.4. Each figure contains two plots.

```python
# Simple3Dplot.py: matplotlib 3D plot, rotate & scale with mouse

import matplotlib.pylab as p
from mpl_toolkits.mplot3d import Axes3D

print ("Please be patient while I do importing & plotting")
delta = 0.1
x = p.arange( -3., 3., delta )
y = p.arange( -3., 3., delta )
X, Y = p.meshgrid(x, y)
Z = p.sin(X) * p.cos(Y)                     # Surface height
fig = p.figure()                            # Create figure
ax = Axes3D(fig)                            # Plots axes
ax.plot_surface(X, Y, Z)                    # Surface
ax.plot_wireframe(X, Y, Z, color = 'r')     # Add wireframe
ax.set_xlabel('X')
ax.set_ylabel('Y')
ax.set_zlabel('Z')
p.show()                                    # Output figure
```

Listing 1.6. **Simple3Dplot.py** produces the Matplotlib 3-D surface plots in Figure 1.5.

```python
#   PondMatPlot.py: Monte-Carlo integration via von Neumann ←
        rejection

import numpy as np, matplotlib.pyplot as plt

N, Npts = 100,    5000;
analyt = np.pi**2
x1 = np.arange(0, 2*np.pi+2*np.pi/N,2*np.pi/N)
xi = [];   yi = [];   xo = [];   yo = []
fig,ax = plt.subplots()
y1 = x1 * np.sin(x1)**2                         # Integrand
ax.plot(x1, y1, 'c', linewidth=4)
ax.set_xlim ((0, 2*np.pi))
ax.set_ylim((0, 5))
ax.set_xticks([0, np.pi, 2*np.pi])
ax.set_xticklabels(['0', '$\pi$','2$\pi$'])
ax.set_ylabel('$f(x) = x\,\sin^2 x$', fontsize=20)
ax.set_xlabel('x',fontsize=20)
fig.patch.set_visible(False)

def fx(x):    return x*np.sin(x)**2            # Integrand

j = 0                                  # Inside curve counter
xx = 2.* np.pi * np.random.rand(Npts)       # 0 =< x <= 2pi
yy = 5*np.random.rand(Npts)                 # 0 =< y <= 5
```

```
boxarea = 2. * np.pi *5.                              # Box area
for i in range(1,Npts):
    #plt.pause(0.000001)
    if (yy[i] <= fx(xx[i])):                          # Below curve
        xi.append(xx[i])
        yi.append(yy[i])
        j +=1
    else:
        yo.append(yy[i])
        xo.append(xx[i])

    area = boxarea*j/(Npts-1)                          # Area under curve
ax.plot(xo,yo,'bo',markersize=1)
ax.plot(xi,yi,'ro',markersize=1)
ax.set_title('Answers: Analytic = %5.3f, MC = %5.3f'%(analyt,area))
plt.show()
```

Listing 1.7. **PondMatPlot.py** produces the scatter plot and the curve in Figure 1.6 left.

```
# Scatter3dPlot.py: Matplotlib scatter plot example

import numpy as np
from mpl_toolkits.mplot3d import Axes3D
import matplotlib.pyplot as plt

def randrange(n, vmin, vmax):
    return (vmax-vmin)*np.random.rand(n) + vmin

fig = plt.figure()
ax = fig.add_subplot(111, projection='3d')
n = 100
for c, m, zl, zh in [('r', 'o', -50, -25), ('b', '^', -30, -5)]:
    xs = randrange(n, 23, 32)
    ys = randrange(n, 0, 100)
    zs = randrange(n, zl, zh)
    ax.scatter(xs, ys, zs, c=c, marker=m)
ax.set_xlabel('X Label')
ax.set_ylabel('Y Label')
ax.set_zlabel('Z Label')
plt.show()
```

Listing 1.8. **Scatter3dPlot.py** produces a 3-D scatter plot using Matplotlib 3D tools.

```
# TwoForces.py Forces on two moving strings

from visual.graph import *

posy=100; Lcord=250  # basic height, cord length
Hweight=50; W = 10   # cylinder height, weight

scene=display(height=600,width=600,range=380)
alt=curve(pos=[(-300,posy,0),(300,posy,0)])
divi=curve(pos=[(0,-150,0),(0,posy,0)])
```

```
kilogr=cylinder(pos=(0,posy-Lcord,0),radius=20,axis=(0,-Hweight,0),
                color=color.red) # kg as a cylinder
cord1=cylinder(pos=(0,posy,0),axis=(0,-Lcord,0),color=color.yellow,
               radius=2)
cord2=cylinder(pos=(0,posy,0),axis=(0,-Lcord,0),color=color.yellow,
               radius=2)

arrow1=arrow(pos=(0,posy,0), color=color.orange) # Tension cord 1
arrow2=arrow(pos=(0,posy,0), color=color.orange) # Tension cord 2

magF=W/2.0              # initial force of each student
v=2.0                  # (m/s) velocity of each student
x1=0.0                 # initial position student 1
anglabel=label(pos=(0,240,0), text='angle (deg)',box=0)
angultext=label(pos=(20,210,0),box=0)
Flabel1=label(pos=(200,240,0), text='Force',box=0)
Ftext1=label(pos=(200,210,0),box=0)
Flabel2=label(pos=(-200,240,0), text='Force',box=0)
Ftext2=label(pos=(-200,210,0),box=0)
local_light(pos=(-10,0,20), color=color.yellow)   # light

for t in arange(0.,100.0,0.2):
    rate(50)                    # slow motion
    x1=v*t                      # 1 to right, 2 to left
    theta=asin(x1/Lcord)        # angle cord
    poscil=posy-Lcord*cos(theta)  # cylinder height
    kilogr.pos=(0,poscil,0)     # y-position kilogram
    magF=W/(2.*cos(theta))      # Cord tension
    angle=180.*theta/pi
    cord1.pos=(x1,posy,0)           # position cord end
    cord1.axis=(-Lcord*sin(theta),-Lcord*cos(theta),0)
    cord2.pos=(-x1,posy,0)              # position end cord
    cord2.axis=(Lcord*sin(theta),-Lcord*cos(theta),0)
    arrow1.pos=cord1.pos            # axis arrow
    arrow1.axis=(8*magF*sin(theta),8*magF*cos(theta),0)
    arrow2.pos=cord2.pos
    arrow2.axis=(-8*magF*sin(theta),8*magF*cos(theta),0)
    angultext.text='%4.2f'%angle
    force=magF
    Ftext1.text='%8.2f'%force       # Tension
    Ftext2.text='%8.2f'%force
```

Listing 1.9. **TwoForces.py** produces a 3-D animation showing the forces on two strings as the students pulling on each mass move apart. Almost all of the code deals with creating and moving the various graphical objects.

```
# SlidingBox.py: 3-D animation of forces on a beam as box slides

from vpython import *

Hsupport,d =30, 100                      # height, distance supports
Lbeam=500; Wbeam=80;   thickness=10 # beam dimensions
W =200;   WeightBox=400                   # weight of table,box
Lbox=60; Wbox=60; Hbox=60                 # Box Dimensions

# Graphics
scene=canvas(width=750, height=500,range=300)
```

```
scene.forward=vector(0.5,-0.2,-1)   # to change point of view
support1=cone(pos=vector(-d,0,0),axis=vector(0,Hsupport,0),
   color=color.yellow, radius=20)
support2=cone(pos=vector(d,0,0),axis=vector(0,Hsupport,0),
   color=color.yellow, radius=20)
beam=box(pos=vector(0,Hsupport+thickness/2,0),color=color.orange,\
   length=Lbeam,width=Wbeam,height=thickness)
cube=box(pos=vector(-d,Hsupport+Hbox/2+thickness,0),length=Lbox,
   width=Wbox,height=Hbox)
piso=curve(pos=[(-300,0,0),(300,0,0)],color=color.green, radius=1)
arrowcube=arrow(color=color.orange,axis=vector(0,-0.15*Wbox,0))
arrowbeam=arrow(color=color.orange,axis=vector(0,-0.15*W,0))
arrowbeam.pos=vector(0,Hsupport+thickness/2,0)

v=4.0                           # box speed
x=-d                            # box initial position
Mg=WeightBox+W                  # weight box+beam
Fl=(2*Wbox+W)/2.0
arrowFl = arrow(color=color.red,
   pos=vector(-d,Hsupport+thickness/2,0), axis=vector(0,0.15*Fl,0))
Fr = Mg-Fl                      # right force
arrowFr=arrow(color=color.red,pos=vector(d,Hsupport+thickness/2,0),
   axis=vector(0,0.15*Fr,0))
anglabel=label(pos=vector(-100,150,0), text='Fl=',box=0)
Ftext1=label(pos=vector(-50,153,0),box=0)
anglabel2=label(pos=vector(100,150,0), text='Fr=',box=0)
Ftext2=label(pos=vector(150,153,0),box=0)
rate(4)                         # to slow motion

for t in arange(0.0,65.0,0.5):
    rate(10)
    x = -d+v*t
    cube.pos=vector(x, Hsupport+Hbox/2+10,0) # position cube
    arrowcube.pos=vector(x,Hsupport+5,0)
    if Fl>0:
        Fl=(d*Mg-x*WeightBox)/(2.0*d)
        Fr=Mg-Fl
        cube.pos=vector(x, Hsupport+Hbox/2+10,0)
        arrowcube.pos=vector(x,Hsupport+thickness/2,0)
        arrowFl.axis=vector(0,0.15*Fl,0)
        arrowFr.axis=vector(0,0.15*Fr,0)
        Ftext1.text='%8.2f'%Fl                # Left force
        Ftext2.text='%8.2f'%Fr                # Right force
    elif Fl==0:
        x=300
        beam.rotate(angle=-0.2,axis=vector(0,0,1),\
            origin=vector(d,Hsupport+thickness/2,0))
        cube.pos=vector(300,Hsupport,0)
        arrowcube.pos=vector(300,0,0)
        break

rate(5)
arrowFl.axis=vector(0,0.15*0.5*W,0)   # return beam
arrowFr.axis=arrowFl.axis
beam.rotate(angle=0.2,axis=vector(0,0,1), ↵
    origin=vector(d,Hsupport+thickness/2,0))
Fl=100.0
Ftext1.text='%8.2f'%Fl
Ftext2.text='%8.2f'%Fl
```

Listing 1.10. SlidingBox.py produces a 3-D animation showing the forces on a sliding box.

```
# CentralValue.py: Gaussian distribution from sum of randoms

import random, matplotlib.mlab as mlab
import numpy as np, matplotlib.pyplot as plt

N = 1000; NR = 10000    # Sum N variables, distribution of sums
SumList = []  # empty list

def SumRandoms():  # Sum N randoms in [0,1]
    sum = 0.0
    for i in range(0,N):  sum = sum + random.random()
    return sum

def normal_dist_param():
    add = sum2 =0
    for i in range(0,NR):  add = add + SumList[i]
    mu = add/NR                      # Average distribution
    for i in range(0,NR): sum2 = sum2 + (SumList[i]-mu)**2
    sigma = np.sqrt(sum2/NR)
    return mu,sigma

for i in range(0,NR):
    dist =SumRandoms()
    SumList.append(dist)                  # Fill list with NR sums
plt.hist(SumList, bins=50, color='white', normed=True) # True: ↵
    normalize
mu, sigma = normal_dist_param()
x = np.arange(450,550)
rho = np.exp(-(x-mu)**2/(2*sigma**2))/(np.sqrt(2*np.pi*sigma**2))
plt.plot( x,rho, 'g-', linewidth=3.0)           # Normal distrib
plt.xlabel('Random Number x 1000')
plt.ylabel('Average of Random Numbers')
plt.title('Generated vs Analytic Normal Distribution')
plt.show()
```

Listing 1.11. CentralValue.py takes the average of a large number of uniformly distributed random numbers and compares it to a normal distribution.

```
# rk4Call.py: 4th-O Runge-Kutta that calls rk4Algor
#             Here for ODE y" = -100y-2y'+ 100 sin(3t)

from rk4Algor import rk4Algor
from numpy import *
import numpy as np, matplotlib.pyplot as plt

Tstart = 0.;   Tend = 10.;   Nsteps = 100        # Initialization
tt=[];   yy=[];   yv=[];   y = zeros((2), float)
y[0] = 3.;    y[1] = -5.              # Initial position & velocity
t = Tstart;          h = (Tend-Tstart)/Nsteps;

def f(t, y):                              # Force (RHS) function
    fvector = zeros((2), float)
    fvector[0] = y[1]
    fvector[1] = -100.*y[0]-2.*y[1] + 10.*sin(3.*t)
    return fvector

while (t < Tend):
```

```
        tt.append(t)                           # Time loop
        if ((t + h) > Tend):   h = Tend - t    # Last step
        y = rk4Algor(t, h, 2, y, f)
        yy.append(y[0])
        yv.append(y[1])
        t = t + h
fig=plt.figure()
plt.subplot(111)
plt.plot(tt,yy,'r')
plt.title('Position versus ')
plt.xlabel('t')
plt.ylabel('y')
fig1=plt.figure()
plt.subplot(111)
plt.plot(tt,yv)
plt.title('Velocity versus time')
plt.xlabel('t')
plt.ylabel('y')
plt.show
```

Listing 1.12. rk4Call.py solves an ordinary differential equation using the fourth-order Runge-Kutta algorithm in Python and plots the results with VPython. The details of the rk4 algorithm, not to be modified, are in **rk4Algor.py**

```
# rk4Duffing.py solve ODE for Duffing Osc via rk4 & Matplotlib

import numpy as np, matplotlib.pylab as plt
from rk4Algor import rk4Algor

tt =[];   yy = []; vy = []
y = np.zeros((2),float)
a = 0.5;   b = -0.5;   g = 0.02;
A, w, h = 0.0008, 1., 0.01
y[0] = 0.09; y[1] =  0.              # Initial x, velocity

def f(t,y):
    rhs = np.zeros((2))
    rhs[0] = y[1]
    rhs[1] = -2*g*y[1] - a*y[0] - b*y[0]**3 + A*np.cos(w*t)
    return rhs
f(0,y)

for t in np.arange(0,40,h):                     # Time Loop
    y = rk4Algor(t,h,2,y,f)                      # Call rk4
    tt.append( t)
    yy.append(y[0])                                         # x(t)
    vy.append(y[1])                                         # v(t)

fig, axes = plt.subplots(nrows=1, ncols=2,figsize=(12,5) )
axes[0].plot(tt[1000:],yy[1000:])     # 1000 avoids transients
axes[0].grid()                                         # x(t)
axes[0].set_title('Duffing Oscillator x(t)')
axes[0].set_xlabel('t')
axes[0].set_ylabel('x(t)')
axes[1].plot(yy[1000:],vy[1000:])
axes[1].grid()
axes[1].set_title('Phase Space Orbits, Duffing Oscillator')
```

```
axes [1]. set__xlabel('x(t)')
axes [1]. set__ylabel('v(t)')
plt.show()
```

Listing 1.13. **rk4Duffing.py** uses rk4 to solve the ODE for a Duffing oscillator $\ddot{x} = -2g\dot{x} - ax - bx^3 + F\cos wt$, and plots results with Matplotlib. The rk4 algorithm in **rk4Algor.py** is called but not modified.

```
# rk4Algor.py: algorithm for Delta y, input y, f; do NOT modify

import numpy as np

def rk4Algor(t, h, N, y, f):
    k1=np.zeros(N); k2=np.zeros(N); k3=np.zeros(N); k4=np.zeros(N)
    k1 = h*f(t,y)
    k2 = h*f(t+h/2.,y+k1/2.)
    k3 = h*f(t+h/2.,y+k2/2.)
    k4 = h*f(t+h,y+k3)
    y=y+(k1+2*(k2+k3)+k4)/6.
    return y
```

Listing 1.14. **rk4Algor.py** is the detailed part of the rk4 algorithm that should not be modified by the user. It is called by a user's program using rk4 to solve an ODE.

```
# TrapMethods.py: trapezoid integrnt, a<x<b, N pts, N-1 intervals

from numpy import *

def func(x):
    return 5*(sin(8*x))**2*exp(-x*x)-13*cos(3*x)

def trapezoid(A,B,N):
    h = (B - A)/(N - 1)                    # step size
    sum = (func(A)+func(B))/2              # (1st + last)/2
    for i in range(1, N-1):
        sum += func(A+i*h)
    return h*sum
A = 0.5
B = 2.3
N = 1200
print(trapezoid(A,B,N-1))
```

Listing 1.15. **TrapMethods.py** integrates a function f(y) with the trapezoid rule. Note that the step size h depends upon the size of interval.

```
# IntegGaussCall.py: N point Gaussian quadrature \int [a,b] f(x)dx

from numpy import *;   from GaussPoints import GaussPoints

Npts = 10; Ans = 0;   a = 0.;   b = 1.;   eps = 3.E−14
w = zeros(2001, float);   x = zeros(2001, float)          # Arrays

def f(x):   return exp(x)                                 # Integrand

GaussPoints(Npts, a, b, x, w, eps)        # eps: precison of pts
for i in range(0,Npts): Ans += f(x[i])*w[i]   # Sum integrands
print ('\n Npts =', Npts, ',   Ans =', Ans)
print (' eps =',eps, ',  Error =', Ans−(exp(1)−1) )
```

Listing 1.16. **IntegGaussCall.py**, Gaussian quadrature using points and weights generated in GaussPoints.py with precision eps.

```
# GaussPoints.py: N point Gaussian quadrature pts & Wts generation

import math
from numpy import *

def GaussPoints(Npts, a, b, x, w, eps):
    m = 0; i = 0; j = 0; t = 0.; t1 = 0.; pp = 0.
    p1 = 0.; p2 = 0.; p3 = 0.
    m = int((Npts+1)/2)
    for i in range(1, m+1):
        t = math.cos(math.pi*(float(i)−0.25)/(float(Npts)+0.5))
        t1 = 1
        while((abs(t−t1)) >=  eps):
            p1 = 1. ;   p2 = 0.
            for j in range(1, Npts + 1):
                p3 = p2;    p2 = p1
                p1 = ((2.*float(j)−1)*t*p2 − ↩
                      (float(j)−1.)*p3)/(float(j))
            pp = Npts*(t*p1 − p2)/(t*t − 1.)
            t1 = t
            t = t1 − p1/pp
        x[i−1] = −t
        x[Npts−i] = t
        w[i−1] = 2./( (1.−t*t)*pp*pp)
        w[Npts−i] = w[i−1]

    for j in range(0, Npts):                    # Scale [−1,+1] to [a,b]
        x[j] = x[j]*(b−a)/2. + (b+a)/2.
        w[j] = w[j]*(b−a)/2.
```

Listing 1.17. **GaussPoints.py** generates points and weights for Gaussian quadrature. Called by IntegGaussCall.py.

2

Data Analytics for Physics

2.1 Chapter Overview

We start this chapter with two methods for finding roots via trial-and-error searching. This is a widely-used computational tool which we apply in §2.3 to least-squares fitting. We then discuss and give problems dealing with Fourier analysis and its implementation as the discrete Fourier transform (DFT) algorithm. In §2.5 we discuss the industrial-strength version of DFT known as the fast Fourier transform (FFT), which we consider optional due to its intricacies. We then show two methods that use Fourier analysis to reduce noise in signals. In §2.7.1 we go on to extend the Fourier analysis to nonstationary signals, first by use of short-term Fourier transforms, and then by wavelet analysis (§2.7.2). This leads naturally to the industrial-strength multi-resolution wavelet analysis in §2.7.3, which we also consider optional due to its intricacies. In §2.8 we introduce principal components analysis (PCA), a powerful tool for analyzing complex and large data sets, and for extracting space-time correlations. We conclude the chapter with the determination of the fractal dimension of an object.

2.2 Root Finding

Many problems require finding the root or zero of a function:

$$\mathbf{f}(\mathbf{x}) \simeq 0, \tag{2.1}$$

where $\mathbf{f}(\mathbf{x})$ may be vector of equations. Numerical root finders employ a trial-and-error technique that starts with a guess for x (the "trial"), substitutes the guess into $f(x)$, sees how far $f(x_{guess})$ is from zero (the "error"), and then makes a better guess for x based on the error found. The procedure continues until $f(x) \simeq 0$ to some desired level of precision, or until the changes in x are insignificant, or until the search seems endless.

The most elementary trial-and-error technique is the *bisection algorithm*. It is reliable, slow, and bound to work if you know that $f(x)$ changes sign within some

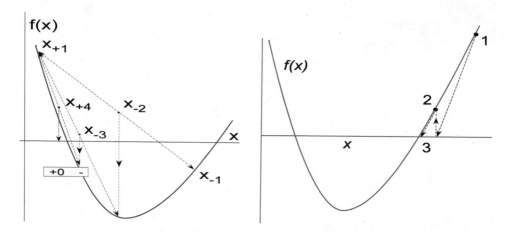

Figure 2.1. *Left:* Some of the steps involved in solving for a zero of $f(x)$ using the bisection algorithm. The algorithm progressively takes the midpoint of the interval as the new guess for x, which progressively reduces the interval size by one-half. *Right:* Some of the steps involved in solving for a zero of $f(x)$ using the Newton-Raphson method. The method takes the new guess as the zero of the straight line tangent to $f(x)$ at the x value of the old guess.

interval. As shown on the left of Figure 2.1, we start with two values x_- and x_+ between which a zero occurs:

$$f(x_-) < 0, \qquad f(x_+) > 0, \tag{2.2}$$

where x_+ may be less than x_- in value. As the next guess, pick a value of x halfway between x_+ and x_-, and then change x_+ or x_- to this new x based upon the value for $f(x)$:

```
# Pseudocode for Bisection.py:  Find x for f(x)=0 via Bisection algorithm
Define function f(x) = 2 cos(x) - x
Define function bisection(x-, x+, N, eps)
    Repeat N times
        x = (x+ + x-)/2
        if f(x+)f(x)  > 0 then x+=x
        else x- = x
        if abs(f(x)) < eps then quit
    return x
Call bisection(x_min, x_max, N, eps)
```

Our code `Bisection.py` is given in Listing 2.6, where you will note that, without being modified, the method `Bisection` can be called from your program.

The Newton-Raphson algorithm (Figure 2.1 right) starts with a guess x_0 for the root, assumes that we are in a regime where $f(x)$ is linear, and then computes a correction Δx based on where the zero of the linear function occurs:

$$f(x = x_0 + \Delta x) \simeq f(x_0) + \frac{df}{dx}\bigg|_{x_0} \Delta x \simeq 0, \tag{2.3}$$

$$\Rightarrow \quad \Delta x \simeq -\frac{f(x_0)}{df/dx|_{x_0}}. \tag{2.4}$$

The procedure is then repeated. If a guess is in a region where $f(x)$ is nearly linear, then the convergence is more rapid than the bisection algorithm; however if the guess is in a region where $f(x)$ has local minima or maxima $(df/dx \simeq 0)$, the method may fail.

If you have an analytic expression for the derivative, you can use that in the algorithm. Otherwise, or for complicated functions, it may be easier and less error-prone to use a numerical forward-difference or central-difference approximation to the derivative:

$$\frac{df}{dx}\bigg|_{FD} \simeq \frac{f(x + \delta x) - f(x)}{\delta x}, \qquad \frac{df}{dx}\bigg|_{CD} \simeq \frac{f(x + \frac{\delta x}{2}) - f(x - \frac{\delta x}{2})}{x}, \tag{2.5}$$

where $\delta x \neq \Delta x$ is some small change in x. (Once you have found a root, the approximations made to get there are irrelevant.)

In Listing 2.7 we give a program `NewtonCall.py` that implements a Newton-Raphson search with the central-difference derivative. Note that you should supply your own integrand function `f(x)`, but leave the function `NewtonR` as is.

1. A standard problem in elementary quantum mechanics is to find the energies of the bound states within a square well potential. The energies $E = -E_B < 0$ of the bound states are solutions of the transcendental equations [Gott]

$$\sqrt{10 - E_B} \tan\left(\sqrt{10 - E_B}\right) = \sqrt{E_B} \quad \text{(even)}, \tag{2.6}$$

$$\sqrt{10 - E_B} \, \cotan\left(\sqrt{10 - E_B}\right) = \sqrt{E_B} \quad \text{(odd)}, \tag{2.7}$$

where even and odd refer to the symmetry of the wave function.

a. A good first step in a search is to get some idea of what your function looks like. For this purpose plot the LHS-RHS of (2.6) or (2.7) versus E_B.
b. Use the bisection algorithm to find several solutions of (2.6) and (2.7).
c. Use the Newton-Raphson algorithm to find some solutions of (2.6) and (2.7) to the same level of precision as demanded of the bisection algorithm, and compare the speed of the two methods.

2. *Warning:* Because the tan function has singularities, numerical procedures may become inaccurate near singularities. One cure is to use a different, though equivalent, form of the equation. Show that an equivalent form of (2.6) is

$$\sqrt{E} \cot(\sqrt{10 - E}) - \sqrt{10 - E} = 0. \tag{2.8}$$

    a. Make a plot of (2.8) and note how the singularities are at different places.
    b. Compare the roots you find with those given by Maple or Mathematica.
    c. The "10" in (2.6) is proportional to the potential's depth. Verify that making the potential deeper, say, by changing the 10 to a 20 or a 30, produces a larger number of, and deeper, bound states.

3. Nonlinear equations are often challenging to solve analytically. Find the solutions of the nonlinear, simultaneous equations:

$$x^2 - 4y^2 = 1, \qquad x - 8y^3 = 3.$$

4. Find the x values at which the function $10\,x^4 \sin(k\,x)$ and its first two derivatives vanish.

2.3 Least-Squares Fitting

Least-square fitting is the preferred approach for fitting a formula ("theory") to data that contain statistical uncertainties. We start with N_D data values of some independent variable y as a function of dependent variable x:

$$(x_i, y_i \pm \sigma_i), \quad i = 1, N_D. \tag{2.9}$$

The function also contains M_P parameters $\{a_1, a_2, \ldots, a_{M_P}\}$:

$$g(x) = g(x; \{a_1, a_2, \ldots, a_{M_P}\}) = g(x; \{a_m\}). \tag{2.10}$$

Our aim is to fit these data to the function $g(x)$; specifically, to determine the values of $\{a_m\}$ that produce a best fit to the data.

The fitting is based on finding the minimum value of the chi-square measure of goodness of fit [Bevington & Robinson(02)]:

$$\chi^2 \stackrel{\text{def}}{=} \sum_{i=1}^{N_D} \left(\frac{y_i - g(x_i; \{a_m\})}{\sigma_i} \right)^2 . \tag{2.11}$$

The sum here is over the N_D data points, where the $1/\sigma_i^2$ weighting ensures that measurements with greater uncertainties contribute proportionally less to χ^2. For M_P parameters, minimization of χ^2 leads to M_P equations to solve:

$$\frac{\partial \chi^2}{\partial a_m} = 0, \qquad \Rightarrow \qquad \sum_{i=1}^{N_D} \frac{[y_i - g(x_i)]}{\sigma_i^2} \frac{\partial g(x_i)}{\partial a_m} = 0, \qquad m = 1, M_P. \tag{2.12}$$

A necessary but not necessarily sufficient condition for a solution to exist is that the number of data points N_D is equal to or greater than the number of parameters.

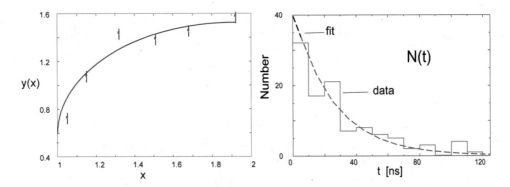

Figure 2.2. *Left:* A linear least-squares best fit of a parabola to data. Here we see that the fit misses approximately one-third of the points, as expected from the statistics for a good fit. *Right:* The number of decays per unit time of π mesons as a function of time since their creation. The curve is a least-squares fit to the $\log N(t)$.

2.3.1 Linear Least-Square Fitting

The M_P simultaneous equations (2.12) simplify considerably if the functions $g(x; \{a_m\})$ depend *linearly* on the parameters a_i, e.g.,

$$g(x; \{a_1, a_2\}) = a_1 + a_2 x. \tag{2.13}$$

Notice that even though there are only two parameters to determine, there is still an arbitrary number N_D of data points to fit. For a linear dependence on the a_i's, the solution can be written as [Press et al.(94)]:

$$\frac{\partial g(x_i)}{\partial a_1} = 1, \quad \frac{\partial g(x_i)}{\partial a_2} = x_i, \quad \Rightarrow \quad a_1 = \frac{S_{xx}S_y - S_x S_{xy}}{\Delta}, \quad a_2 = \frac{SS_{xy} - S_x S_y}{\Delta},$$

$$S = \sum_{i=1}^{N_D} \frac{1}{\sigma_i^2}, \quad S_x = \sum_{i=1}^{N_D} \frac{x_i}{\sigma_i^2}, \quad S_y = \sum_{i=1}^{N_D} \frac{y_i}{\sigma_i^2}, \tag{2.14}$$

$$S_{xx} = \sum_{i=1}^{N_D} \frac{x_i^2}{\sigma_i^2}, \quad S_{xy} = \sum_{i=1}^{N_D} \frac{x_i y_i}{\sigma_i^2}, \quad \Delta = SS_{xx} - S_x^2. \tag{2.15}$$

The function being fitted does not have to be linear in the independent variable, rather only in the parameters. As a case in point, imagine fitting a parabola

$$g(x) = a_1 + a_2 x + a_3 x^2 \tag{2.16}$$

to data (Figure 2.2 left). In this case (2.12) leads to three simultaneous, linear equations for a_1, a_2, and a_3:

$$\sum_{i=1}^{N_D} \frac{[y_i - g(x_i)]}{\sigma_i^2} = 0, \quad \sum_{i=1}^{N_D} \frac{[y_i - g(x_i)]}{\sigma_i^2} x = 0, \quad \sum_{i=1}^{N_D} \frac{[y_i - g(x_i)]}{\sigma_i^2} x^2 = 0. \tag{2.17}$$

Table 2.1. Temperature versus distance along a metal rod.

x_i (cm)	2.0	4.0	6.0	84.0	10.0	12.0	14.0	16.0	18.0
T_i (C)	7.3	9.3	18.3	15.4	30	30	31.1	39.7	49.8

These equations can be rewritten as simple extensions of those used in (2.14)–(2.15):

$$A\mathbf{x} = \mathbf{b}, \tag{2.18}$$

$$A = \begin{bmatrix} S & S_x & S_{xx} \\ S_x & S_{xx} & S_{xxx} \\ S_{xx} & S_{xxx} & S_{xxxx} \end{bmatrix}, \quad \mathbf{x} = \begin{bmatrix} a_1 \\ a_2 \\ a_3 \end{bmatrix}, \quad \mathbf{b} = \begin{bmatrix} S_y \\ S_{xy} \\ S_{xxy} \end{bmatrix}, \tag{2.19}$$

where the matrix form is convenient for computation. In Listing 2.8 we present the code `Fit.py` that performs such a fit using a linear algebra subroutine library to solve for the parameter vector $\mathbf{a}$. The results are shown in Figure 2.2. A pseudocode is

```
# Fit.py: Linear least-squares fit via matrix solution
Import packages
Define constants
Define arrays with data values
for i in range(0, Nd)
    calculate s's
Set up A matrix, b vector
Solve for x = A^{-1} b
Plot data and fit
```

1. Table 2.1 gives the temperature T along a metal rod as a function of the distance x along the rod.

   a. Plot the data in Table 2.2 and, accordingly, confirm the linear relation

   $$T(x) \simeq a + bx. \tag{2.20}$$

   b. Perform a least-squares straight-line fit to these data.
   c. Plot your fit $a + bx$ on the same graph as the data.
   d. Verify that your fit produce a minimum in χ^2.

2. Show that (2.17) can be written as (2.19).

3. Unstable particles often decay spontaneously and stochastically. In the limit of large number of particles, the decay rate can be approximated with an exponential function:

   $$\frac{dN(t)}{dt} \simeq \frac{dN(0)}{dt} e^{-t/\tau}. \tag{2.21}$$

   Fit the exponential decay law (2.21) to the data of [Stetz et al.(73)] in Figure 2.2 and thus deduce a values for the pion lifetime τ.

Table 2.2. Distance r in megaparsecs versus radial velocity for 24 extragalactic nebulae.

Object	r	v (km/s)	Object	r	v (km/s)	Object	r	v (km/s)
	0.032	170	3627	0.9	650		0.034	290
4826	0.9	150	6822	0.214	-130	4649	2.0	1090
5236	0.9	500	598	0.263	-70	1068	1.0	920
221	0.275	-185	5055	1.1	450	224	0.275	-220
7331	1.1	500	5457	0.45	200	4258	1.4	500
4736	0.5	290	4141	1.7	960	5194	0.5	270
4382	2.0	500	4449	0.63	200	4472	2.0	850
4214	0.8	300	4486	2.0	800	3031	0.9	-30

a. Read from the figure approximate values for $[\Delta N_i/\Delta t, t_i]$, picking the middle of each bin as the time value.

b. Deduce an approximate value for the error σ_i for each point equal to how much the histogram values appear to fluctuate about a smooth curve.

c. Although the exponential is a nonlinear function of the parameter τ, the logarithm $\ln[-dN/dt(t)] \simeq \ln[-dN(0)/dt] + t/\tau$ is expected to be linear in both t and $1/\tau$.

d. Compare your deduction to the tabulated pion lifetime of $\tau = 2.6 \times 10^{-8}$ s.

e. Verify that your fit is a minimum in χ^2.

f. Repeat the exercise without taking the logarithms, and compare to the deduced lifetime obtained previously. (*Hint: See Problem 7 for a sample approach to use.*)

4. In 1929 Edwin Hubble examined the data in Table 2.2 relating the radial velocity v of 24 extra galactic nebulae to their distance r from our galaxy [Hubble(29)]. Although there was considerable scatter in the data, he fit them with a straight line:

$$v = Hr, \tag{2.22}$$

where H is now called the Hubble constant.

a. Plot the data to verify the appropriateness of a linear relation

$$v(r) \simeq a + Hr. \tag{2.23}$$

b. Deduce a value for the error σ_i for each point as how much the histogram values appear to fluctuate about a smooth curve.

c. Compute a least-squares straight-line fit to these data.

d. Plot your fit on the curve with the data.

e. After fitting the data, compute the variance and verify that approximately one-third of the points miss the σ error band (that's what is expected for a random distribution of errors).

f. Determine the χ^2 of the fit and comment on its value.

Table 2.3. Measured total cross sections in millibarns of neutron-carbon scattering as a function of neutron energy in KeVs.

i	1	2	3	4	5	6	7	8	9
E_i (keV)	0	13	25	38	50	63	75	87	100
σ_T (mb)	10.6	16.0	45.0	83.5	52.8	19.9	10.8	8.25	4.7
Error (mb)	9.34	17.9	41.5	85.5	51.5	21.5	10.8	6.29	4.14

5. **Linear Quadratic Fit Assessment** Fit the quadratic (2.16) to the four following data sets. In each case indicate the values found for the a's, the number of degrees of freedom, *and* the value of χ^2.

   a. $(0, 2)$
   b. $(0, 2), (2, 6)$
   c. $(0, 2), (2, 6), (4, 14)$
   d. $(0, 2), (2, 6), (4, 14), (6, 30)$

6. Find a fit to the last set of data to the function

$$y = Ae^{-bx^2}. \qquad (2.24)$$

   Hint: A judicious change of variables will permit you to convert this to a linear fit. Does a minimum χ^2 still have the usual meaning here?

7. **Nonlinear Fitting to a Resonance** Table 2.3 gives total cross section data σ_T for low energy neutron scattering from carbon nuclei as a function of energy E. Determine if these data may be fit by the Breit-Wigner resonance formula:

$$\sigma_T(E) = \frac{\sigma_0}{(E - E_r)^2 + \Gamma^2/4}, \qquad (2.25)$$

   where E_r is the resonance energy and Γ is the full width at half maxima.

   a. Plot these data so you have some idea as to the function you wish to fit.
   b. Show that for this problem the parameters and function are:

$$g(x) = \frac{a_1}{(x - a_2)^2 + a_3}, \quad a_1 = \sigma_0, \quad a_2 = E_R, \quad a_3 = \Gamma^2/4, \quad x = E. \ (2.26)$$

   c. Verify that the three derivatives required by the best fit equations (2.12) are:

$$\frac{\partial g}{\partial a_1} = \frac{1}{(x - a_2)^2 + a_3}, \quad \frac{\partial g}{\partial a_2} = \frac{-2a_1(x - a_2)}{[(x - a_2)^2 + a_3]^2}, \quad \frac{\partial g}{\partial a_3} = \frac{-a_1}{[(x - a_2)^2 + a_3]^2}.$$

   d. Show that minimum χ^2 leads to three equations nonlinear in the a_i's:

$$\sum_{i=1}^{9} \frac{[y_i - g(x_i)]}{\sigma_i^2} \frac{\partial g(x_i)}{\partial a_m} = 0, \qquad (m = 1, 3). \qquad (2.27)$$

e. Determine the best fit values for the parameters E_r, σ_0, and Γ to the data in Table 2.3. Write a program that finds the three roots of (2.27). We recommend a Newton-Raphson search procedure with the equations written in matrix form.

f. Compare the deduced values of the parameters to estimates based on a casual inspection of a plot of these data.

2.4 Discrete Fourier Transforms (DFT)

A function is said to be periodic in the variable t, which does not have to be time, if it keeps repeating after each interval T:

$$y(t + T) = y(t). \tag{2.28}$$

Here T is called the *fundamental period* if it is the smallest number for which the function repeats; smallest, because there's also repetition after $2T$, $3T$, ... Equivalently, we can say that the function has a *fundamental frequency* ω, where

$$\omega \equiv \omega_1 = \frac{2\pi}{T}. \tag{2.29}$$

Given a set $\{y_i\}$ of N signal measurements at times t_i, $i = 1, \ldots N$, *Fourier's theorem* states that any single valued, periodic functions with at most a finite number of discontinuities can be approximated as the infinite series:

$$y(t) \simeq \frac{a_0}{2} + \sum_{n=1}^{\infty} \left(a_n \cos n\omega t + b_n \sin n\omega t \right). \tag{2.30}$$

This series (2.30) is a "best or least-square" fit (§2.3) in the sense that it minimizes the squared deviation between theory and measurement, $\sum_i [y(t_i) - y_i]^2$. A Fourier series will miss the function at discontinuities, where it converges to the mean, or at sharp corners, where it has a "Gibbs overshoot".

The coefficients a_n and b_n in (2.30) measure, respectively, the amount of $\cos n\omega t$ and $\sin n\omega t$ present in $y(t)$:

$$\begin{pmatrix} a_n \\ b_n \end{pmatrix} = \frac{2}{T} \int_0^T dt \begin{pmatrix} \cos n\omega t \\ \sin n\omega t \end{pmatrix} y(t). \tag{2.31}$$

If a function is not periodic, then a series is not as accurate a representation as is a *Fourier integral*:

$$y(t) = \int_{-\infty}^{+\infty} d\omega \, Y(\omega) \frac{e^{i\omega t}}{\sqrt{2\pi}}. \tag{2.32}$$

Here the Fourier transform $Y(\omega)$ is analogous to the Fourier coefficients (a_n, b_n):

$$Y(\omega) = \int_{-\infty}^{+\infty} dt \, \frac{e^{-i\omega t}}{\sqrt{2\pi}} y(t), \tag{2.33}$$

where the $1/\sqrt{2\pi}$ factor in (2.32) and (2.33) is a common convention in physics. Likewise, our sign conventions in the exponents are sometimes reversed, which is okay as long as one remains consistent.

In the *discrete Fourier transform (DFT)* algorithm we assume that the signal $y(t)$ is measured or computed at only a finite number N of *uniform* time intervals $\Delta t = h$, and for a total time $T = Nh$:

$$y_k \stackrel{\text{def}}{=} y(t_k), \qquad t_k \stackrel{\text{def}}{=} kh, \qquad k = 0, 1, 2, \ldots, N. \tag{2.34}$$

By its very nature, the DFT is an approximation because the signal is not known for all times and because we will evaluate the integrals approximately. However, the DFT can be used to reconstruct $y(t)$ for any time. The algorithm evaluates the integral in (2.33) using the equally-spaced measurements $\{y_i\}$ and the trapezoid integration rule:

$$Y(\omega_n) \stackrel{\text{def}}{=} \int_{-\infty}^{+\infty} dt \, \frac{e^{-i\omega_n t}}{\sqrt{2\pi}} y(t) \simeq \int_0^T dt \, \frac{e^{-i\omega_n t}}{\sqrt{2\pi}} y(t), \tag{2.35}$$

$$\simeq \sum_{k=1}^N h \, y(t_k) \frac{e^{-i\omega_n t_k}}{\sqrt{2\pi}} = h \sum_{k=1}^N y_k \frac{e^{-2\pi i k n/N}}{\sqrt{2\pi}}. \tag{2.36}$$

To make the notation symmetric, the step size h is factored out from Y:

$$Y_n \stackrel{\text{def}}{=} \frac{1}{h} Y(\omega_n) = \sum_{k=1}^N y_k \frac{e^{-2\pi i k n/N}}{\sqrt{2\pi}}, \qquad n = 0, 1 \ldots, N, \tag{2.37}$$

$$\Rightarrow \quad y(t) \simeq \sum_{n=1}^N \frac{2\pi}{N} \frac{e^{i\omega_n t}}{\sqrt{2\pi}} Y_n, \qquad \omega_n = n\omega_1 = n\frac{2\pi}{T} = n\frac{2\pi}{Nh}. \tag{2.38}$$

Here the extra $n = 0$ value, $\omega_{n=0} = 0$, corresponds to the zero-frequency or DC component of the transform, that is, the part of the signal that does not oscillate. Regardless of the true periodicity of the signal, when we sample the signal over a finite period T, the mathematics produces a $y(t)$ that is periodic with period T,

$$y(t + T) = y(t). \tag{2.39}$$

We build this periodicity into the algorithm by having a fictitious, measurement Y_N at time Nh, that is equals to the first signal measurement:

$$y_N = y_0. \tag{2.40}$$

This does not change the fact that there are just N independent measurements spanning one period.

We see from (2.38) that the larger we make the time $T = Nh$ over which we sample the function, the smaller will be the frequency steps or resolution. Accordingly, if you want a smooth frequency spectrum, you need to have a small frequency step $2\pi/T$, which means a longer observation time T.

The DFT algorithm can be expressed succinctly by introducing a complex variable Z for the exponential, and then raising Z to various powers:

$$y_k = \frac{\sqrt{2\pi}}{N} \sum_{n=1}^{N} Z^{-nk} Y_n, \qquad Y_n = \frac{1}{\sqrt{2\pi}} \sum_{k=1}^{N} Z^{nk} y_k, \qquad Z = e^{-2\pi i/N}. \qquad (2.41)$$

With this formulation, the computer needs to compute only powers of Z. Our DFT code `DFTcomplex.py` is given in Listing 2.1. Here is its pseudocode:

```
# DFTcomplex.py:  Discrete Fourier Transform
Import packages
Define constant
Declare arrays
Set up plots
  Function Signal(y)              # Generates a sample signal y[i]
Function DFT(Ycomplex}    #  Computes DFT, returns complex Y
    Y = Sum y_k  exp(- i 2 pi k n/N)
Call Signal
Call DFT
```

1. It is always a good idea to perform simple checks before applying your own or packaged Fourier tools. And so, sample the mixed-symmetry signal

$$y(t) = 5\sin(\omega t) + 2\cos(3\omega t) + \sin(5\omega t). \qquad (2.42)$$

   a. Decompose this into its components.
   b. Check that the components are essentially real and in the ratio 5:2:1.
   c. Verify that the frequencies have the expected values.
   d. Verify that the summed transform values reproduce the input signal.

2. Sum the Fourier series for the *sawtooth function* up to order $n = 2, 4, 10, 20$, and plot the results over two periods.

   a. Check that in each case the series gives the mean value of the function *at* the points of discontinuity.
   b. Check that in each case the series *overshoots* by about 9% the value of the function on either side of the discontinuity (the *Gibbs overshoot*).
   c. Experiment on the effects of picking different values of the step size h and of enlarging the measurement period $T = Nh$.

3. An electron initially localized at $x = 5$ with momentum k_0 is described by the wave packet ($\hbar = 1$):

$$\psi(x, t = 0) = \exp\left[-\frac{1}{2}\left(\frac{x - 5.0}{\sigma_0}\right)^2\right] e^{ik_0 x}. \qquad (2.43)$$

Determine and plot the momenta components in this wave packet by evaluating the Fourier transform

$$\psi(p) = \int_{-\infty}^{+\infty} dx \frac{e^{ipx}}{\sqrt{2\pi}} \psi(x). \qquad (2.44)$$

4. For your signal of choice, take the output from `DFTcomplex.py` and inverse-transform it back to signal space. Compare it to your input.

5. The sampling of a signal by DFT for only a finite number of times both limits the accuracy of the deduced high-frequency components, and contaminates the deduced low-frequency components (called *aliasing*). Consider the two functions $\sin(\pi t/2)$ and $\sin(2\pi t)$ for $0 \le t \le 8$.

   a. Make graphs of both functions on the same plot.
   b. Perform a DFT on both functions.
   c. Sample at times $t = 0, 2, 4, 6, 8, \ldots$ and draw conclusions.
   d. Sample at times $t = 0, 12/10, 4/3, \ldots$ and draw conclusions about the high-frequency components (*Hint:* they may be *aliased* by the low-frequency components).
   e. The *Nyquist criterion* states that when a signal containing frequency f is sampled at a rate of $s = N/T$ measurements per unit time, with $s \le f/2$, then aliasing occurs. Verify specifically that the frequencies f and $f - 2s$ yield the same DFT.

6. Perform a Fourier analysis of the chirp signal $y(t) = \sin(60t^2)$. As seen in Figure 2.11, this signal is not truly periodic, and is better analyzed with methods soon to be discussed.

7. Consider the following wave packets:

$$y_1(t) = e^{-t^2/2}, \quad y_2(t) = \sin(8t)e^{-t^2/2}, \quad y_3(t) = (1 - t^2)e^{-t^2/2}. \qquad (2.45)$$

   For each wave packet:

   a. Estimate $\Delta t =$ the *full width at half-maxima* (FWHM) of $|y(t)|$.
   b. Compute and plot the Fourier transform $Y(\omega)$ for each wave packet, going out to large enough ω's to see periodicity. Make *both* linear and semilog plots (small components may be important however not evident in linear plots).
   c. What are the units for $Y(\omega)$ and ω in your DFT?
   d. For each wave packet, estimate $\Delta\omega =$ the *full width at half-maxima* of $|Y(\omega)|$.
   e. For each wave packet compute approximate value for the constant C in the uncertainty principle:

$$\Delta t \, \Delta\omega \ge 2\pi C. \qquad (2.46)$$

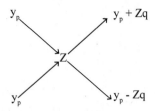

Figure 2.3. The basic butterfly operation in which input data y_p and y_q on the left are transformed into $y_p + Z\,y_q$ and $y_p - Z\,y_q$ on the right.

2.5 Fast Fourier Transforms (FFT)⊙

Equation (2.41) presents the discrete Fourier transform in the compact form

$$Y_n = \frac{1}{\sqrt{2\pi}} \sum_{k=1}^{N} Z^{nk}\, y_k, \quad Z = e^{-2\pi i/N}, \quad n = 0, 1, \ldots, N-1. \qquad (2.47)$$

As n and k range over their N integer values, the multiplications of complex numbers, $(Z^n)^k y_k$, lead to long computation times. Yet [Cooley & Tukey(65), Donnelly & Rust(05)] showed that these N^2 multiplications can be reduced to $N \log_2 N$ steps, which can lead to order-of-magnitude speedup in processing. Because of its widespread use (such as in cell phones), this fast Fourier transform (FFT) is considered one of the 10 most important algorithms of all time.

The FFT's economy arises from the computationally expensive complex factor $Z^{nk} = [(Z)^n]^k$ having values that are repeated as the integers n and k vary sequentially. Here we have you work out problems that lead you through the basis of the FFT. Although realistic application might have millions of observations, you only have to consider a signal that has been observed for $N = 8$ times: $y_0, y_1, \ldots, y_7$.

1. Write the eight equations for $Y_0, Y_1, \ldots, Y_7$ in terms of the y_i's and powers of the eight Z^i's in (2.47).

2. Show that these eight equations contain 64 elements, each of which contains several multiplications.

3. Show that the equations contain only four independent Z values, $Z^{0,\ldots,3}$, and, accordingly, only 64 complex number multiplications.

4. Show that the transform values can be expressed in terms of sums and differences

of the y's:

$$Y_0 = Z^0(y_0 + y_4) + Z^0(y_1 + y_5) + Z^0(y_2 + y_6) + Z^0(y_3 + y_7),$$
$$Y_1 = Z^0(y_0 - y_4) + Z^1(y_1 - y_5) + Z^2(y_2 - y_6) + Z^3(y_3 - y_7),$$
$$Y_2 = Z^0(y_0 + y_4) + Z^2(y_1 + y_5) - Z^0(y_2 + y_6) - Z^2(y_3 + y_7),$$
$$Y_3 = Z^0(y_0 - y_4) + Z^3(y_1 - y_5) - Z^2(y_2 - y_6) + Z^1(y_3 - y_7),$$
$$Y_4 = Z^0(y_0 + y_4) - Z^0(y_1 + y_5) + Z^0(y_2 + y_6) - Z^0(y_3 + y_7),$$
$$Y_5 = Z^0(y_0 - y_4) - Z^1(y_1 - y_5) + Z^2(y_2 - y_6) - Z^3(y_3 - y_7),$$
$$Y_6 = Z^0(y_0 + y_4) - Z^2(y_1 + y_5) - Z^0(y_2 + y_6) + Z^2(y_3 + y_7),$$
$$Y_7 = Z^0(y_0 - y_4) - Z^3(y_1 - y_5) - Z^2(y_2 - y_6) - Z^1(y_3 - y_7),$$
$$Y_8 = Y_0.$$

5. Display graphically, either on the computer or on a piece of paper, the symmetry in these equations, namely the relation among the repeating $y_p \pm y_q$ factors.

6. The symmetries of the FFT can be incorporated into the *butterfly operation* (Figure 2.3) that takes the y_p and y_q elements from the left wing and converts them into the $y_p + Zy_q$ elements on the right. Write down (again) your eight equations for Y_{0-7}, and show how butterfly operations connect the pairs (y_0, y_4), (y_1, y_5), (y_2, y_6), and (y_3, y_7).

7. Prove that the number of multiplications of complex numbers has been reduced, namely from the 64 required in the original DFT (2.47) to 24 made in four butterflies.

8. Look again at the equations for Y_{0-7}. Indicate how starting with 8 data elements in the order 0–7, the butterfly operators leads to transforms in the order 0, 4, 2, 6, 1, 5, 3, 7.

9. Show that the numbers 0, 4, 2, 6, 1, 5, 3, 7 correspond to the bit-reversed order of 0–7. This means further speedup in the FFT can be obtained by reshuffling the input data into bit-reversed order, in which case the output transforms are ordered correctly.

10. Show that in order for the FFT to work, *the number of input data elements must always be a power of two*, in our case, $8 = 2^3$.

11. Compare your analysis with the scheme in Figure 2.4.

Our FFT program `FFTmod.py` is given in Listing 2.3. Here's its pseudocode:

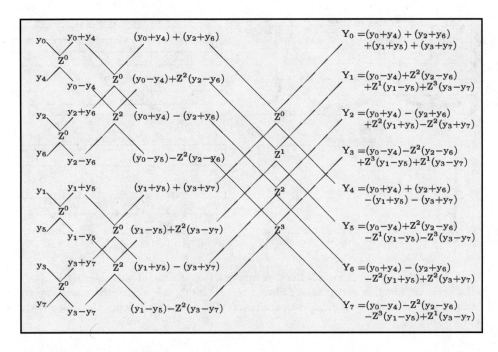

Figure 2.4. A modified FFT in which the eight input data on the left are transformed into eight transforms on the right. Note how the output transforms are now in numerical order.

```
# FFTmod.py:   FFT for complex numbers in dtr[][2], returned in dtr
Assign parameters: isign = -1, 1 (TF, inverse TF), N = 2^n
Declare arrays
Function FFT(N, isign)
    Separate even and odd input
    Place in bit-reversed order
    Print reordered data
    Evaluate transforms
        Loop through data
        Evaluate Z's
        Sum data y*Z's
    Replace input data with processed data
Call FFT
Output results
```

1. Compile and execute FFTmod.py. Make sure you understand the input and output.

2. Take the output from FFTmod.py, inverse-transform it back to signal space, and compare it to your input. [Checking that the double transform is proportional to itself is adequate, although the normalization factors in (2.41) should make the two equal.]

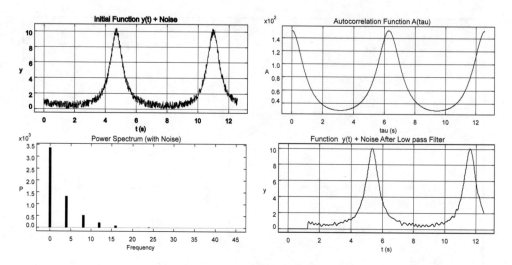

Figure 2.5. *From top left to right bottom:* A function that is a signal plus noise $s(t)+n(t)$; the autocorrelation function versus time deduced by processing this signal; the power spectrum obtained from autocorrelation function; the signal plus noise after passage through a lowpass filter.

3. Compare the transforms obtained with a FFT to those obtained with a DFT. Make sure to compare both precision and execution times.

2.6   Noise Reduction

2.6.1   Noise Reduction via Autocorrelation Function

All measured signals $y(t)$ contain noise $n(t)$. If the true signal is $s(t)$, and if the noise is random and just adds to it as

$$y(t) = s(t) + n(t), \tag{2.48}$$

then we can use Fourier transforms to "average out" some of the noise. To do that, we introduce the *autocorrelation function* $A(\tau)$, which folds or *convolutes* the measured signal onto itself, thereby measuring the correlation of a signal with itself:

$$A(\tau) \stackrel{\text{def}}{=} \int_{-\infty}^{+\infty} dt\, y^*(t)\, y(t+\tau) \equiv \int_{-\infty}^{+\infty} dt\, y(t)\, y^*(t-\tau). \tag{2.49}$$

Here τ is called the *lag time*, and $A(t)$ tends to look like $y(t)^2$.

To see how this folding removes noise from a signal, we start with the Fourier transform of $y(t) = s(t) + n(t)$:

$$Y(\omega) = S(\omega) + N(\omega), \tag{2.50}$$

$$S(\omega) = \int_{-\infty}^{+\infty} dt\, s(t)\frac{e^{-i\omega t}}{\sqrt{2\pi}}, \qquad N(\omega) = \int_{-\infty}^{+\infty} dt\, n(t)\frac{e^{-i\omega t}}{\sqrt{2\pi}}. \tag{2.51}$$

We evaluate the autocorrelation function (2.49) of the signal $y(t) = s(t) + n(t)$:

$$A_y(\tau) = \int_{-\infty}^{+\infty} dt\,[s(t)s(t+\tau) + s(t)n(t+\tau) + n(t)n(t+\tau)]. \tag{2.52}$$

If the noise $n(t)$ is random and uncorrelated at times t and $t+\tau$, then it should average out to zero, leaving an approximate autocorrelation function of the pure signal:

$$A_y(\tau) \simeq \int_{-\infty}^{+\infty} dt\, s(t)\, s(t+\tau) = A_s(\tau). \tag{2.53}$$

Yet the convolution theorem tells us that the Fourier transform of a convolution is proportional to the product of the two functions being convoluted, and so:

$$A_s(\tau) \int_{-\infty}^{+\infty} dt s(t+\tau)s(t) \quad \Rightarrow \quad A(\omega) \simeq \sqrt{2\pi}\,|S(\omega)|^2. \tag{2.54}$$

The function $|S(\omega)|^2$ is called the *power spectrum* of the true signal, and is proportional to the squared modulus of the Fourier transform. Often the power spectrum itself provides all that we need to know about the components in a signal.

For example, in Figure 2.5 we see a noisy signal (upper left), the autocorrelation function (upper right), which clearly is smoother than the signal, and, lastly, the deduced power spectrum (lower left). Notice that broadband high-frequency components, characteristic of noise, are absent from the power spectrum, and for this reason $|S(\omega)|^2$ provides a clean indication of the components in the true signal.

1. Modify the program `DFTcomplex.py` in Listing 2.1 so that it computes the autocorrelation function $A(\tau)$ and then extract the power spectrum from $A(\omega)$.

2. Consider the noiseless (true) signal

$$s(t) = \frac{1}{1 - 0.9\sin t}. \tag{2.55}$$

   a. Compute the DFT $S(\omega)$ of the true signal ensuring that you have good sensitivity to the high-frequency components.
   b. Create a semilog plot of the power spectrum $|S(\omega)|^2$ of the true signal.
   c. Compute the autocorrelation function $A(\tau)$ of the input signal $s(t)$.
   d. Compute the power spectrum by computing a DFT of the autocorrelation function.

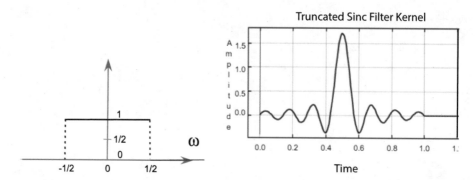

Figure 2.6. *Left:* The rectangle function rect(ω) that is constant for a finite frequency interval. *Right:* The sinc filter, the Fourier transform of the rectangular function.

    e. Compare the extracted power spectrum to that obtained by computing $|S(\omega)|^2$ directly.

3. Use a random number generator to add noise to the true signal:

$$y(t_i) = s(t_i) + \alpha(2r_i - 1), \qquad 0 \leq r_i \leq 1, \qquad (2.56)$$

where α is an adjustable parameter and the t_i's cover an entire period in detail.

    a. Try a range of α values, from small ones that just add fuzz to the signal, to large ones that nearly hide the signal.

    b. Plot the noisy signal, its Fourier transform, and its power spectrum.

    c. Compute the autocorrelation function $A(\tau)$ of the nosy signal and its Fourier transform $A(\omega)$.

    d. Compute the DFT of $A(\tau)$ and compare to the true power spectrum. Comment on the effectiveness of reducing noise by use of the autocorrelation function.

    e. For what value of α do you essentially lose all the information in the input?

2.6.2   Noise Reduction via Digital Filters

A filter converts an input signal $f(t)$ to an output signal $g(t)$ with a desired specific property. An *analog filter* does this via a convolution [Hartmann(98)]:

$$g(t) = \int_{-\infty}^{+\infty} d\tau\, f(\tau)\, h(t - \tau) \overset{\text{def}}{=} f(t) * h(t), \qquad (2.57)$$

where $h(t)$ is the filter's *transfer function* and τ is the *lag time*. The *convolution theorem* tells us that filtering is a multiplication of the signal's transform with the filter's transform:

$$G(\omega) = \sqrt{2\pi}\, F(\omega)\, H(\omega). \qquad (2.58)$$

Figure 2.7. *Left:* The original photograph. *Center:* With noise added. *Right:* The noisy photograph after filtering.

Filters that decrease high-frequency components are called *lowpass* filters, and those that filter out the low frequencies are called *highpass filters*.

An ideal low-frequency filter look like the rectangular pulse rect(ω) shown on the left of Figure 2.6. The Fourier transform of this rectangular pulse in frequency is the *sinc function* in the time domain [Smith(99)]:

$$\int_{-\infty}^{+\infty} d\omega\, e^{-i\omega t}\mathrm{rect}(\omega) \;=\; \mathrm{sinc}\left(\frac{t}{2}\right) \;\overset{\mathrm{def}}{=}\; \frac{\sin(\pi t/2)}{\pi t/2}. \tag{2.59}$$

A popular digital lowpass filter is the *windowed sinc filter* [Smith(99)]. By decreasing the high frequencies in a signal it tends to remove some of the noise from the signal. For example, the graph in the lower right corner of Figure 2.5 was obtained by passing the noisy signal in the upper left corner through a sinc filter using the program `SincFilter.py`. In practice, filtering with a pure sinc function may lead to excessive *Gibbs overshoot*, that is, rounded corners and oscillations beyond the corner. This is usually corrected by multiplying the sinc function by a smoothly tapered curve such as the *Hamming window function*:

$$w[i] = 0.54 - 0.46\ \cos(2\pi i/M). \tag{2.60}$$

The resulting truncated sinc filter is shown on the right of Figure 2.6. In terms of discrete times i, the filter's kernel is now

$$h[i] = \frac{\sin[2\pi\omega_c(i - M/2)]}{i - M/2}\ [0.54 - 0.46\ \cos(2\pi i/M)]. \tag{2.61}$$

The cutoff frequency ω_c should be a fraction of the sampling rate, with the time length M determining the *bandwidth* over which the filter changes from 1 to 0.

On the left of Figure 2.7 we show a photograph of Mariana. It is stored in the *plain portable gray map* (`pmg`) digital format in the file `MarianaNoise.pgm` in the `Codes` directory. The picture is constructed from 512 rows and 512 columns of integer numbers, each in the range [0,255]. A 0 represents black, 255 represents white, and intermediate numbers represent gray tones. In the center of Figure 2.7 we show the same photograph to which random noise, `20*(2*random.random()-1)`, has been added to each number in the file. On the right of the figure we show the output after filtering the data with a truncated sinc filter in the program `SincFilter.py`. The filtered image shows reduced noise, though also with reduced contrast.

1. Your problem is either to repeat the process we have just gone through with `Mariana.pgm`, or, better yet, repeat the process with one of your own digital photographs that you have converted to `plain pgm` format.

   a. Read the disk file `Mariana.dat` containing 512×512 integers in range $0 \leq N \leq 255$. These are the gray scale tones. Fill the array `fg[512,512]` with the integers.

   b. Write a disk file `Mariana.pgm`. The first entries in the file define the format:
   ```
   P2
   512 512
   255
   ```
   After this definition place the 512 lines of 512 elements from `Mariana.dat`.

   c. View the `.pgm` file with a photo viewing program such as `irfanview` or Gimp.

   d. Add noise (`20 (2*random.random()-1)`) to each integer in the file, and output the numbers to `MarianaNoise.pgm` maintaining 512 lines of 512 elements.

   e. To clean up the image, form an array of 100 entries to sample with parameters $w_c = 0.7$, for M=100:
   ```
   for i in range(0,100):      # Calculate low-pass filter kernel
       if ((i-(m//2)) == 0):       h[i] = 2*math.pi*fc
       if ((i-(m//2)) != 0):       h[i] = sin(2*math.pi*fc*(i-m/2))/(i-m/2)
       h[i] = h[i]*(0.54 - 0.46*cos(2*math.pi*i/m))    # Hamming window
   ```
   Normalize the values by summing all of the elements of `h[i]`, and then dividing each element by the sum.

   f. Compute the convolution integral of the filter and data. Output the result to the file `MarianaFiltered.pgn`, still maintaining the 512 rows of 512 elements.

   g. Compare the images from the input file, the noisy file, and the filtered file.

   h. See what happens if you repeat the filtering several times.

2.7   Spectral Analysis of Nonstationary Signals

Consider analyzing the signal in Figure 2.8 containing an increasing number of frequencies as time increases. Imagine that you actually measured this signal and that no one told you that it derives from the analytic form

$$y(t) = \begin{cases} \sin 2\pi t, & \text{for } 0 \leq t \leq 2, \\ 5\sin 2\pi t + 10\sin 4\pi t, & \text{for } 2 \leq t \leq 8, \\ 2.5\sin 2\pi t + 6\sin 4\pi t + 10\sin 6\pi t, & \text{for } 8 \leq t \leq 12. \end{cases} \quad (2.62)$$

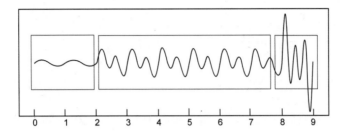

Figure 2.8. A signal containing additional frequencies as time increases. The boxes are possible placements of windows for short-time Fourier transforms.

2.7.1   Short-Time Fourier Transforms

A limitation of the Fourier series (2.30) for analyzing nonstationary signals such as Figure 2.8 is the absence of any time dependence of a_n and b_n, and the infinite extent in time of $\sin \omega_n t$ and $\cos \omega_n t$. This leads to considerable overlap and correlation among the components, which in turn can lead to excessively large data storage. An extension of Fourier analysis for nonstationary signals is the *short-time Fourier transform* that "chops " the signal $y(t)$ into segments for differing times, and then analyzes each segment separately. For instance, we show three such segments as the boxes of Figure 2.8, and these would lead to the Fourier transforms $\{Y_{\tau_1}^{(\mathrm{ST})}, Y_{\tau_2}^{(\mathrm{ST})}, Y_{\tau_3}^{(\mathrm{ST})}\}$, where the superscript $^{(\mathrm{ST})}$ indicates short time.

Rather than chopping up a signal by hand, we can do it mathematically by translating a *window $w(t)$* of finite length (a box in the figure) over the signal:

$$Y^{(\mathrm{ST})}(\omega, \tau) = \int_{-\infty}^{+\infty} dt\, \frac{e^{i\omega t}}{\sqrt{2\pi}}\, w(t - \tau)\, y(t). \qquad (2.63)$$

Here different values for the translation time τ correspond to different locations of the window w over the signal. The price paid for this extension of Fourier analysis is that the transform is now a function of two variables, ω and τ.

1. Use a short-time Fourier transform to analyze the signal (2.62).

   a. Modify your Fourier transform code to include an arbitrary window function.
   b. Choose a smooth window function such as a Gaussian. A rectangular window is simple, though the sharp corners may lead to spurious least-squares components.
   c. Does your transform exhibit three distinct frequencies, with a progressive increase in time?
   d. Invert the computed short-time transform and compare to input signal.

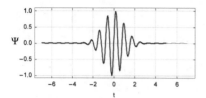 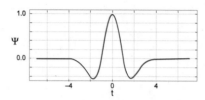

Figure 2.9. *Left:* The real part of the Morlet mother wavelet. *Right:* The Mexican hat mother wavelet.

2.7.2 Wavelet Analysis

Wavelet analysis extends the short-time Fourier transform idea by using basis functions that oscillate for only a short period of time (*localized* in both time and frequency). The complete set of bases functions are called *wavelets* or wave packets, with each individual wavelet centered at a different time. For example, Figure 2.9 shows two such wavelets:

$$\Psi_{Morlet}(t) = e^{2\pi it}e^{-t^2/2\sigma^2}, \qquad \Psi_{hat}(t) = (1 - \frac{t^2}{\sigma^2})e^{-t^2/2\sigma^2}. \qquad (2.64)$$

The wavelet transform is similar in notation to the short-time Fourier transform (2.63):

$$Y(s,\tau) = \int_{-\infty}^{+\infty} dt\, \psi_{s,\tau}^*(t)\, y(t), \qquad s = \frac{2\pi}{\omega}. \qquad (2.65)$$

Here the τ variable indicates the time portion of the signal being decomposed, while the s variable is proportional to the inverse of frequency present during that time (small s corresponds to high-frequencies). Because each wavelet is localized in time, each acts as its own window function. Furthermore, because each wavelet is oscillatory, each contains its own small range of frequencies.

The wavelet transform is not restricted to any particular basis. You start with a *mother* wavelet $\Psi(t)$, and then generate the *daughter* wavelets $\psi_{s,\tau}$ that form the basis from it. For example, as displayed in Figure 2.10, the mother wavelet

$$\Psi(t) = \sin(8t)e^{-t^2/2} \qquad (2.66)$$

is scaled and translated to form the daughters:

$$\psi_{s,\tau}(t) \stackrel{\text{def}}{=} \frac{1}{\sqrt{s}}\Psi\left(\frac{t-\tau}{s}\right) = \frac{1}{\sqrt{s}}\sin\left[\frac{8(t-\tau)}{s}\right]e^{-(t-\tau)^2/2s^2}, \qquad (2.67)$$

where $\sqrt{s}$ is a normalization factor. We see that larger or smaller values of s, respectively, expand or contract the mother wavelet, while different values of τ shift the center of the wavelet. Because the wavelets are oscillatory, the scaling leads to

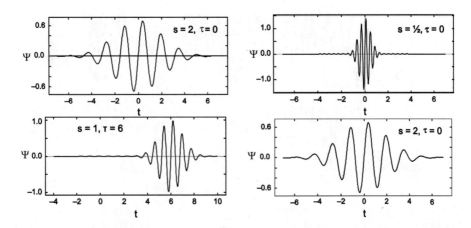

Figure 2.10. Four wavelet basis functions (daughters) generated by scaling (s value) and translating (τ value) an oscillating Gaussian mother wavelet. *Clockwise from top:* ($s = 1$, $\tau = 0$), ($s = 1/2$, $\tau = 0$), ($s = 1$, $\tau = 6$), and ($s = 2$, $\tau = 60$). Note how $s < 1$ is a wavelet with higher frequency, while $s > 1$ has a lower frequency than the $s = 1$ mother. Likewise, the $\tau = 6$ wavelet is a translated and compressed version of the $\tau = 0$ one directly above it.

the same number of oscillations occurring in different time spans, which is equivalent to having basis states with differing frequencies. Values of $s < 1$ produce higher-frequency wavelets, while $s > 1$ produces lower-frequency ones, both of the same shape. After substituting in the definition of daughters, the wavelet transform (2.65) and its inverse become

$$Y(s,\tau) = \frac{1}{\sqrt{s}} \int_{-\infty}^{+\infty} dt\, \Psi^* \left(\frac{t - \tau}{s} \right) y(t), \tag{2.68}$$

$$y(t) = \frac{1}{C} \int_{-\infty}^{+\infty} d\tau \int_{0}^{+\infty} ds \frac{\psi_{s,\tau}^*(t)}{s^{3/2}} Y(s,\tau), \tag{2.69}$$

where the normalization constant C depends on the wavelet used.

As an illustration of how the s and τ degrees of freedom in a wavelet transform are used, consider the analysis of a chirp signal $y(t) = \sin(60t^2)$ (Figure 2.11). We see that a segment at the beginning of the signal is compared to the first basis function using a narrow version of the wavelet, that is, a low scale one. The comparison at this scale continues with the next signal segment, and eventually ends when the entire signal has been covered (the top row in Figure 2.11). Then in the second row the wavelet is expanded to larger s values, and comparisons are repeated. Eventually, the data are processed at all scales and at all time intervals. As used in the wavelet-based PEG compression of digital images, transform values from the narrow wavelets provide high-resolution reconstruction, while the broad wavelets provide the overall shape. As the scales get larger, fewer details of the time signal remain visible, though the gross

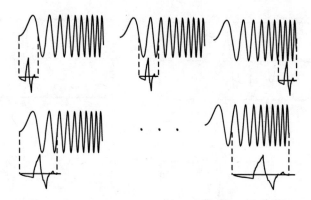

Figure 2.11. A schematic representation of the steps followed in performing a wavelet transformation over all time displacements and scales. The upper signal is first analyzed by evaluating its overlap with a narrow wavelet at the signal's beginning. The wavelet is successively shifted over the length of the signal and the overlaps are successively evaluated. After the entire signal is covered, the wavelet is expanded and the entire analysis is repeated at a larger scale.

features of the signal are preserved.

1. Modify the program used for the Fourier transform so that it now computes the wavelet transform.

2. Examine the effect of using different wavelets by running the computation with these other mother wavelets:

   a. a Morlet wavelet (2.64),
   b. a Mexican hat wavelet (2.64).

3. Transform the following input signals and comment on the results:

   a. A pure sine wave $y(t) = \sin 2\pi t$,
   b. A sum of sine waves $y(t) = 2.5 \sin 2\pi t + 6 \sin 4\pi t + 10 \sin 6\pi t$,
   c. The nonstationary signal of (2.62)

$$y(t) = \begin{cases} \sin 2\pi t, & \text{for } 0 \le t \le 2, \\ 5 \sin 2\pi t + 10 \sin 4\pi t, & \text{for } 2 \le t \le 8, \\ 2.5 \sin 2\pi t + 6 \sin 4\pi t + 10 \sin 6\pi t, & \text{for } 8 \le t \le 12. \end{cases} \qquad (2.70)$$

   d. The half-wave function

$$y(t) = \begin{cases} \sin \omega t, & \text{for } 0 < t < T/2, \\ 0, & \text{for } T/2 < t < T. \end{cases} \qquad (2.71)$$

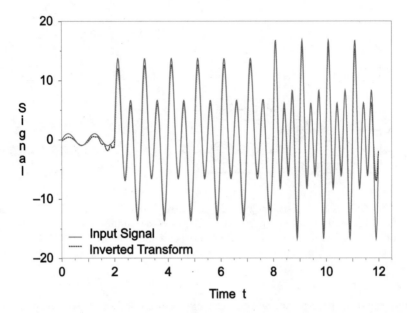

Figure 2.12. Comparison of an input and reconstituted signal (2.70) using Morlet wavelets (the curves overlap nearly perfectly), as computed with the program `CWT.py`. As occurs for Fourier transforms, the reconstruction is least accurate near the endpoints.

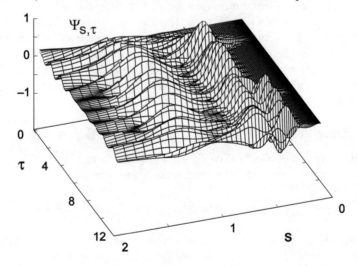

Figure 2.13. The continuous wavelet spectrum obtained by analyzing the input signal (2.62) with Morelet wavelets. Observe how at small values of time τ there is predominantly one frequency present, how a second, higher-frequency (smaller-scale) component enters at intermediate times, and how at larger times still higher-frequency components enter. (Figure courtesy of Z. Dimcovic.)

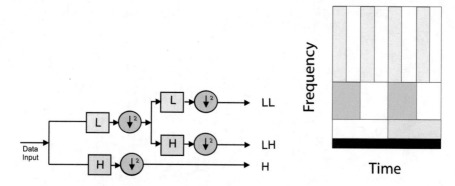

Figure 2.14. *Left:* A multifrequency dyadic filter tree used for discrete wavelet transformations. The L boxes represent lowpass filters, the H boxes highpass filters, and the $\downarrow 2$ filter out half of their input. *Right:* The relation between time and frequency resolutions (uncertainty relation). Each box contains the same area but with different proportions of time and frequency.

4. Use (2.69) to invert your wavelet transform and compare the reconstructed signal to the input signal (feel free to adjust the normalization). In Figure 2.12 we show our reconstruction.

In Listing 2.4 we give our *continuous wavelet transformation* `CWT.py` [Lang & Forinash(98)]. Because wavelets, with their transforms in two variables, are somewhat hard to grasp at first, we suggest that you write your own code and include a portion that does the inverse transform as a check. Figure 2.13 shows a surface plot of the spectrum produced for the input signal (2.62) of Figure 2.8. We see predominantly one frequency at short times, two frequencies at intermediate times, and three frequencies at longer times.

2.7.3 Discrete Wavelet Transforms, Multi-Resolution Analysis⊙

The *discrete wavelet transform* (DWT) evaluates the transforms with discrete values for the scaling parameter s and the time translation parameter τ:

$$\psi_{j,k}(t) = \frac{\Psi\left(t/2^j - k\right)}{\sqrt{2^j}} \qquad s = 2^j, \qquad \tau = \frac{k}{2^j}, \qquad k,\, j = 0, 1, \ldots, \qquad (2.72)$$

where the times are scaled so that the total interval $T = 1$. This choice of s and τ based on powers of 2 is called a *dyadic grid* arrangement and will automatically perform the scalings and translations at the different time scales inherent in wavelet analysis. The discrete wavelet transform and its inverse now become the sums

$$Y_{j,k} \simeq \sum_m \psi_{j,k}(t_m)y(t_m)h, \qquad y(t) = \sum_{j,\,k=-\infty}^{+\infty} Y_{j,k}\,\psi_{j,k}(t). \qquad (2.73)$$

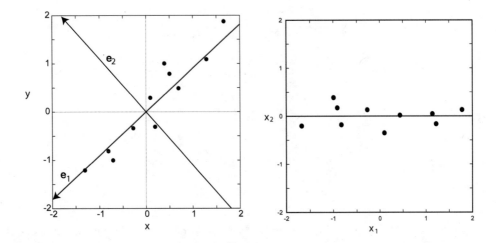

Figure 2.15. *Left:* The normalized data and eigenvectors of covariance matrix. *Right:* The normalized data using the PCA eigenvectors as basis.

These are the basic equations for the DWT. Like with the FFT, a significant speedup is possible by being more systematic, though at the expense of extra complications. As indicated in Figure 2.14, the speedup follows by using filters instead of explicit integrations and storing only the minimum number of components needed for a faithful reproduction [LPB(15)]. The key is that the low-frequency components provide the gross or *smooth* outline of a signal which, being smooth, require only a small number of components. In contrast, the high-frequency components provide the fine details of the signal over a short time interval, and so require many components in order to reproduce these details with high resolution. Listing 2.5 gives our program for performing a DWT on the chirp signal $y(t) = \sin(60t^2)$.

1. Use a digital wavelet transform to analyze the chirp signal $y(t) = \sin(60t^2)$.

   a. Invert the transform and compare the output to the original signal.

2. Use a digital wavelet transform to analyze the input signal (2.62) and compare to the results of the continuous wavelet transform.

   a. Invert the transform and compare the output to the original signal.

2.8   Principal Components Analysis (PCA)

Principal components analysis (PCA) is a powerful tool for deducing and ordering the dominate functions in complicated signals, such as those arising from nonlinear and

Table 2.4. PCA data.

Data		Zero Mean Data		In PCA Basis	
x	y	x	y	x_1	x_2
2.5	2.4	0.69	0.49	-0.828	-0.175
0.5	0.7	-1.31	-1.21	1.78	0.143
2.2	2.9	0.39	0.99	-0.992	0.484
1.9	2.2	0.09	0.29	-0.274	0.130
3.1	3.0	1.29	1.09	-1.68	-0.209
2.3	2.7	0.49	0.79	0.913	0.175
2	1.6	0.19	-0.31	0.0991	-0.350
1.0	1.1	-0.81	-0.81	1.14	0.464
1.6	1.6	-0.31	-0.31	0.438	0.0178
1.1	0.9	-0.71	-1.01	1.22	-0.163

multiple variable systems.[1] PCA views a data set as elements in a multi-dimensional *data space*, and then finds the basis vectors for this space [Jackson(91), Jolliffe(01), Smith(02)]. This is analogous to the principal axes theorem of mechanics, in which the description of the rotation of a solid object is greatly simplified if moments of inertia relative to the principal axes are used.

To understand how PCA works, we present a demonstration, following [Smith(02)]. We assume that the data have two dimensions, which we call x and y, although they need not be related to spatial positions.

1. Input Data Input first two columns of Table 2.4 into a matrix.

2. Subtract the Mean PCA theory assumes that the data in each dimension has zero mean. Accordingly, columns three and four in Table 2.4 each have zero mean.

3. Calculate the Covariance Matrix The *covariance* is a measure of how much the variance of one variable from the mean is correlated with the deviation of another variable from the mean:

$$\text{cov}(x,y) = \text{cov}(y,x) = \frac{1}{N-1}\sum_{i=1}^{N}(x_i - \bar{x})(y_i - \bar{y}). \tag{2.74}$$

A positive covariance indicates that the x and y variables tend to change together in the same direction. Place the covariance values into the covariance matrix:

$$C = \begin{bmatrix} \text{cov}(x,x) & \text{cov}(x,y) \\ \text{cov}(y,x) & \text{cov}(y,y) \end{bmatrix} = \begin{bmatrix} 0.6166 & 0.6154 \\ 0.6154 & 0.7166 \end{bmatrix}. \tag{2.75}$$

---

[1]The PCA approach [Wiki(14)] is used in many fields, sometimes with names such as the Karhunen-Loève transform, the Hotelling transform, the proper orthogonal decomposition, singular value decomposition, factor analysis, empirical orthogonal functions, empirical component analysis, empirical modal analysis, and so forth.

4. Compute Unit Eigenvector and Eigenvalues of C (Easy with NumPy):

$$\lambda_1 = 1.284, \qquad \lambda_2 = 0.4908, \qquad \mathbf{e}_1 = \begin{bmatrix} -0.6779 \\ -0.7352 \end{bmatrix}, \qquad \mathbf{e}_2 = \begin{bmatrix} -0.7352 \\ +0.6789 \end{bmatrix}, \quad (2.76)$$

where the largest eigenvalue is placed first. In Figure 2.15 we show the normalized data and the two eigenvectors. The first eigenvector corresponds to the major component in the data, with $\mathbf{e}_1$ looking very much like a best straight-line fit to the data. The second eigenvector $\mathbf{e}_2$ is orthogonal to (and, accordingly, independent of) $\mathbf{e}_1$, and contains much less of the signal strength.

5. Express Data in Terms of Principal Components: Form the *feature vector* F from the eigenvectors. For simplicity's sake, those eigenvectors of minimal signal strength can be discarded.

$$F_2 = \begin{bmatrix} -0.6779 & -0.7352 \\ -0.7352 & 0.6779 \end{bmatrix}, \qquad F_1 = \begin{bmatrix} -0.6779 \\ -0.7352 \end{bmatrix}, \qquad (2.77)$$

where F_1 keeps just one principal component, while F_2 keeps two. Form the transpose of the feature matrix and of the adjusted data matrix:

$$F_2^T = \begin{bmatrix} -0.6779 & -0.7352 \\ -0.7352 & 0.6779 \end{bmatrix}, \qquad (2.78)$$

$$X^T = \begin{bmatrix} .69 & -1.31 & .39 & .09 & 1.29 & .49 & .19 & -.81 & -.31 & -.71 \\ .49 & -1.21 & .99 & .29 & 1.09 & .79 & -.31 & -.81 & -.31 & -1.01 \end{bmatrix}. \quad (2.79)$$

Express the data in terms of the principal components by multiplying the transposed feature matrix by the transposed-adjusted data matrix:

$$X^{PCA} = F_2^T X^T = \begin{bmatrix} .82 & 1.8 & -.99 & -.27 & -1.7 & -.91 & .10 & 1.2 & .44 & 1.2 \\ -.18 & .14 & .38 & .13 & -.21 & .18 & -.35 & .46 & .18 & -.16 \end{bmatrix}$$

In columns five and six of Table 2.4 we place the transformed data elements back into standard form, and on the right of Figure 2.15 we plot the normalized data using the eigenvectors $\mathbf{e}_1$ and $\mathbf{e}_2$ as basis. The plot shows just where each datum point sits relative to the trend in the data. If we use only the principal component, we would have all of the data on a straight line (we leave that as an exercise). Of course our data are so simple that this example does not show the power of the technique. On the other hand if there are large numbers of data, it is valuable to be able to categorize them in terms of just a few components.

1. Use just the principal eigenvector to perform the PCA analysis just completed with two eigenvectors.

2. Store data from ten cycles of the chaotic pendulum studied in Chapter 3, but do not include transients. Perform a PCA of these data and plot the results using principal component axes.

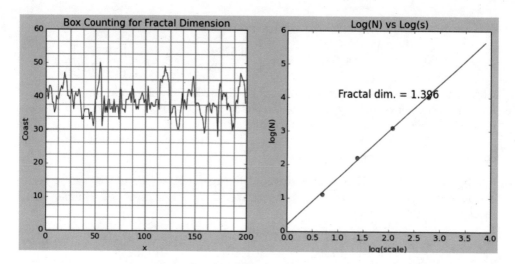

Figure 2.16. *Left:* A simulated coastline of Britain. *Right:* A linear fit to the results of box counting leading to a deduced fractal dimension.

2.9   Fractal Dimension Determination

Mandelbrot asked the classic question "How Long Is the Coast of Britain?." The answer depends upon the length of the ruler you lay down to measure the coastline, and whether the coastline has a geometric or fractal shape. If the coastline is geometric, then a finite number of measurements will suffice, though if the coastline is fractal, then it looks similar, regardless of how small a scale it is observed at, and an infinite number of measurements is needed, which would yield an infinite length.

The rule we ask you to apply to determine the fractal dimension of an object, is simply to count the number of spheres or cubes of successively diminishing size that are needed to cover the object. If it takes N little spheres or cubes of side $r \to 0$, then the fractal dimension d_f is deduced as

$$N(r) = C \left(\frac{1}{r}\right)^{d_f} = C' s^{d_f} \quad \text{(as } r \to 0\text{)}, \tag{2.80}$$

$$\Rightarrow \quad d_f = -\lim_{r \to 0} \frac{\Delta \log N(r)}{\Delta \log r}. \tag{2.81}$$

Here $s \propto 1/r$ is called the *scale*, so $r \to 0$ corresponds to an infinite scale. Once we have a value for d_f, we determine a value for the perimeter P of an object as

$$P(r) = \lim_{r \to 0} M r^{1-d_f}, \tag{2.82}$$

where M is an empirical parameter. For a geometric figure $d_f = 1$, and the perimeter approaches a constant as $r \to 0$. Yet for a fractal with $d_f > 1$, $L \to \infty$ as $r \to 0$.

1. Use box counting to determine the fractal dimension of a square.

2. Use box counting to determine the fractal dimension of a straight line.

3. Use box counting to determine the fractal dimension of a model of the British coastline.

   a. Take the output from the random deposition program (top of Figure 7.6 left) and use it as your hypothetical coastline.

   b. Print your coastline graph with the same physical scale (*aspect ratio*) for the vertical and horizontal axes.

   c. Place a piece of graph paper over your printout and look through the graph paper at your coastline. Count the number of boxes containing any part of the coastline (Figures 7.6 right). If you prefer, you can do this digitally as well, but the paper version is quick and sort of quaint.

   d. If the vertical height of your printout is 17 cm, and the largest boxes on your graph paper is 1 cm, then your lowest scale is 1:17, or $s = 17$.

   e. With our largest boxes of $1\,\text{cm} \times 1\,\text{cm}$, we found that the coastline passed through $N = 24$ large boxes, that is, that 24 large boxes covered the coastline at $s = 17$. Determine how many of the largest boxes (lowest scale) are needed to cover your coastline.

   f. Determine and record how many of the largest boxes (lowest scale) of your graph paper are needed to cover the coastline.

   g. Determine and record how many midsize boxes (midrange scale) are needed to cover your coastline.

   h. Determine and record how many of the smallest boxes (highest scale) are needed to cover your coastline.

   i. Equation (2.81) tells us that as the box sizes get progressively smaller,

   $$\log N \simeq \log A + d_f \log s, \tag{2.83}$$

   $$\Rightarrow \quad d_f \simeq \frac{\Delta \log N}{\Delta \log s} = \frac{\log N_2 - \log N_1}{\log s_2 - \log s_1} = \frac{\log(N_2/N_1)}{\log(s_2/s_1)}. \tag{2.84}$$

   Plot $\log N$ versus $\log s$ and determine the slope of the resulting straight line. Although only two points are needed to determine the slope, use your lowest scale point as a check.

4. Does your fractal imply an infinite coastline? Does it make sense that a small island like Britain, which you can walk around, has an infinite perimeter?

5. Figure 8.2 is a bifurcation graph from the problems involving the logistics map. Take that graph, or one of the bifurcation graphs you produced, and determine the fractal dimension of different parts of the graph by using the same technique that was applied to the coastline of Britain.

Our program CoastDimension.py in Listing 2.9 determines the dimension of a simulated coastline, and produces the output shown in Figure 2.16.

2.10   Code Listings

```
# DFTcomplex.py:   Discrete Fourier Transform wi Python complex math

from vpython import *
from numpy import *
import cmath                                          # Complex math

N = 100;  twopi = 2.*pi;  h = twopi/N;   sq2pi = 1./sqrt(twopi)
y = zeros(N+1, float); Ycomplex = zeros(N, complex)          # Arrays

SignalGraph = graph(x=0, y=0, width=600, height=250,
   title='Signal y(t)', xtitle='x', ytitle='y(t)',
   xmax=2*math.pi, xmin=0, ymax=250, ymin=-250)
SignalCurve = gcurve(color=color.red)
TransformGraph = graph(x=0, y=250, width=600, height=250,
   title='Im Y(omega)', xtitle = 'x', ytitle='Im Y(omega)',
   xmax=10., xmin=-1, ymax=100, ymin=-250)
TransformCurve = gvbars(delta = 0.05, color = color.red)

def Signal(y):                                        # Signal
   h = twopi/N;          x = 0.
   for i in range(0, N+1):
      y[i] = 30*cos(x) + 60*sin(2*x) + 120*sin(3*x)
      SignalCurve.plot(pos = (x, y[i]))               # Plot
      x += h

def DFT(Ycomplex):                                    # DFT
   for n in range(0, N):
     zsum = complex(0.0, 0.0)
     for  k in range(0, N):
        zexpo = complex(0, twopi*k*n/N)    # Complex exponent
        zsum += y[k]*exp(-zexpo)
     Ycomplex[n] = zsum * sq2pi
     if Ycomplex[n].imag != 0:
           TransformCurve.plot(pos=(n,Ycomplex[n].imag))

Signal(y)                                  # Generate signal
DFT(Ycomplex)                              # Transform signal
```

Listing 2.1. **DFTcomplex.py** uses the built-in complex numbers of Python to compute the discrete Fourier transform for the signal in method **f(signal)**.

```
# SincFilter.py: VPthon, Noise reduction by filtering

from vpython import *;   from numpy import *
import random

max = 4000
array = zeros((max),float)
ps = zeros((max),float)
step = 2*math.pi/1000
graph1 = graph(x=0,y=0,width=600,height=250,title='Pure Signal',
   xtitle='t (s)',ytitle='f(t)',xmin=0,xmax=25,ymin=0,ymax=10)
```

```
funct1 = gcurve(color=color.red)
graph2 = graph(x=0, y=250 ,width=600, height=250,
   title='Signal+Noise', xtitle='t (s)', ytitle='y(t)',
   xmin=0, xmax=25, ymin=0, ymax=10)
funct2 = gcurve(color=color.red)
graph3 = graph(x=0, y=500, width=600, height=250,
   title='Filtered Input', xtitle='t (s)', ytitle='y(t)',
   xmin=0, xmax=25, ymin=0, ymax=10)
funct3 = gcurve(color=color.red)

def function(array,max):
    f = zeros((max + 1),float)
    step = 2*pi/1000; x = 0.
    for i in range(0,max):
        f[i] = 1/(1. - 0.9*sin(x))                  # Function, then noise
        array[i] = (1/(1-0.9*sin(x)))+0.5*(2.*random.random()-1)
        funct1.plot(pos=(x,f[i]))
        funct2.plot(pos=(x,array[i]))
        x += step

def filter():                              # Low-pass windowed sinc filter
    y = zeros((max),float); h = zeros((max),float)
    step = 2*pi/1000
    m = 100                                     # Set filter length
    fc = .07
    for i in range(0,100):                           # Low-pass filter
        if ((i-(m/2)) == 0):  h[i] = 2*pi*fc
        if ((i-(m/2))!= 0):  h[i] = sin(2*pi*fc*(i-m/2))/(i-m/2)
        h[i] = h[i]*(0.54 - 0.46*cos(2*pi*i/m))     # Hamming window
    sum = 0.                                 # Normalize low-pass filter
    for i in range(0,100):   sum = sum + h[i]
    for i in range(0,100):   h[i] = h[i] / sum
    for j in range(100,max-1):              # Convolute input + filter
        y[j] = 0.
        for i in range(0,100): y[j] = y[j] + array[j-i] * h[i]
    for j in range(0,max-1):     funct3.plot(pos=(j*step,y[j]))

function(array, max)
filter()
```

Listing 2.2. **SincFilter.py** uses a windowed sinc filter to remove noise from an input digital image.

```
# FFTmod.py:  FFT for complex numbers in dtr[][2], returned in dtr

import numpy as np    # from sys import version
max = 2100;   points = 1026;  N = 16               # Power of 2
isign = -1                                  # -1, 1 TF, inverse TF
data = np.zeros((max));   dtr  = np.zeros((points,2))

def FFT(N, isign):                            # FFT of dtr[n,2]
    n = 2*N
    for i in range(0,N):
        j = 2*i+1
        data[j] = dtr[i,0]              # Real dtr, odd data[j]
        data[j+1] = dtr[i,1]           # Imag dtr, even data[j+1]
        print('data   ',data[j],data[j+1])
    j = 1                              # Place reverse order
```

```
    for i in range(1,n+2, 2):
        if (i-j) < 0 :
            tempr = data[j]
            tempi = data[j+1]
            data[j] = data[i]
            data[j+1] = data[i+1]
            data[i] = tempr
            data[i+1] = tempi
        m = n//2;   # NB divide to remain integer
        while (m-2 > 0):
            if  (j-m) <= 0 :   break
            j = j-m
            m = m//2 # integer part
        j = j + m;
    print("\n Bit-Reversed Input Data ")
    for i in range(1,n+1,2):  print("%2d data[%2d] %9.5f ↵
        "%(i,i,data[i]))
    mmax = 2
    while (mmax-n) < 0 :                              # Begin transform
        istep = 2*mmax
        theta = 6.2831853/(isign*mmax)
        sinth = sin(theta/2.)
        wstpr = -2.0*sinth**2
        wstpi = sin(theta)
        wr = 1.0
        wi = 0.0
        for m in range(1,mmax +1,2):
            for i in range(m,n+1,istep):
                j = i + mmax
                tempr = wr*data[j]    - wi *data[j+1]
                tempi = wr*data[j+1] + wi *data[j]
                data[j]   = data[i]    - tempr
                data[j+1] = data[i+1] - tempi
                data[i]   = data[i]    + tempr
                data[i+1] = data[i+1] + tempi
            tempr = wr
            wr = wr*wstpr - wi*wstpi + wr
            wi = wi*wstpr + tempr*wstpi + wi;
        mmax = istep
    for i in range(0,N):
        j = 2*i+1
        dtr[i,0] = data[j]
        dtr[i,1] = data[j+1]

print('\n          Input')
print("  i   Re part    Im part")
for i in range(0,N ):                                # Form array
    dtr[i,0] = 1.0*i                                 # Real part
    dtr[i,1] = 1.0*i                                 # Im part
    print(" %2d %9.5f %9.5f" %(i,dtr[i,0],dtr[i,1]))
FFT(N, isign)                           # Call FFT, use global dtr[][]
print('\n     Fourier Transform \n  i       Re      Im     ')
for i in range(0,N): print(" %2d %9.5f %9.5f " ↵
    %(i,dtr[i,0],dtr[i,1])  )
```

Listing 2.3. **FFTmod.py** computes the FFT or inverse transform depending upon the sign of **isign**.

```
# CWT.py    Continuous  Wavelet  TF,  a  la  Zlatko  Dimcovic

import matplotlib.pylab as p, numpy as np
from mpl_toolkits.mplot3d import Axes3D
from vpython import *

originalsignal=graph(x=0, y=0, width=600, height=200, \
        title='Input Signal',xmin=0,xmax=12,ymin=-20,ymax=20)
orsigraph = gcurve(color=color.red)
invtrgr = graph(x=0, y=200, width=600, height=200, title=
 'Inverted Transform',xmin=0,xmax=12,ymin=-20,ymax=20)
invtr = gcurve(color=color.green)
iT =  0.0;           fT =   12.0;        W = fT - iT;
N =  240;            h = W/N
noPtsSig = N;        noS = 20;           noTau =  90;
iTau = 0.;           iS = 0.1;           tau = iTau;       s =  iS


dTau = W/noTau; dS = (W/iS)**(1./noS); # Tiny s for hi freq
maxY =  0.001;      sig = np.zeros((noPtsSig), float)      # Signal

def signal(noPtsSig, y):                        # Signal function
    t = 0.0;        hs = W/noPtsSig;      t1 = W/6.;    t2 = 4.*W/6.
    for i in range(0, noPtsSig):
        if  t >= iT  and t <= t1: y[i] = sin(2*pi*t)
        elif t >= t1 and t <= t2:
          y[i] = 5*sin(2*pi*t)+10*sin(4*pi*t);
        elif t >= t2 and t <= fT:
            y[i] = 2.5*sin(2*pi*t) + 6.*sin(4*pi*t) + ↩
            10.*sin(6*pi*t)
        else:
            print("In signal(...) : t out of range.")
            sys.exit(1)
        yy=y[i]
        orsigraph.plot(pos=(t,yy))
        t += hs

signal(noPtsSig, sig)                            # Form signal
Yn = np.zeros( (noS+1, noTau+1), float)          # Transform

def morlet(t, s, tau):                           # Mother
    T = (t - tau)/s
    return sin(8*T) * exp( - T*T/2. )

def transform(s, tau, sig):                      # Find wavelet TF
    integral = 0.
    t = iT;
    for i in range(0, len(sig) ):
        t += h
        integral += sig[i]*morlet(t, s, tau)*h
    return integral / sqrt(s)

def invTransform(t, Yn):                   # Compute inverse
    s = iS                                 # Transform
    tau = iTau
    recSig_t = 0
    for i in range (0, noS):
        s *= dS                            # Scale graph
        tau = iTau
        for j in range (0, noTau):
            tau += dTau
            recSig_t += dTau*dS*(s**(-1.5))*Yn[i,j]*morlet(t,s,tau)
    return recSig_t
```

```
    print("Working at finding transform, count 20")
    for i in range( 0, noS):
        s *= dS                                      # Scaling
        tau = iT
        print(i)
        for j in range(0, noTau):
            tau += dTau                              # Translate
            Yn[i, j] = transform(s, tau, sig)
    print("transform found")
    for i in range( 0, noS):
        for j in range( 0, noTau):
            if Yn[i,j] > maxY or Yn[i,j] < -maxY: maxY=abs(Yn[i,j])
    tau = iT
    s = iS
    print("normalize")
    for i in range( 0, noS):
        s *= dS
        for j in range( 0, noTau):
            tau +=   dTau                            # Transform
            Yn[i, j] = Yn[i, j]/maxY
        tau = iT
    print("finding inverse transform")              # Inverse TF
    recSigData =  "recSig.dat"
    recSig = np.zeros(len(sig) )
    t =   0.0;
    print("count to 10")
    kco = 0;              j = 0;              Yinv =  Yn
    for rs in range(0, len(recSig) ):
        recSig[rs] = invTransform(t, Yinv)          # Find input signal
        xx=rs/20
        yy=4.6*recSig[rs]
        invtr.plot(pos=(xx,yy))
        t += h
        if kco %24 == 0:
            j += 1
            print(j)
        kco += 1
    x = list(range(1, noS + 1))
    y = list(range(1, noTau + 1))
    X,Y = p.meshgrid(x, y)

    def functz(Yn):                                 # Transform function
        z = Yn[X, Y]
        return z

    Z = functz(Yn)
    fig = p.figure()
    ax = Axes3D(fig)
    ax.plot_wireframe(X, Y, Z, color = 'r')
    ax.set_xlabel('s: scale')
    ax.set_ylabel('Tau')
    ax.set_zlabel('Transform')
    p.show()
    print("Done")
```

Listing 2.4. **CWT.py** computes a normalized continuous wavelet transform of the signal data in **input** (here assigned as a sum of sine functions) using Morlet wavelets (courtesy of Z. Dimcovic). The discrete wavelet transform (DWT) is faster and yields a compressed transform, though it is less transparent.

```
# DWT.py:   Discrete Wavelet Transform, Daubechies type

from vpython import *;   import numpy as np

sq3 = sqrt(3);    fsq2 = 4.0*sqrt(2);    N = 1024    # N = 2^n
c0 = (1+sq3)/fsq2;    c1 = (3+sq3)/fsq2    # Daubechies 4 coeff
c2 = (3-sq3)/fsq2;    c3 = (1-sq3)/fsq2
transfgr1 = None                         # Display indicator
transfgr1 = None

def chirp( xi):                          # Chirp signal
    y = sin(60.0*xi**2);
    return y;

def daube4(f, n, sign):      # DWT if sign >= 0, inverse if < 0
    global transfgr1, transfgr2
    tr = np.zeros( (n + 1), float)           # Temporary
    if n < 4 : return
    mp = n//2
    mp1 = mp + 1                             # midpoint + 1
    if sign >= 0:                            # DWT
        j = 1
        i = 1
        maxx = n/2
        if n > 128:                          # Scale
            maxy = 3.0
            miny = - 3.0
            Maxy = 0.2
            Miny = - 0.2
            speed = 50                       # Fast rate
        else:
            maxy = 10.0
            miny = - 5.0
            Maxy = 7.5
            Miny = - 7.5
            speed = 8                        # Lower rate
        if transfgr1:
            transfgr1.visible = False
            transfgr2.visible = False
            del transfgr1
            del transfgr2
        transfgr1 = graph(x=0, y=0, width=600, height=400,
        title='Wavelet TF, down sample + low pass',
        xmax=maxx, xmin=0, ymax=maxy, ymin=miny)
        transf = gvbars(delta=2.*n/N, color=color.cyan)
        transfgr2 = graph(x=0, y=400, width=600, height=400,
    title='Wavelet TF, down sample + high pass',
        xmax=2*maxx, xmin=0, ymax=Maxy, ymin=Miny)
        transf2 = gvbars(delta=2.*n/N, color=color.cyan)
        while j <= n - 3:
            rate(speed)
            tr[i] = c0*f[j] + c1*f[j+1] + c2*f[j+2] + c3*f[j+3]
            transf.plot(pos = (i, tr[i]) )       # c coefficents
            tr[i+mp] = c3*f[j] - c2*f[j+1] + c1*f[j+2] - c0*f[j+3]
            transf2.plot(pos = (i + mp, tr[i + mp]))
            i += 1                             # d coefficents
            j += 2                             # downsampling
        tr[i] = c0*f[n-1] + c1*f[n] + c2*f[1] + c3*f[2] # low-pass
        transf.plot(pos = (i, tr[i]) )         # c coefficients
        tr[i+mp] = c3*f[n-1] - c2*f[n] + c1*f[1] - c0*f[2] # hi
        transf2.plot(pos = (i+mp, tr[i+mp]) )
    else:                                      # inverse DWT
        tr[1] = c2*f[mp] + c1*f[n] + c0*f[1] + c3*f[mp1] # low-pass
```

```
        tr[2] = c3*f[mp] − c0*f[n] + c1*f[1] − c2*f[mp1]  # hi−pass
        j = 3
        for i in range (1, mp):
            tr[j] = c2*f[i] + c1*f[i+mp] + c0*f[i+1] + c3*f[i+mp1]
            j += 1                                         # upsample
            tr[j] = c3*f[i] − c0*f[i+mp] + c1*f[i+1] − c2*f[i+mp1]
            j += 1;                                        # upsampling
    for i in range(1, n+1):
        f[i] = tr[i]                              # copy TF to array

def pyram(f, n, sign):                            # f −> TF
    if (n < 4): return                            # too few data
    nend = 4                                      # when to stop
    if sign >= 0 :                                # Transform
        nd = n
        while nd >= nend:                 # Downsample filtering
            daube4(f, nd, sign)
            nd //= 2
    else:                                         # Inverse TF
        while nd <= n:        # Upsampling fix, thanks Pavel Snopok
            daube4(f, nd, sign)
            nd *= 2

f = np.zeros( (N + 1), float)                 # data vector
inxi = 1.0/N                                  # for chirp signal
xi = 0.0
for i in range(1, N + 1):
    f[i] = chirp(xi)                          # Function to TF
    xi += inxi;
n = N                                         # must be 2^m
pyram(f, n, 1)                                # TF
# pyram(f, n, − 1)                            # Inverse TF
```

Listing 2.5. **DWT.py** performs a discrete wavelet transform on the chirp signal.

```
# Bisection.py: Matplotlib, 0 of f(x) via Bisection algorithm

from numpy import *
eps = 1e−3;  Nmax = 100;  a = 0.0;  b = 7.0     # Precision, [a,b]

def f(x): return 2*math.cos(x) − x             # Your function here

def Bisection(Xminus, Xplus, Nmax, eps):               # Do not change
    for it in range(0, Nmax):
        x = (Xplus + Xminus)/2.
        print(" it =", it, " x = ", x, " f(x) =", f(x))
        if (f(Xplus)*f(x) > 0.): Xplus = x         # Change x+ to x
        else: Xminus = x                           # Change x− to x
        if (abs(f(x) ) < eps):                     # Converged?
            print("\n Root found with precision eps = ", eps)
            break
        if it == Nmax−1: print ("\n No root after N iterations\n")
    return x

root = Bisection(a, b, Nmax, eps)
print(" Root =", root)
```

Listing 2.6. Bisection.py is a simple implementation of the bisection algorithm for finding a zero of a function, in this case, $2\cos x - x$.

```python
# NewtonCall.py      Newton Search with central difference

from math import cos

x0 = 1111.;   dx = 3.e-4; eps = 0.002; Nmax = 100;   # Parameters

def f(x):   return 2*cos(x) - x # Function

def NewtonR(x, dx, eps, Nmax):
    for it in range(0, Nmax + 1):
        F = f(x)
        if (abs(F) <= eps):                           # Converged?
            print("\n Root found, f(root) =", F, ", eps = " , eps)
            break
        print("Iteration # = ", it, " x = ", x, " f(x) = ", F)
        df = (f(x+dx/2)  -  f(x-dx/2))/dx             # Central diff
        dx = - F/df
        x  += dx                                      # New guess
    if it == Nmax+1: print("\n Newton Failed for Nmax =", Nmax)
    return x

NewtonR(x0,dx,eps,Nmax)
```

Listing 2.7. NewtonCall.py uses the Newton-Raphson method to search for a zero of the function $f(x)$. The method `NewtonR` does not get changed by the user.

```python
# Fit.py: Linear least-squares fit via matrix solution

import pylab as p
from numpy import*; from numpy.linalg import inv, solve

Nd = 7
A = zeros( (3,3), float );   bvec = zeros((3,1), float)  # Declare
ss= sx = sxx = sy = sxxx = sxxxx = sxy = sxy = sxxy = 0.
x = array([1., 1.1, 1.24, 1.35, 1.451, 1.5, 1.92])
y = array([0.52, 0.8, 0.7, 1.8, 2.9, 2.9, 3.6])
sig = array([0.1, 0.1, 0.2, 0.3, 0.2, 0.1, 0.1])        # Error bars
xRange = arange(1.0, 2.0, 0.1)                          # For plots
p.plot(x, y, 'bo')                                      # Blue data
p.errorbar(x,y,sig)
p.title('Least-Squares Fit of Parabola to Blue Data')
p.xlabel('x'); p.ylabel('y');   p.grid(True)            # Plot grid

for i in range(0, Nd):
        sig2 = sig[i] * sig[i]
        ss += 1. / sig2; sx += x[i]/sig2; sy += y[i]/sig2
        rhl = x[i] * x[i]; sxx += rhl/sig2;   sxxy += rhl*y[i]/sig2
```

```
            sxy += x[i]*y[i]/sig2; sxxx +=rhl*x[i]/sig2
            sxxxx +=rhl*rhl/sig2

A      = array([  [ss,sx,sxx], [sx,sxx,sxxx], [sxx,sxxx,sxxxx] ])
bvec = array([sy,  sxy,  sxxy])
xvec = multiply(inv(A), bvec)                    # Invert matrix
print('\n x via Inverse A\n', xvec, '\n' )
xvec = solve(A, bvec)                         # Solve via elimination
print('\n x via Elimination \n', xvec, '\n Fit to Parabola\n')
print('y(x) = a0+a1 x+a2 x^2\n a0 =', x[0],'a1 =',x[1],'a2=',x[2])
print('\n i   xi      yi      yfit   ')
for i in range(0, Nd):
    s = xvec[0] + xvec[1]*x[i] + xvec[2]*x[i]*x[i]
    print(" %d %5.3f   %5.3f   %8.7f"  %(i,  x[i],  y[i],  s))
# red line is the fit, red dots the fits at y[i]m
curve  = xvec[0] + xvec[1]*xRange + xvec[2]*xRange**2
points = xvec[0] + xvec[1]*x + xvec[2]*x**2
p.plot(xRange, curve,'r', x, points, 'ro')
p.show()
```

Listing 2.8. Fit.py performs a least-squares fit of a parabola to data using the NumPy linalg package to solve the set of linear equations $S\vec{a} = \vec{s}$.

```
# CoastDimension.py: fractal dimension analysis of coastline

import random, numpy as np
import matplotlib.pyplot as plt, matplotlib.mlab as mlab

MinX = 0;  MaxX = 200;  MinY = 0;  MaxY = 60
coast = np.zeros((200)); nboxes = [];  scales = []
for i in range(0,5000):
    spot = int(200*random.random())
    if (spot == 0):                  # Hitting edge = filling hole
        if (coast[spot] < coast[spot+1]): coast[spot]=coast[spot+1]
        else: coast[spot] =  coast[spot] + 1
    else:
        if (spot == 199):                        # Extreme right
            if (coast[spot] < coast[spot-1]):  ↵
                coast[spot]=coast[spot-1]
            else: coast[spot] = coast[spot] + 1
        else:                                  # All other cases
            if ((coast[spot]<coast[spot - 1]) and
               (coast[spot]<coast[spot + 1])):
                if (coast[spot-1] > coast[spot+1]):
                    coast[spot]=coast[spot-1]
                else: coast[spot] = coast[spot + 1]
            else: coast[spot] = coast[spot] + 1
i =   range(0,200)
fig, axes = plt.subplots(nrows=1, ncols=2,figsize=(12,5) )
axes[0].plot( i,coast)
p = [0,200]
sc = 1
countbox = 0
M = 1                                    # To plot divisions
for k in range(0,4):                     # 4 scales and boxes
  M *= 2
  cajas = 0
  for j in range (0,M):
```

```
        dx = 200/M                              # Horizontal/M
        dy = MaxY/M
        q = [MaxY−dy*j,MaxY−dy*j]    # Plot horiz & vert lines
        axes[0].plot(p,q,'r-')
        r = [dx*(j+1),dx*(j+1)]
        ver = [0,60]
        axes[0].plot(r,ver,'r-')                    # Vert lines
        plt.pause(0.1)                           # Delay while plot
        for m in range(0,M):
            for i in range(int(m*dx),int((m+1)*dx)):
                if coast[i]<60−dy*j and coast[i]>60−dy*(j+1):
                    countbox = 1        # Occupied, count & break
                    if countbox ==1:
                        cajas = cajas + 1
                        countbox=0
                    break                      # Continue counting
        nboxes.append(np.log(cajas))               # Log N
        sc = sc*2                              # Multiply scale
        scales.append(np.log(sc))    # Scale = number boxes
        axes[1].plot(scales[k],nboxes[k],'ro')  # Plot logs
axes[0].set_xlabel('x')
axes[0].set_ylabel('Coast')
axes[0].set_title('Box Counting for Fractal Dimension')
m,b=np.polyfit(scales,nboxes,1)             # Linear fit
x= np.arange(0,4,0.1)                        # Plot line
axes[1].plot(x, m*x+ b)
axes[1].set_title('Log(N) vs Log(s)')
axes[1].set_xlabel('log(scale)')
axes[1].set_ylabel('log(N)')
fd="%6.3f"%m                 # Convert dimension to string
axes[1].text(1,4,'Fractal dim. ='+ fd,fontsize=15,)
plt.show()
```

Listing 2.9. **CoastDimension.py** determines the fractal dimension of a simulated coastline by box counting.

3

Classical & Nonlinear Dynamics

3.1 Chapter Overview

The first half of this chapter focuses on nonlinear oscillations. The prime tool is the numerical solution of ordinary differential equations (§1.7), which is both easy and precise. We look at a variety of systems, with emphases on behaviors in phase-space plots and bifurcation diagrams. In addition, the solution for the realistic pendulum gets new life since we can actually evaluate elliptic integrals using the integration techniques of §1.5. We then analyze the output from the simulations using the discrete Fourier transform of §2.4. The second half of the chapter examines projectile motion, bound states of three-body systems, and Coulomb and chaotic scattering. We also look at some of the unusual behavior of billiards, which are a mix of scattering and bound states. (The quantum version of these same billiards is examined in §6.8.4.) Problems related to Lagrangian and Hamiltonian dynamics then follow, with the actual computation of Hamilton's principle. Finally, we end the chapter with the problem of several weights connected by strings: a simple problem that requires a complex solution involving both a derivative algorithm and a search algorithm, as discussed in Chapters 1 and 2.

Note that Chapter 8 contains a number of problems dealing with the several discrete maps that lead to chaotic behavior in biological systems. These materials, as well as the development of the predator-prey models in that chapter, might well be included in a study of classical dynamics.

3.2 Oscillators

3.2.1 First a Linear Oscillator

1. Consider the 1-D harmonic (linear) oscillator with viscous friction:

$$\frac{d^2x}{dt^2} + \kappa\frac{dx}{dt} + \omega_0^2 x = 0. \tag{3.1}$$

a. Verify by hand or by using a symbolic manipulation package (§3.2.7) that

$$x(t) = e^{at}\left[x_0 \cos\omega t + (p_0/m\omega)\sin\omega t\right] \tag{3.2}$$

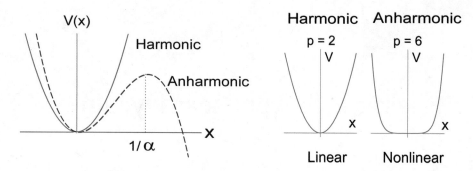

Figure 3.1. *Left:* The potentials of an harmonic oscillator (solid curve) and of an anharmonic oscillator (dashed curve). If the amplitude becomes too large for the anharmonic oscillator, the motion becomes unbound. *Right:* The shapes of the potential energy function $V(x) \propto |x|^p$ for $p = 2$ and $p = 6$. The "linear" and "nonlinear" labels refer to the restoring force derived from these potentials.

is a solution of (3.1).

    b. Determine the constants ω, x_0, and p_0 in (3.2) in terms of initial conditions.

    c. Plot the phase-space portrait $[x(t), p(t)]$ for $\omega_0 = 0.8$ and several values of $p(0)$. (Phase space portraits are discussed in §3.3.3.)

2. Do a number of things to check that your ODE solver is working well and that you know the proper integration step size needed for high precision.

    a. Choose initial conditions corresponding to a frictionless oscillator initially at rest, for which the analytic solution is:

$$x(t) = A \sin(\omega_0 t), \quad v = \omega_0 A \cos(\omega_0 t), \quad \omega_0 = \sqrt{k/m}. \qquad (3.3)$$

    b. Pick values of k and m such that the period $T = 2\pi/\omega = 10$.

    c. Start with a time step size $h \simeq T/5$ and make h smaller until the solution looks smooth, has a period that remains constant for a large number of cycles, and agrees with the analytic result. As a general rule of thumb, we suggest that you start with $h \simeq T/100$, where T is a characteristic time for the problem at hand. You should start with a large h so that you can see a bad solution turn good.

    d. Make sure that you have exactly the same initial conditions for the analytic and numerical solutions (zero displacement, nonzero velocity) and then plot the two solutions together. Also make a plot of their difference versus time since graphical agreement may show only 2–3 places of sensitivity.

    e. Try different initial velocities and verify that a *harmonic* oscillator is *isochronous*, that is, that its period does *not* change as the amplitude is made large.

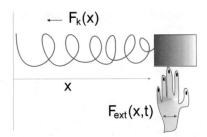

Figure 3.2. A mass m (the block) attached to a spring with restoring force $F_k(x)$ driven by an external time-dependent driving force (the hand).

3.2.2 Nonlinear Oscillators

Figure 3.2 shows a mass m attached to a spring that exerts a restoring force $F_k(x)$ toward the origin, as well as a hand that exerts a time-dependent external force $F_{ext}(x,t)$ on the mass. The motion is constrained to one dimension and so Newton's second law provides the equation of motion

$$F_k(x) + F_{ext}(x,t) = m\frac{d^2x}{dt^2}. \tag{3.4}$$

Consider two models for a nonlinear oscillator:

$$V(x) \simeq \frac{1}{2}kx^2\left(1 - \frac{2}{3}\alpha x\right), \qquad \text{Model 1}, \tag{3.5}$$

$$V(x) = \frac{1}{p}kx^p, \qquad \text{Model 2 (p even)}. \tag{3.6}$$

Model 1's potential is quadratic for small displacements x, but also contains a perturbation that introduces a nonlinear term to the force for large x values: If $\alpha x \ll 1$, we would expect harmonic motion, though as $x \to 1/\alpha$ the anharmonic effects should increase. Model 2's potential is proportional to an arbitrary p of the displacement x from equilibrium, with the power p being even for this to be a restoring force. Some characteristics of both potentials can be seen in Figure 3.1.

1. Modify your harmonic oscillator program to study anharmonic oscillations for strengths in the range $0 \le \alpha x \le 2$. Do *not* include any explicit time-dependent forces yet.

2. Test that for $\alpha = 0$ you obtain simple harmonic motion.

3. Check that the solution remains periodic as long as $x_{max} < 1/\alpha$ in model 1 and for all initial conditions in model 2.

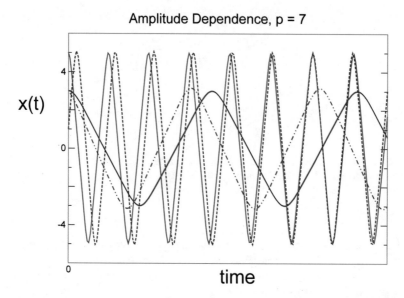

Figure 3.3. The position versus time for oscillations within the potential $V \propto x^7$ for four different initial amplitudes. Each is seen to have a different period.

4. Check that the maximum speeds always occur at $x = 0$ and that the velocity vanishes at the maximum x's.

5. Verify that nonharmonic oscillators are *nonisochronous*, that is, that vibrations with different amplitudes have different periods (Figure 3.3).

6. Describe how the shapes of the oscillations change for different α or p values.

7. In Model 2, for what values of p and x will the potential begin to look like a square well? Note that for large values of p, the forces and accelerations get large near the turning points, and so you may need a smaller time step h to track the rapid changes in motion.

8. Devise an algorithm to determine the period T of the oscillation by recording times at which the mass passes through the origin. Note that because the motion may be asymmetric, you must record at least three times to deduce the period.

9. Verify that the oscillators are *nonisochronous*, that is, that vibrations with different amplitudes have different periods.

10. Construct a graph of the deduced period as a function of initial amplitude.

11. Verify that the motion is oscillatory, though not harmonic, as the energy approaches $k/6\alpha^2$, or for $p \neq 2$.

12. Verify that for oscillations with energy $E = k/6\alpha^2$, the motion in potential 1 changes from oscillatory to translational.

13. For Model 1, see how close you can get to the *separatrix* where a single oscillation takes an infinite amount of time.

3.2.3   Assessing Precision via Energy Conservation

It is important to test the precision and reliability of a numerical solution. For the present cases, as long as there is no friction and no external forces, we expect energy to be conserved. Energy conservation, which follows from the mathematics and not the algorithm, is hence an independent test of our algorithm.

1. Plot for 50 periods the potential energy $\text{PE}(t) = V[x(t)]$, the kinetic energy $\text{KE}(t) = mv^2(t)/2$, and the total energy $E(t) = \text{KE}(t) + \text{PE}(t)$.

2. Check the long-term *stability* of your solution by plotting

$$-\log_{10} \left| \frac{E(t) - E(t=0)}{E(t=0)} \right| \simeq \text{number of places of precision} \qquad (3.7)$$

for a large number of periods. Because $E(t)$ should be independent of time, the numerator is the absolute error in your solution, and when divided by $E(0)$, becomes the relative error. If you cannot achieve 11 or more places, then you need to decrease the value of h or debug.

3. Because $(x < 1)^p$ is a small number, a particle bound by a large-p oscillator is essentially "free" most of the time, and so you should observe that the average of its kinetic energy over time exceeds the average of its potential energy. This is actually the physics behind the Virial theorem for a power-law potential [Marion & Thornton(03)]:

$$\langle \text{KE} \rangle = \frac{p}{2} \langle \text{PE} \rangle. \qquad (3.8)$$

Verify that your solution satisfies the Virial theorem and in doing so compute the effective value of p.

3.2.4   Models of Friction

Three simple models for frictional force are *static, kinetic,* and *viscous* friction:

$$F_f^{(\text{static})} \leq -\mu_s N, \qquad F_f^{(\text{kinetic})} = -\mu_k N \frac{v}{|v|}, \qquad F_f^{(\text{viscous})} = -bv, \qquad (3.9)$$

where N is the *normal force* on the object under consideration, μ and b are parameters, and v is the velocity.[1]

---
[1] The effect of air resistance on projectile motion is studied §3.6.

1. Extend your harmonic oscillator code to include the three types of friction in (3.9) and observe how the motion differs for each.

   a. For the simulation with static plus kinetic friction, each time the oscillator has $v = 0$ you need to check that the restoring force exceeds the static frictional force. If not, the oscillation must end at that instant. Check that your simulation terminates at nonzero x values.

   b. For your simulations with viscous friction, investigate the qualitative changes that occur for increasing b values:

Underdamped: $b < 2m\omega_0$ Oscillate within decaying envelope
Critically: $b = 2m\omega_0$ Nonoscillatory, finite decay time
Overdamped: $b > 2m\omega_0$ Nonoscillatory, infinite decay time

3.2.5   Linear & Nonlinear Resonances

A periodic external force of frequency ω_f is applied to an oscillatory system with natural frequency ω_0. As the frequency of the external force passes through ω_0, a *resonance* may occur. If the oscillator and the driving force remain in phase over time, the amplitude of oscillation will increase continuously unless there is some mechanism, such as friction or nonlinearity, that limits the growth. If the frequency of the driving force is close to, though not exactly equal to ω_0, a related phenomena, *beating*, may occur in which there is interference between the natural vibration and the driven vibrations:

$$x \simeq x_0 \sin \omega_f t + x_0 \sin \omega_0 t = \left(2x_0 \cos \frac{\omega_f - \omega_0}{2} t \right) \sin \frac{\omega_f + \omega_0}{2} t. \qquad (3.10)$$

The resulting motion resembles the natural oscillation of the system at the average frequency $(\omega_f + \omega_0)/2$, however with an amplitude $2x_0 \cos(\omega_f - \omega_0)/2t$ that varies slowly with the *beat frequency* $(\omega_f - \omega_0)/2$.

1. Include the time-dependent external force $F \cos(\omega_f t)$ in your rk4 ODE solver. You can modify the `rk4Call.py` program given earlier in Listing 1.12 which uses VPython, or `ForcedOscillate.py` in Listing 3.1 which uses *Matplotlib*.

2. Start with a harmonic oscillator with these parameters and the initial conditions:

$$p = 2, \quad k = 1, \quad m = 1, \quad \mu = 0.001, \quad \omega_f = 2, \quad y^{(0)}(0) = 0.1, \quad y^{(1)}(0) = 0.3.$$

3. Starting with a large value for the magnitude of the driving force F_0 should lead to *mode locking* in which the system is overwhelmed by the driving force, and, after the transients die out, will oscillate in phase with the driver. See if you can reproduce a behavior like that found on the left of Figure 3.4.

4. Why don't the oscillations appear damped?

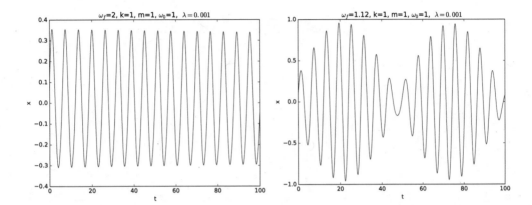

Figure 3.4. *Left:* Position versus time for an oscillator with $p = 2$, $k = 1$, $m = 1$, $\mu = 0.001$, $\omega_f = 2$, and $\omega_0 = 1$. *Right:* Position versus time for $k = 1$, $m = 1$, $\lambda = 0.001$, $\omega_f = 1.12$, and $\omega_0 = 1$.

5. With the same constants as before, change to $w_f = 1.12$, which is close to the natural frequency $\omega_0 = 1$. You should obtain oscillations similar to that on the right of Figure 3.4.

6. Verify that for $p = 2$, the beat frequency, that is, the number of variations in intensity per unit time, equals the frequency difference in cycles per second $(\omega_f - \omega_0)/2\pi$. With the same constants as in Figure 3.4, see the effects of a large viscosity.

7. Return to parameters that gave you distinct beating and make a series of runs in which you progressively increase the frequency of the driving force in the range $\omega_0/10 \le \omega_f \le 10\omega_0$. Plot up the response for a number of interesting runs.

8. Make a plot of the maximum amplitude of oscillation versus the driver's ω_f. This should exhibit a resonance peak at ω_0.

9. Explore what happens when you drive a nonlinear system. Start with a system being close to harmonic, and verify that you obtain beating in place of the blowup that occurs for the linear system.

10. Explain the origin of the beating in nonlinear resonances.

11. Investigate how the inclusion of viscous friction modifies the curve of maximum amplitude versus driver frequency. You should find that friction broadens the curve.

12. Explain how the character of the resonance changes as the exponent p in model 2 is made progressively larger. You should find that at large p the mass effectively

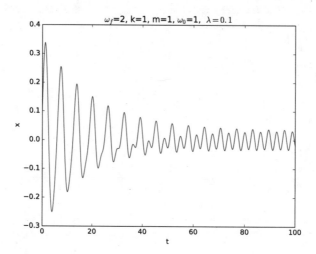

Figure 3.5. Position versus time for an oscillator with $p = 2$, $k = 1$, $m = 1$, $\mu = 0.1$, $\omega_f = 2$, and $\omega_0 = 1$.

"hits" the wall and falls out of phase with the driver, thereby making the driver less effective at pumping in energy.

3.2.6   Famous Nonlinear Oscillators

1. The nonlinear behavior in once-common objects such as vacuum tubes and metronomes is described by the **van der Pool Equation**,

$$\frac{d^2x}{dt^2} + \mu(x^2 - x_0^2)\frac{dx}{dt} + \omega_0^2 x = 0. \tag{3.11}$$

   a. Explain why you can think of (3.11) as describing an oscillator with x-dependent damping.
   b. Create some phase space plots of the solutions to this equation, that is, plots of $\dot{x}(t)$ versus $x(t)$.
   c. Verify that this equation produces a limit cycle in phase space, that is orbits internal to the limit cycle spiral out until they reach the limit cycle, and those external to it spiral in to it.

The **Duffing oscillator** is another example of a damped, driven nonlinear oscillator. It is described by the differential equation [Kov(11), Enns(01)]:

$$\frac{d^2x}{dt^2} = -2\gamma\frac{dx}{dt} - \alpha x - \beta x^3 + F\cos\omega t. \tag{3.12}$$

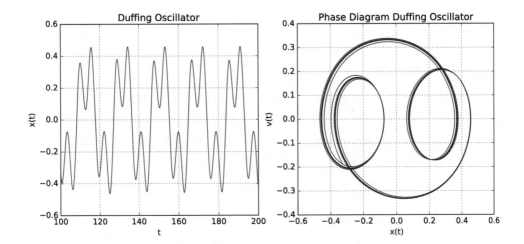

Figure 3.6. A period three solution for a forced Duffing oscillator. *Left*: $x(t)$, *Right*: $v(t)$ versus $x(t)$.

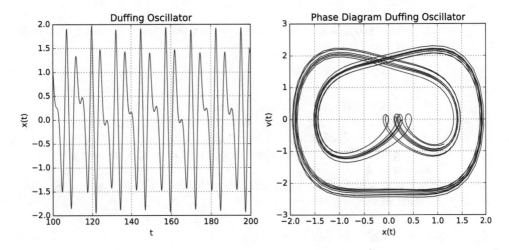

Figure 3.7. The Ueda oscillator *Left*: $x(t)$, *Right*: $v(t)$ versus $x(t)$.

In Listing 1.13 we gave as an example the code `rk4Duffing.py` that solves a simple form of this equation.

1. Modify your ODE solver program to solve (3.12).

2. First choose parameter values corresponding to a simple harmonic oscillator and verify that you obtain sinusoidal behavior for $x(t)$ and a closed elliptical phase

space plot.

3. Include a driving force, wait 100 cycles in order to eliminate transients, and then create a phase space plot. We used $\alpha = 1.0$, $\beta = 0.2$, $\gamma = 0.2$, $\omega = 1.$, $F = 4.0$, $x(0) = 0.009$, and $\dot{x}(0) = 0$.

4. Search for a period-three solution like those in Figure 3.6, where we used $\alpha = 0.0$, $\beta = 1.$, $\gamma = 0.04$, $\omega = 1.$, and $F = 0.2$.

5. Change your parameters to $\omega=1$ and $\alpha=0$ in order to model an *Ueda* oscillator. Your solution should be similar to Figure 3.7.

6. Consider a nonlinear perturbed harmonic oscillator with friction:

$$p = \dot{q}, \qquad \dot{p} = q - q^3 - p. \tag{3.13}$$

   a. Create several phase space portraits for this system.
   b. Determine analytically the Liapunov coefficients and from these the position and nature of the critical points.
   c. Does your analytic analysis agree with your computed results?

7. Investigate the simplified version of the Lorenz attractors developed by **Roessler** [Becker(86)]:

$$\dot{x} = -y - z \quad \dot{y} = x + ay \quad \dot{z} = b + xz - cz \quad (a, b, c) = (0.2, 0.2, 5.7). \tag{3.14}$$

   a. Compute and plot $x(t)$, $y(t)$, and $z(t)$ as functions of time.
   b. Plot projections of your solutions onto the (x, y) and $(x, \dot{x})$ planes.
   c. Make a Poincaré mapping of the transverse section $\dot{x} = 0$. (A Poincaré mapping is the intersection of a periodic orbit in the phase space with a lower-dimensional subspace.)
   d. When $\dot{x} = 0$, x has an extremum. Plot the value of the extrema x_{i+1} as a function of the previous extremum x_i.

3.2.7   Solution via Symbolic Computing

1. Repeat the study of the damped and driven harmonic oscillator using a symbolic manipulation package. Listing 3.2 presents a direct solution of the differential equation using SymPy (see Chapter 1 for discussion of Python packages) and produces the output

```
ODE to be solved:
Eq(kap*Derivative(f(t), t) + w0**2*f(t) + Derivative(f(t), t, t), 0)
 Solution of ODE:
Eq(f(t), C1*exp(t*(-kap - sqrt(kap**2 - 4*w0**2))/2)
          + C2*exp(t*(-kap + sqrt(kap**2 - 4*w0**2))/2))
```

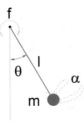

Figure 3.8. A pendulum of length l driven through resistive air (dotted arcs) by an external sinusoidal torque (semicircle). The strength of the external torque is given by f and that of air resistance by α.

In turn, Listing 3.2 presents a determination of the parameters in the solution by evaluating the initial conditions, and produces the output

```
Soltn:
(x0*cos(t*w) + p0*sin(t*w)/(m*w))*exp(alf*t)
Derivatives:
kap*(alf*(x0*cos(t*w) + p0*sin(t*w)/(m*w))*exp(alf*t) + (-w*x0*sin(t*w)
    + p0*cos(t*w)/m)*exp(alf*t)) + w0**2*(x0*cos(t*w)
    + p0*sin(t*w)/(m*w))*exp(alf*t) + (alf**2*(x0*cos(t*w) + p0*sin(t*w)/(m*w))
    - 2*alf*(w*x0*sin(t*w) - p0*cos(t*w)/m) - w*(w*x0*cos(t*w)
    + p0*sin(t*w)/m))*exp(alf*t)
Initial value y2:
y2 = alf**2*x0 + 2*alf*p0/m + kap*(alf*x0 + p0/m) - w**2*x0 + w0**2*x0
  Coefficients of p0/m, A =  {-kap/2}
  W =  {-sqrt(alf**2 + alf*kap + w0**2), sqrt(alf**2 + alf*kap + w0**2)}
  frequency w =  {-sqrt(-kap**2 + 4*w0**2)/2, sqrt(-kap**2 + 4*w0**2)/2}
```

2. As you can see from the output, the analytic solution is output as exponentials. How do you reconcile these results compared with the previous ones in terms of sines and cosines?

3. Use a symbolic manipulation package to solve the equations of motion for a nonlinear oscillator.

3.3   Realistic Pendula

We call a pendulum without a small angle approximation "realistic" or "nonlinear", and a realistic pendulum with a periodic driving torque "chaotic." The chaotic pendulum in Figure 3.8 is described by the ordinary differential equation,

$$\frac{d^2\theta}{dt^2} = -\omega_0^2 \sin\theta - \alpha\frac{d\theta}{dt} + f\cos\omega t, \qquad \omega_0 = \frac{mgl}{I}, \qquad \alpha = \frac{\beta}{I}, \qquad f = \frac{\tau_0}{I}. \quad (3.15)$$

Here ω_0 is the natural frequency, the α term arises from friction, and the f term measures the strength of the driving torque. The difficulty with the computer study of this system is that the four parameters space $(\omega_0, \alpha, f, \omega)$ is immense, and the results may be hypersensitive to the exact values used for them. So you may have to adjust somewhat the suggested parameter values to obtain the predicted behaviors.

1. Consider the ODE for an undriven realistic pendulum without friction:

$$\frac{d^2\theta}{dt^2} = -\omega_0^2 \sin\theta. \tag{3.16}$$

   a. Use conservation of energy to show that the pendulum's velocity as a function of position is

$$\frac{d\theta}{dt}(\theta) = 2\sqrt{\frac{g}{l}} \left[\sin^2(\theta_0/2) - \sin^2(\theta/2)\right]^{1/2}, \tag{3.17}$$

   where θ_0 is the angle of displacement for a pendulum released from rest.
   b. Solve this equation for dt and then integrate analytically to obtain the integral expression for the period of oscillation as a function of θ_0:

$$\frac{T}{4} = \frac{T_0}{4\pi} \int_0^{\theta_m} \frac{d\theta}{\left[\sin^2(\theta_m/2) - \sin^2(\theta/2)\right]^{1/2}} = 4\sqrt{\frac{L}{g}} K(\sin^2 \frac{\theta_0}{2}). \tag{3.18}$$

   The K function in (3.18) is an *elliptic integral* of the first kind, and in §3.3.1 we discuss its numerical evaluation.

2. Again consider the ODE for an undriven realistic pendulum without friction (3.16), though now solve it numerically.

   a. To ensure that you can solve the ODE for the realistic pendulum with high accuracy, start by plotting the total energy $E(t)$ as a function of time. Adjust (decrease) the integration step size in rk4 until the relative energy of your solution $E(t)/E(0)$ varies by less than 10^{-6}, even for exceedingly large times.
   b. It may be easiest to start the pendulum at $\theta = 0$ with $\dot{\theta}(0) \neq 0$, and gradually increase $\dot{\theta}(0)$ to increase the energy of the pendulum. Check that for all initial conditions your solution is periodic with unchanging amplitude.
   c. Verify that as the initial KE approaches $2mgl$, the motion remains oscillatory but with ever-increasing period.
   d. At $E = 2\,\mathrm{mgl}$ (the *separatrix*), the motion changes from oscillatory to rotational ("over the top" or "running"). See how close you can get your solution to the separatrix and hence to an infinite period.
   e. Convert your different oscillations to sound and hear the difference between harmonic motion (boring) and anharmonic motion containing overtones (interesting). Some ways to do this are discussed in §3.4.2.

3.3.1 Elliptic Integrals

Conservation of energy permits us to solve for the period of a realistic pendulum released from rest with initial displacement of θ_0:

$$T = 4\sqrt{\frac{L}{g}} K(\sin^2 \frac{\theta_0}{2}), \tag{3.19}$$

$$\simeq T_0 \left[1 + \left(\frac{1}{2}\right)^2 \sin^2 \frac{\theta_m}{2} + \left(\frac{1 \cdot 3}{2 \cdot 4}\right)^2 \sin^4 \frac{\theta_m}{2} + \cdots \right]. \tag{3.20}$$

Here K is the incomplete elliptic integral of the first kind,

$$K(m) = \int_0^1 \frac{dt}{\sqrt{(1 - t^2)(1 - mt^2)}} = \int_0^{\pi/2} \frac{d\theta}{\sqrt{1 - m \sin^2 \theta}}. \tag{3.21}$$

Tabulated values for elliptic integrals are available, or they can be evaluated directly using, for instance, Gaussian quadrature. In a mathematical sense, an infinite power series provides an exact representation of a function. However, it is often not good as an algorithm because it may converge slowly and because round–off error may dominate when there are many terms summed or when there is significant cancellation of terms. On the other hand, a polynomial approximation, such as [Abramowitz & Stegun(72)]

$$K(m) \simeq a_0 + a_1 m_1 + a_2 m_1^2 - [b_0 + b_1 m_1 + b_2 m_1^2] \ln m_1 + \epsilon(m),$$

$$m_1 = 1 - m, \quad 0 \le m \le 1, \quad |\epsilon(m)| \le 3 \times 10^{-5}, \tag{3.22}$$

$$\begin{array}{lll} a_0 = 1.38629\ 44 & a_1 = 0.11197\ 23 & a_2 = 0.07252\ 96 \\ b_0 = 0.5 & b_1 = 0.12134\ 78 & b_2 = 0.02887\ 29 \end{array},$$

provides an approximation of known precision with only a few terms, and is often very useful in its own right or as a check on numerical quadrature.

1. Compute $K(m)$ by evaluating its integral representation numerically. Tune your evaluation until you obtain agreement at the $\le 3 \times 10^{-5}$ level with the polynomial approximation.

2. Use numerical quadrature to determine the ratio T/T_0 for five values of θ_m between 0 and π. Show that you have attained at least four places of accuracy by progressively increasing the number of integration points until changes occur only in the fifth place, or beyond.

3. Now use the power series (3.20) to determine the ratio T/T_0. Continue summing terms until changes in the sum occur only in the fifth place, or beyond and note the number of terms needed.

4. Plot the values you obtain for T/T_0 versus θ_m for both the integral and power series solutions. Note that any departure from 1 indicates breakdown of the familiar small-angle approximation for the pendulum.

3.3.2   Period Algorithm

1. Devise an algorithm to determine the period T of a realistic pendulum or of a nonlinear oscillator by recording times at which the pendulum passes through the origin $\theta = 0$. Because the oscillations may be asymmetric, you will need to record at least three times to deduce the period.

2. Verify that realistic pendula are *nonisochronous*, that is, that oscillations with different initial amplitudes have different periods.

3. Construct a graph of the deduced period versus initial amplitude.

4. Compare the graph of the deduced period versus initial amplitude deduced from your simulation to that predicted by numerical evaluation of the elliptic integral in (3.18).

3.3.3   Phase Space Orbits

1. Plot $[\theta(t), d\theta/dt]$ for a large range of time t values for the linear and for the nonlinear pendulum (no torque, no friction). The geometric figures obtained are called phase space portraits or orbits.

   a. Obtain orbits for small energies as well as for energies large enough for the pendulum to go over the top (Figure 3.9).
   b. Indicate the hyperbolic points, that is, points through which trajectories flow in and out.
   c. Plot the gravitational torque on the pendulum as a function of θ and relate it to your phase-space plot. Align vertically the phase-space plot with a plot of the torque versus angle so that both have the same abscissa.

2. Use your numerical solution to produce the phase-space orbits of the nonlinear pendulum with friction, though no driving force, and compare to frictionless solutions.

3. Determine analytically the value of f for which the *average* energy dissipated by friction during one cycle is balanced by the energy put in by the driving force during that cycle. This is a stable configuration.

4. Show by computation that when the above condition is met, there arises a *limit cycle* near the phase space origin.

5. Sometimes you may have position data for a dynamical system, but not velocity data (or the positions are too chaotic to attempt forming derivatives). In cases like these, you can produce an alternative phase-space plot by plotting $q(t + \tau)$ versus $q(t)$, where t is the time and τ is a convenient lag time chosen as some fraction of a characteristic time for the system [Abarbanel et al.(93)]. Create

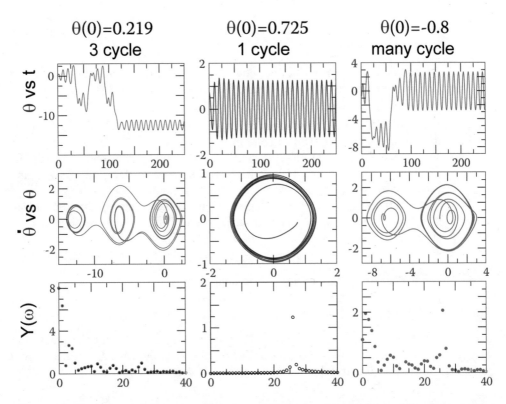

Figure 3.9. From top to bottom, position versus time, phase space plot, and Fourier spectrum for a chaotic pendulum with $\omega_0 = 1$, $\alpha = 0.2$, $f = 0.52$, and $\omega = 0.666$ and, from left to right, three different initial conditions. The leftmost column displays three dominant cycles, the center column only one, while the rightmost column has multiple cycles.

a phase space plot from the output of your realistic or chaotic pendulum by plotting $\theta(t + \tau)$ versus $\theta(t)$ for a large range of t values.

    a. Explore how the graphs change for different values of the lag time τ.

    b. Compare your results to the conventional phase space plots you obtained previously for the same parameters.

6. Extend your ODE solution to the chaotic pendulum with parameters

$$f = 0.52, \quad \alpha = 0.2, \quad \omega_0 = 1. \tag{3.23}$$

    a. Using your previously tested ODE solver, create phase-space orbits by plotting $[\theta(t), d\theta/dt(t)]$ for long time intervals (Figure 3.9).

    b. Indicate which parts of the orbits are transients.

    c. Correlate phase-space structures with the behavior of $\theta(t)$ by also plotting θ versus t (preferably next to $d\theta/dt$ versus θ).

  d. Gain some physical intuition about the flow in phase space by watching how it builds up with time.

7. For the second part of the chaotic pendulum study, use the same parameters as in first part, though now sweep through a range of ω values.

  a. Use initial conditions: $d\theta(0)/dt = 0.8$, and $\theta(0) = -0.0888$.
  b. Verify that for $\omega \simeq 0.6873$ there is a period-three limit cycle where the pendulum jumps between three orbits in phase space.
  c. Verify that for $\omega \simeq 0.694 - 0.695$ there are running solutions in which the pendulum goes over the top multiple times. Try to determine how many rotations are made before the pendulum settles down.
  d. For $\omega \simeq 0.686$ and long times, the solutions for very slightly different initial conditions tend to fill in bands in phase space. If needed, decrease your time step and try to determine how the bands get filled, in particular, just how small a difference in ω values separates the regular and the chaotic behaviors.

8. Create a Poincaré map for the chaotic pendulum.

3.3.4 Vibrating Pivot Pendulum

As an alternative to what we have called the chaotic pendulum, repeat the pendulum analysis for another version of the chaotic pendulum, this one with a vibrating pivot point (in contrast to our usual sinusoidal external torque):

$$\frac{d^2\theta}{dt^2} = -\alpha\frac{d\theta}{dt} - \left(\omega_0^2 + f\cos\omega t\right)\sin\theta. \tag{3.24}$$

Essentially, the acceleration of the pivot is equivalent to a sinusoidal variation of g or ω_0^2 [Landau & Lifshitz(77), DeJong(92), Gould et al.(06)]. The scatterplot in Figure 3.10 displays a sampling of $\dot\theta$ as a function of the magnitude of the vibrating pivot point.

3.4 Fourier Analysis of Oscillations

1. Consider a particle oscillating in the nonharmonic potential of (3.6):

$$V(x) = \frac{1}{p}k|x|^p, \qquad p \neq 2. \tag{3.25}$$

While nonforced oscillations in this potential are always periodic, they are not sinusoidal.

  a. For $p = 12$, decompose the solution $x(t)$ into its Fourier components.
  b. Determine the number of components that contribute at least 10%.
  c. Check that resuming the components reproduces the input $x(t)$.

2. Recall the perturbed harmonic oscillator (3.5):

$$V(x) = \frac{1}{2}kx^2\left(1 - \frac{2}{3}\alpha x\right) \qquad \Rightarrow \qquad F(x) = -kx(1 - \alpha x). \qquad (3.26)$$

For small oscillations $(x \ll 1/\alpha)$, $x(t)$ should be well approximated by solely the first term of the Fourier series.

   a. Fix your value of α and the maximum amplitude of oscillation x_{max} so that $\alpha x_{max} \simeq 10\%$. Plot up resulting $x(t)$ along with a pure sine wave.

   b. Decompose your numerical solution into a discrete Fourier spectrum.

   c. Make a semilog plot of the power spectrum $|Y(\omega)|^2$ as a function of x_{max}. Because power spectra often vary over several orders of magnitude, a semi-log plot is needed to display the smaller components.

   d. As always, check that summation of your transform reproduces the signal.

3. For cases in which there are one-, three-, and five-cycle structures in phase space (Figure 3.9), store your post-transients solutions for the chaotic pendulum, or for the double pendulum.

4. Perform a Fourier analysis of $x(t)$. Does it verify the statement that "the number of cycles in the phase-space plots corresponds to the number of major frequencies contained in $x(t)$"?

5. See if you can deduce a relation among the Fourier components, the natural frequency ω_0, and the driving frequency ω.

6. Examine your system for parameters that give chaotic behavior and plot the power spectrum in a semi-logarithmic plot. Does this verify the statement that "a classic signal of chaos is a broad Fourier spectrum"?

3.4.1   Pendulum Bifurcations

Fourier analysis and phase-space plots indicate that a chaotic system contains a number of dominant frequencies, and that the system tends to "jump" from one frequency to another. In contrast to a linear system in which the Fourier components occur simultaneously, in nonlinear systems the dominant frequencies may occur sequentially. Thus a sampling of the instantaneous angular velocity $\dot{\theta} = d\theta/dt$ of the chaotic pendulum for a large number of times indicates the frequencies to which the system is *attracted*, and, accordingly, should be related to the system's Fourier components.

1. Make a scatter plot of the sampled $\dot{\theta}$s for many times as a function of the magnitude of the driving torque.

2. For each value of f, wait 150 periods of the driver before sampling to permit transients to die off. Then sample $\dot{\theta}$ for 150 times at the instant the driving force passes through zero.

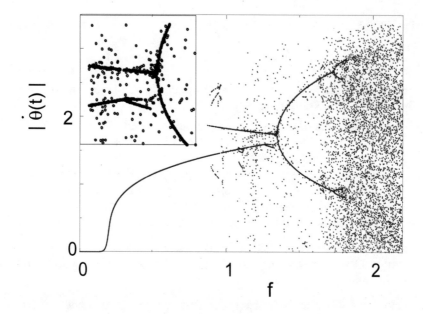

Figure 3.10. A bifurcation diagram for the damped pendulum with a vibrating pivot (see also the similar diagram for a double pendulum, Figure 3.12 right). The ordinate is $|d\theta/dt|$, the absolute value of the instantaneous angular velocity at the beginning of the period of the driver, and the abscissa is the magnitude of the driving torque f. Note that the heavy line results from the overlapping of points, not from connecting the points (see enlargement in the inset).

3. Plot values of $|\dot{\theta}|$ versus f as unconnected dots.

3.4.2 Sonification

Human ears respond to air vibrations in the approximate frequency range 20–20,000Hz, which our brain interprets as sound. So to create a sonification of an oscillation we need to map the oscillation's frequencies into the range of human hearing. Python and MATLAB have utilities to create sound files from arrays, as do third party software packages such as *Audacity*. In Python, the function `scipy.io.wavfile` is used to create a `wav` file, for example,

```
import numpy as np
from scipy.io.wavfile import write
data = np.random.uniform(-1,1,44100)          # Random samples  -1 < r < 1
scaled = np.int16(data/np.max(np.abs(data)) * 32767)
write('test.wav', 44100, scaled)
```

Sonify an harmonic and a nonharmonic oscillation, and listen to the differences. The overtones (higher harmonics) in the nonharmonic oscillation should make it sound

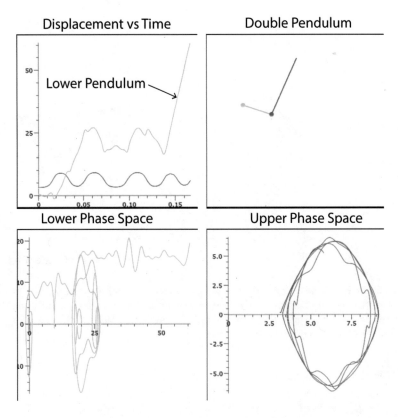

Figure 3.11. A large-angle, chaotic double pendulum (lower bob started at top).

more interesting.

3.5   The Double Pendulum

Repeat the preceding study of the chaotic pendulum but now do it for the double pendulum with no small angle approximations. This is a compound system in which a lower pendulum is attached to the bottom of an upper pendulum (Figure 3.11 top right). On the left of Figure 3.12 we see a phase space diagram for one bob undergoing fairly regular motion. Because each pendulum acts as a driving force for the other, we need not include an external driving torque to produce a chaotic system. In Figure 3.11 (and the animation DoublePend.mp4) we show visualizations for a chaotic double pendulum in which the lower pendulum started off upright.

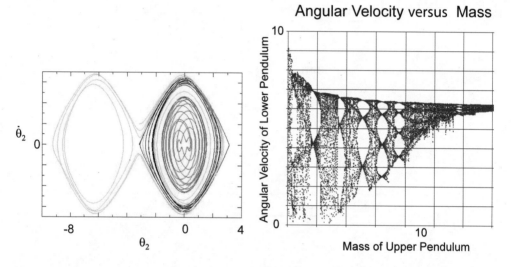

Figure 3.12. *Left:* Phase space trajectories for a double pendulum with $m_1 = 10m_2$ and with two dominant attractors. *Right:* A bifurcation diagram for the double pendulum displaying the instantaneous velocity of the lower pendulum as a function of the mass of the upper pendulum. (Both plots courtesy of J. Danielson.)

1. Show that the Lagrangian for the double pendulum is:

$$L = \text{KE} - \text{PE} = \frac{1}{2}(m_1 + m_2)l_1^2\dot{\theta}_1{}^2 + \frac{1}{2}m_2 l_2^2 \dot{\theta}_2{}^2 \qquad (3.27)$$
$$+ m_2 l_1 l_2 \dot{\theta}_1 \dot{\theta}_2 \cos(\theta_1 - \theta_2) + (m_1 + m_2)gl_1 \cos\theta_1 + m_2 gl_2 \cos\theta_2.$$

2. Use this Lagrangian to show that the equations of motion are

$$(m_1 + m_2)L_1\ddot{\theta}_1 + m_2 L_2 \ddot{\theta}_2 \cos(\theta_2 - \theta_1) = m_2 L_2 \dot{\theta}_2^2 \sin(\theta_2 - \theta_1) - (m_1 + m_2)g\sin\theta_1$$
$$L_2\ddot{\theta}_2 + L_1\ddot{\theta}_1 \cos(\theta_2 - \theta_1) = -L_1\dot{\theta}_1{}^2 \sin(\theta_2 - \theta_1) - g\sin\theta_2. \quad (3.28)$$

3. Deduce the equations of motion for small displacement of each pendulum from its equilibrium position (usually what is found in textbooks).

4. Deduce analytically the frequencies of slow and fast modes for the small angle oscillations.

5. Solve the equations of motion numerically without making any small angle approximations.

6. Verify that your numerical solution has a slow mode in which $\theta_1 = \theta_2$ and a fast mode in which θ_1 and θ_2 have opposite signs.

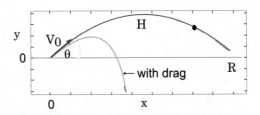

Figure 3.13. Schematics of the trajectories of a projectile fired with initial velocity V_0 in the θ direction. The lower curve includes air resistance.

7. Reproduce the phase space plots on the left in Figure 3.12 describing the motion of the lower pendulum for $m_1 = 10m_2$. When given enough initial kinetic energy to go over the top, the trajectories are seen to flow between two major attractors with energy being transferred back and forth between the pendula.

8. Reproduce the bifurcation diagram for the double pendulum shown on the right in Figure 3.12. This is created by sampling the set of instantaneous angular velocity $\dot{\theta}_2$ of the lower pendulum as it passes through its equilibrium position, and plotting the set as a function of the mass of the upper pendulum. The resulting structure is fractal with the bifurcations indicative of the dominant Fourier components.

9. Compute the Fourier spectrum for the double pendulum with the same parameters used for the bifurcation plot. Do the two plots correlate?

3.6   Realistic Projectile Motion

Figure 3.13 shows trajectories for a projectile shot at inclination θ and with an initial velocity V_0. If we ignore air resistance, the projectile has only the force of gravity acting on it and the trajectory will be a parabola with range $R = 2V_0^2 \sin\theta \cos\theta/g$ and maximum height $H = \frac{1}{2}V_0^2 \sin^2\theta/g$. Because a parabola is symmetric about its midpoint it does not describe what appears to be a sharp, nearly vertical, drop-off of baseballs and golf balls near the end of their trajectories.

1. Investigate several models for the frictional force:

$$\mathbf{F}^{(f)} = -k\,m\,|v|^n\,\frac{\mathbf{v}}{|v|}. \tag{3.29}$$

Here the $-\mathbf{v}/|v|$ factor ensures that the frictional force is always in a direction opposite that of the velocity.

a. Show that for our model of friction, the equations of motion are

$$\frac{d^2x}{dt^2} = -k\,v_x^n\,\frac{v_x}{|v|}, \qquad \frac{d^2y}{dt^2} = -g - k\,v_y^n\,\frac{v_y}{|v|}, \qquad |v| = \sqrt{v_x^2 + v_y^2}. \tag{3.30}$$

b. Consider three values for n, each of which represents a different model for the air resistance: 1) $n = 1$ for low velocities, 2) $n = 3/2$ for medium velocities, and 3) $n = 2$ for high velocities.

c. Modify your `rk4` program so that it solves the simultaneous ODEs for projectile motion (3.30) with friction ($n = 1$). In Listing 3.4 we present the program `ProjectileAir.py` that solves for the trajectory using a form of the velocity–Verlet algorithm accurate to second order in time. Here is its pseudocode:

```
#  Pseudocode for ProjectileAir.py: Projectile motion with air resistance
Import visual library
Initialize variables
Compute analytic R, T, & H
Initialize 2 graphical curves
Define plotNumeric
    loop over times
    compute & output numeric (x,y)
Define plotAnalytic
    loop over times
    compute & analytic numeric (x,y)
Plot up numeric & analytic results
```

d. What conclusions can you draw as to the effects of air resistance on the shape of the projectile?

e. The model (3.29) with $n = 1$ is applicable only for low velocities. Now modify your program to handle $n = 3/2$ (medium-velocity friction) and $n = 2$ (high-velocity friction). Adjust the value of k for the latter two cases such that the initial force of friction $k v_0^n$ is the same for all three cases.

f. Solve the equations of motion for several initial conditions and powers n.

g. How does friction affect the range R and the time aloft T?

h. What conclusion can you draw regarding the observation of balls appearing to fall straight down out of the sky?

3.6.1 Trajectory of Thrown Baton

Classical dynamics describes the motion of the baton as the motion of an imaginary point particle located at the center of mass (CM), plus a rotation about the CM. Because the translational and rotational motions are independent, each may be determined separately. This is made easier in the absence of air resistance, since then there are no torques on the baton and the angular velocity ω about the CM is constant.

1. The baton in Figure 3.14 is let fly with an initial velocity corresponding to a rotation about the center of the lower sphere. Extend the program developed for projectile motion of a point particle to one that computes the motion of the baton as a function of time. Ignore air resistance, assume that the bar is massless, and choose values for all of the parameters.

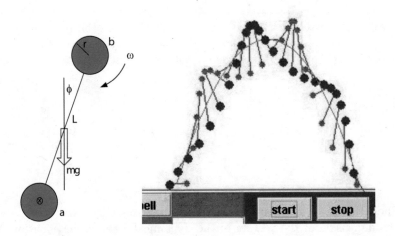

Figure 3.14. *Left:* The initial conditions for a baton as it is thrown. *Right:* The motion of the entire baton as its CM follows a parabola.

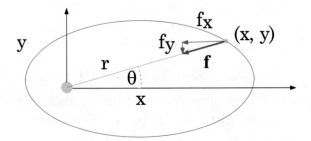

Figure 3.15. The gravitational force on a planet a distance r from the sun. The x and y components of the force are indicated.

   a. Plot the position of each end of the baton as functions of time. Figure 3.14 shows a typical result.

   b. Plot the translational kinetic energy, the rotational kinetic energy, and the potential energy of the baton, all as functions of time.

   c. Solve for the motion of the baton with an additional lead weight at its center, again making separate plots of each end of the baton.

3.7   Bound States

In contrast to the analytic case, the numerical solutions to orbit problems are straightforward. A simple approach is to express the forces and equations of motion in Cartesian coordinates (Figure 3.15), and then to solve the resulting two simultaneous ODEs:

$$\mathbf{F} = m\mathbf{a} = m\frac{d^2\mathbf{x}}{dt^2} \quad \Rightarrow \tag{3.31}$$

$$F_x = F^{(g)}\cos\theta = F^{(g)}\frac{x}{\sqrt{x^2+y^2}}, \qquad F_y = F^{(g)}\sin\theta = F^{(g)}\frac{y}{\sqrt{x^2+y^2}}, \tag{3.32}$$

$$\frac{d^2x}{dt^2} = -GM\frac{x}{(x^2+y^2)^{3/2}}, \qquad \frac{d^2y}{dt^2} = -GM\frac{y}{(x^2+y^2)^{3/2}}. \tag{3.33}$$

1. Show that in order to apply rk4 to simultaneous ODEs, we need only increase the dimension of the vectors in the dynamical form of the equation of motion from two to three:

$$y^{(0)} = x(t), \qquad y^{(1)} = \frac{dx(t)}{dt} = \frac{dy^{(0)}}{dt}, \tag{3.34}$$

$$y^{(2)} = y(t), \qquad y^{(3)} = \frac{dy(t)}{dt} = \frac{dy^{(2)}}{dt}, \tag{3.35}$$

$$\Rightarrow \quad f^{(0)} = y^{(1)}(t), \quad f^{(1)} = \frac{F_x(\mathbf{y})}{m}, \quad f^{(2)} = y^{(3)}(t), \quad f^{(3)} = \frac{F_y(\mathbf{y})}{m}. \tag{3.36}$$

2. What are the explicit expressions for $f^{(2)}$ and $f^{(3)}$ in terms of the $y^{(i)}$s?

3. Modify your ODE solver program to solve the equations of motion.

   a. Assume units such that $GM = 1$ and the initial conditions

$$x(0) = 0.5, \quad y(0) = 0, \quad v_x(0) = 0.0, \quad v_y(0) = 1.63. \tag{3.37}$$

   b. Check that you are using small enough time steps by verifying that the orbits remain closed and fall upon themselves for long periods of time.
   c. Experiment with the initial conditions until you find the ones that produce a circular orbit (a special case of an ellipse).
   d. Progressively increase the initial velocity until the orbits become unbound.
   e. Using the same initial conditions that produce elliptical orbits, investigate the effect of varying continuously the power in Newton's law of gravitation from two. Even small changes should cause the orbits to precess, as predicted by general relativity.

4. Consider the motion of a particle of mass m with angular momentum l in an inverse-square force field subjected to an inverse-cube perturbation:

$$\mathbf{F} = \left(\frac{-k}{r^2} + \frac{C}{r^3}\right)\hat{\mathbf{e}}_r, \tag{3.38}$$

The solutions to the orbit equation fall into three classes depending on the constant C:

$$|C| < l^2/m, \quad |C| = l^2/m, \quad |C| > l^2/m. \tag{3.39}$$

  a. Solve the equations of motion numerically for each of the three conditions given in (3.39).
  b. Indicate which of your solutions are bound, that is, the particle remains localized to some region of space.
  c. What are the conditions for the solutions to look like ellipses with a slow rate of precession?

5. A mass is in a circular orbit about an attractive potential $U(r)$.

  a. Find an analytic expression for the frequency of oscillations for small radial perturbations about the circular orbit.
  b. Consider now the potential

$$U(r) = \frac{-k}{r^{n-1}}, \tag{3.40}$$

  with n an integer. Prove analytically that the angle through which the orbit rotates as r varies from its minimum to maximum value is $\pi/\sqrt{3-n}$.
  c. Solve numerically for the orbits for various values of n and plot your results.
  d. Is it true that the orbit returns to itself only for $n = 2$?
  e. Plot the *phase space portraits* for various combinations of variables for oscillations about a circular orbit, and for various values of n.

6. A particle is confined to a 2-D square well potential of depth W and radius R,

$$V(r) = -W\theta(R - r). \tag{3.41}$$

  a. Solve for the 2-D orbits of a particle of mass $m = 1$ within this square well, either by deriving the appropriate orbit equation, or geometrically by using a ruler, a pencil, and a piece of paper.
  b. Explain why there are problems solving for a square well potential numerically. (*Hint:* Think about derivatives.)
  c. As an approximation to a square well, try the potential

$$V(r) = V_0 r^{10}, \tag{3.42}$$

  which is small for $r < 1$, though gets large rapidly for $r > 1$.
  d. Start by looking at $x(t)$ and $y(t)$ and making sure that they are reasonable (like a free particle for small x or small y, and then like a particle hitting a wall for larger x or y).
  e. Next look at the trajectories $[x(t), y(t)]$ and see if they seem close to what you might expect for a mass reflecting off the walls of a circular cavity.

f. Evaluate the angular momentum $l(t)$ and the total energy of the mass $E(t)$ as functions of time and determine their level of constancy.

7. Solve for the orbits of a particle of mass $m = 1$ confined within a 2-D racetrack shaped (ellipse-like) potential,

$$V(r) = ax^n + by^n, \qquad (3.43)$$

where you are free to choose values for the parameters a, b, and n.

a. Test your program for the two cases $(b = 0, n = 2)$ and $(a = 0, n = 2)$. You should obtain simple harmonic motion with frequency $\omega_0 = \sqrt{a}$ and $\sqrt{b}$.
b. Verify that the orbits are symmetric in the xy plane.
c. Verify that the angular momentum and energy of the particle are constants as functions of time for all $a = b$ values and for all values of n.
d. Plot the orbits for several $a \neq b$ cases.
e. Check that for large n values the orbits look like internal reflections from the racetrack walls.
f. Search for those combinations of (a, b) values for which the orbits close on themselves.
g. Search for those combinations of (a, b) values for which the orbits precess slowly.
h. Evaluate the energy and angular momentum as functions of time for an $a \neq b$ case, and comment on their variability.

3.8   Three-Body Problems: Neptune, Two Suns, Stars

The planet Uranus was discovered in 1781 by William Herschel, and found to have an orbital period of approximately 84 years. Nevertheless, even before it had completed an entire orbit around the sun in the time since its discovery, it was observed that Uranus's orbit was not precisely that predicted by Newton's law of gravity. Accordingly, it was hypothesized that a yet-to-be-discovered and distant planet was perturbing Uranus's orbit. The predicted planet is called Neptune.

Use these data for the calculation:

	Mass (10^{-5} Solar Masses)	Distance (AU)	Orbital Period (Years)	Angular Position (in 1690)
Uranus	4.366244	19.1914	84.0110	~205.64°
Neptune	5.151389	30.0611	164.7901	~288.38°

You may enter these data into your program much as we did:

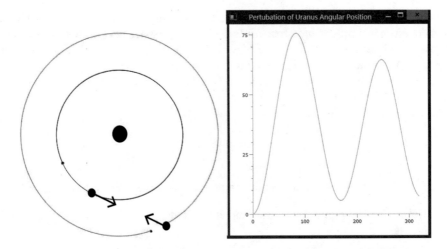

Figure 3.16. A screenshot of an animation from `UranusNeptune.py` showing: *Left:* The orbits of Uranus (inner circle) and of Neptune (outer circle) with the sun in the center. The arrow on the outer circle indicates Neptune's perturbation on Uranus. *Right:* The perturbation on Uranus as a function of angular position.

G = 4*pi*pi (AU, Msun=1)	mu = 4.366244e-5 = Uranus
M = 1.0 = Sun mass	mn = 5.151389e-5 = Neptune
du = 19.1914 = Uran Sun dist	dn = 30.0611 = N sun dist
Tur = 84.0110 = Uran Period	Tnp = 164.7901 = N Period
omeur = 2*pi/Tur = U omega	omennp = 2*pi/Tnp = N omega
omreal = omeur (UA/yr)	urvel = 2*pi*du/Tur = Uran omegal
npvel = 2*pi*dn/Tnp = Nept omega UA/yr	radur = 205.64*pi/180
urx = du*cos(radur) = init x 1690	ury = du*sin(radur) = init y 1690
urvelx = urvel*sin(radur)	urvely = -urvel*cos(radur)
radnp = 288.38*pi/180. = Nept angular pos.	

1. Use rk4 to compute the variation in angular position of Uranus with respect to the Sun as a result of the influence of Neptune during one complete orbit of Neptune. Consider only the forces of the Sun and Neptune on Uranus. Use astronomical units in which $M_s = 1$ and $G = 4\pi^2$.

2. As shown in Figure 3.16, assume that the orbits of Neptune and Uranus are circular and coplanar, and that the initial angular positions with respect to the x axis are as in the table above.

3.8.1   Two Fixed Suns with a Single Planet

The three-body problem in which three particles move via pairwise interactions can be complicated and chaotic. Here we ask you to examine a simple version in which two heavy stars 1 and 2 are kept at a fixed separation along the x axis, while a lighter planet moves about them [Row 2004]. We use natural units $G = 1$ to keep the calculations simpler, and treat all bodies as point particles. It is best to view the output as animations so that you can actually see the planet pass through a number of orbits. A characteristic of this kind of chaotic system is that there are periods with smooth precessions followed by chaotic behavior, and then smooth precession again. This means that we on earth are lucky having only one sun as this makes the year of reliably constant length [Liu(14)].

1. Start with $M_1 = M_2 = 1$ and the planet at $(x, y)_0 = (0.4, 0.5)$ with $(v_x, v_y)_0 = (0, -1)$.

2. Set $M_2 = 2$ and see if the planet remains in a stable orbit about sun 2.

3. Return to the equal mass case and investigate the effect of differing initial velocities.

4. Make M_1 progressively smaller until it acts as just a perturbation on the motion around planet 2, and see if the year now becomes of constant length.

5. What might be the difficulty of having life develop and survive in a two sun system?

6. Explore the effect on the planet of permitting one of the suns to move.

3.8.2   Hénon-Heiles Bound States

The Hénon-Heiles potential

$$V(x, y) = \frac{1}{2}x^2 + \frac{1}{2}y^2 + x^2 y - \frac{1}{3}y^3 \tag{3.44}$$

is used to model the interaction of three close astronomical objects. The potential binds the objects near the origin though releases them if they move far out.

1. Show that the minimum in the potential for $x = 0$ occurs at $y = 1$.

2. Show that the value of the potential at its minimum implies bound orbits occur for energies $0 < E < 1/6$.

3. Show that the equations of motion following from Hamiltonian equations are:

$$\frac{dp_x}{dt} = -x - 2xy, \quad \frac{dp_y}{dt} = -y - x^2 + y^2, \quad \frac{dx}{dt} = p_x, \quad \frac{dy}{dt} = p_y. \tag{3.45}$$

4. Solve for the position $[x(t), y(t)]$ for a number of initial conditions.

5. Verify that the orbits are bounded for energies less than $1/6$.

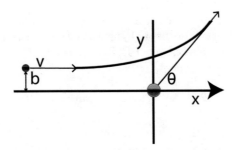

Figure 3.17. A particle with impact parameter b incident along the y axis is scattered by a force center at the origin into a scattering angle θ.

6. As check on your numerics, verify that the Hamiltonian remains constant.

7. Produce a Poincare section by creating a (y, p_y) plot, adding a point each time $x \simeq 0$. With the energy fixed, make several plots for different initial conditions.

8. Isolate a smaller region of phase space and look for unusual structures.

3.9 Scattering

3.9.1 Rutherford Scattering

A particle of mass $m = 1$ and velocity v is scattered by the force center

$$V(r) = \frac{\alpha}{r^2} \tag{3.46}$$

with α positive. As seen in Figure 3.17, the particle starts at the left ($x = -\infty$) with an impact parameter b (distance above the x axis) and is scattered into an angle θ. Because the force center does not recoil, the magnitude of the initial and final velocities are the same, though their directions differ.

1. Calculate and plot the positions $[x(t), y(t)]$ for a range of impact parameters b starting from negative values and proceeding to positive ones, and for a range of velocities v.

2. For fixed angular momentum, for what values of α does the particle make one and two revolutions about the center of force before moving off?

3. See if you can find an orbit that collapses into the origin $r = 0$. This should be possible for angular momentum $l^2 < 2m\alpha$.

4. Use a central-difference approximation to calculate the derivative $d\theta/db$ as a function of θ.

5. Deduce the scattering angle for a trajectory by examining the trajectory of the scattered particle at a large enough distance from the target so that the potential no longer has much effect, say $|PE|/KE \leq 10^{-6}$. The scattering angle is then deduced from the components of velocity,

$$\theta = \tan^{-1}(v_y/v_x) = \texttt{math.atan2(y, x)}. \qquad (3.47)$$

Here `atan2` computes the arctangent in the correct quadrant and eliminates the possibility of division by zero when computing $\tan^{-1}(y/x)$.

6. Calculate the differential scattering cross section from the dependence of the scattering angle upon the classical impact parameter b [Marion & Thornton(03)]:

$$\sigma(\theta) = \left| \frac{d\theta}{db} \right| \frac{b}{\sin\theta(b)}. \qquad (3.48)$$

7. In units for which the Coulomb potential between a target of charge $Z_T e$ and a projectile of charge $Z_P e$ is

$$V(r) = \frac{Z_T Z_P e^2}{r}, \qquad (3.49)$$

the cross section for pure Coulomb scattering is given by the Rutherford formula

$$\sigma(\theta)_R = \left(\frac{Z_T Z_P e^2}{4E \sin^2 \theta/2} \right)^2, \qquad (3.50)$$

where $E = p^2/2\mu$ is essentially the projectile's energy, with p the COM momentum, and μ the reduced target-projectile mass (for a fixed scattering center, the target mass is infinite, and so $\mu = m_P$) [Landau(96)]. Compare the θ dependence of your computed results to that of the Rutherford cross section.

3.9.2   Mott Scattering

Rutherford scattering is appropriate for the scattering of two spinless particles (no magnetic moments). In many important applications, high energy electrons are scattered from the Coulomb field of heavy nuclei and the magnetic moment (spin) of the electron interacts with the Coulomb field. This is called Mott scattering and leads to a multiplicative correction term being affixed to the Rutherford cross section:

$$\sigma(\theta)_{Mott} = \sigma(\theta)_R \left(1 - \frac{v^2}{c^2} \sin^2 \theta/2 \right). \qquad (3.51)$$

1. Compute and plot the Mott and Rutherford differential cross sections for the scattering of 150 MeV electrons from gold nuclei.

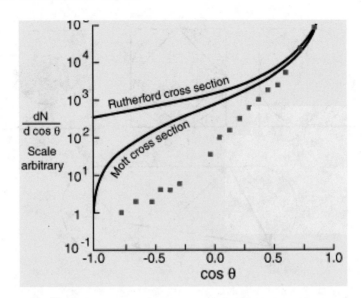

Figure 3.18. Experimental results for high-energy electron scattering from a nucleus showing deviations from assumption of a point nucleus.

2. At what angle does the difference reach 50%?

3. At what angle does the difference reach 10%?

4. It was found experimentally by Hofstadter et al. that the reduction in 150° cross section for 150 MeV electrons scattering was 1000 times greater than that predicted by the Mott formula. Apparently, there must be a reduction in the strength of the Coulomb potential from that given by (3.49). This reduction could arise from the electron penetrating into the nucleus, and , accordingly, not being affected by all of the nucleus's electric charge. Quantum mechanics tells us that the correction for the finite nuclear size is approximately

$$\sigma(\theta)_{finite} \simeq \sigma(\theta)|_{Mott} \left(1 - \frac{q^2 R_{rms}^2}{6\hbar^2}\right)^2, \qquad (3.52)$$

where $q^2 = 2p^2(1 - \sin\theta)$ is the scattered electron's momentum transfer and R_{rms} is the root mean square radius of the nucleus. Based on this reduction in the cross section, what would you estimate as the size of the gold nucleus?

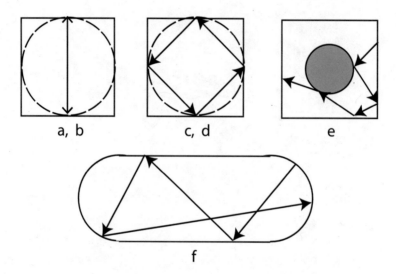

Figure 3.19. Square (a, c), circular (b, d), Sinai (e), and stadium billiards (f). The arrows are trajectories. The stadium billiard has two semicircles on the ends.

3.9.3   Chaotic Scattering

1. One expects the scattering of a projectile from a force center to be a continuous process. Nevertheless in some situations the projectile undergoes multiple internal scatterings and ends up with a final trajectory that seems unrelated to the initial one. Specifically, consider the 2-D potential [Blehel et al.(90)]

$$V(x,y) = \pm x^2 y^2 e^{-(x^2+y^2)}. \tag{3.53}$$

As seen in Figure 3.20, this potential has four circularly symmetric peaks in the xy plane. The two signs produce repulsive and attractive potentials, respectively.

a. Show that the two simultaneous equations of motions are:

$$m\frac{d^2x}{dt^2} = \mp 2y^2 x(1 - x^2)e^{-(x^2+y^2)}, \qquad m\frac{d^2y}{dt^2} = \mp 2x^2 y(1 - y^2)e^{-(x^2+y^2)}. \tag{3.54}$$

b. Show that the force vanishes at the $x = \pm 1$ and $y = \pm 1$ peaks in Figure 3.20, that is, where the maximum value of the potential is $V_{\max} = \pm e^{-2}$. This sets the energy scale for the problem.

c. Apply the `rk4` method to solve the simultaneous second-order ODEs.

d. The initial conditions are 1) an incident particle with only an x component of velocity and 2) an impact parameter b (the initial y value). You do not need to vary the initial x, though it should be large enough such that PE/KE $\leq 10^{-6}$, which means that the KE $\simeq E$.

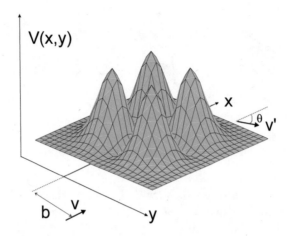

Figure 3.20. Scattering from the potential $V(x,y) = x^2 y^2 e^{-(x^2+y^2)}$ which in some ways models a pinball machine. The incident velocity v is in the y direction, and the impact parameter (y value) is b. The velocity of the scattered particle is v' and its scattering angle is θ.

e. Good parameters to use are $m = 0.5$, $v_y(0) = 0.5$, $v_x(0) = 0.0$, $\Delta b = 0.05$, $-1 \le b \le 1$. You may want to lower the energy and use a finer step size once you have found regions of rapid variation in the cross section.

f. Plot a number of trajectories $[x(t), y(t)]$, some being smooth and others being jumpy. In particular, try to show the apparently discontinuous changes in scattering angle.

g. Plot a number of phase space trajectories $[x(t), \dot{x}(t)]$ and $[y(t), \dot{y}(t)]$. How do these differ from those of bound states?

h. Compute the scattering angle $\theta = \texttt{atan2(Vx,Vy)}$ by determining the velocity components of the scattered particle after it has left the interaction region $(\text{PE/KE} \le 10^{-6})$.

i. Which characteristics of a trajectory lead to discontinuities in $d\theta/db$ and thus in the scattering cross section $\sigma(\theta)$ (3.48)?.

j. Run the simulations for both attractive and repulsive potentials and for a range of energies less than and greater than $V_{max} = \exp(-2)$.

k. Another way to find unusual behavior in scattering is to compute the *time delay* $T(b)$, that is, the increase in the time it takes a particle to travel through the interaction region. Look for highly oscillatory regions in the semilog plot of $T(b)$, and once you find some, repeat the simulation at a finer scale by setting $b \simeq b/10$ and observe the trajectories.

2. Figure 3.21 shows three setups in which one, two, and three hard disks are attached to the surface of a 2-D billiard table. In each case the disks scatter point particles elastically (no energy loss). The disks all have radius R and have

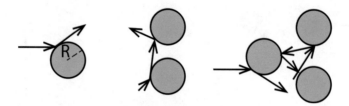

Figure 3.21. One, two, and three stationary disks on a flat billiard table scatter point particles elastically, with some of the internal scattering leading to trapped, periodic orbits.

center-to-center separations of a, with the three disks placed at the vertices of an equilateral triangle. Also shown in the figure are trajectories of particles scattered from the disks, with some of the internal scattering leading to trapped, periodic orbits in which the projectile bounces back and forth endlessly. For the two-disk case there is just a single trapped orbit, but for the three-disk case there are infinitely many, which leads to chaotic scattering. In §6.8.3 we explore the quantum mechanical version of this problem. Note, the quantum version of this problem 3QMdisks.py in Listing 6.17 has a potential subroutine that can be used to model the present disks.

3. Modify the program already developed for the study of scattering from the four-peaked Gaussian (3.53) so that it can be applied to scattering from one, two, or three disks.

4. Repeat the study of scattering from the four-peak Gaussian in §3.9 though now for the one-, two-, and three-disk case.

   a. Since infinite potential values would cause numerical problems, pick instead a very large value for the potential and verify that increasing the strength of the potential even further has no significant effect on the scattering.

   b. Explore different geometries to find some that produce chaos. This should be near $a/R \simeq 6$ for the three hard disks.

3.10   Billiards

Deriving its name from the once-popular parlor game, a mathematical *billiard* describes a dynamical system in which a particle moves freely in a straight line until it hits a boundary wall, at which point it undergoes specular reflection (equal incident and reflected angles), and then continues on in a straight line until the next collision. The confining billiard table can be square, rectangular, circular, polygonial, or a combination of the preceding, and can be three-dimensional or in curved spaces. These are Hamiltonian systems in which there is no loss of energy, in which the motion continues endlessly, and which often display chaos.

In Figure 3.19 we show square (a, c), circular (b, d), Sinai (e), and stadium billiards
(f), with the arrows indicating possible trajectories. Note how right-angle collisions
lead to two-point periodic orbits for both square and circular billiards, while in (c)
and (d) we see how $45°$ collisions leads to four-point periodic orbits. Figures (e) and
(d) show nonperiodic trajectories that are ergodic, that is, orbits that in time will pass
with equal likelihood through all points in the allowed space. These latter orbits lead
to chaotic behaviors.

1. The problems for this section are for you to write programs that compute the
   trajectories for any or all of these billiards and for a range of initial conditions.
   This is straightforward computationally (it's just geometry) and is an excellent
   way to study chaos. Preferably, your programs should produce animated output
   so you can view the trajectories as they occur. In Listing 3.3 we give a sample
   program for a square billiard that produces an animation using the VPython
   package (we do the quantum version of this problem in §6.8.3).

2. Have your program compute, keep track of, and plot the distance between successive collision points as a function of collision number. The plot should be
   simple for periodic motion, but show irregular behavior as the motion becomes
   chaotic. *Hint:* 20–30 collisions typically occur before chaos sets in.

3. Keep in mind that not all initial conditions lead to chaos, especially for circles,
   and so you may need to do some scanning of initial conditions.

4. Keep track of how many collisions occur before chaos sets in. You need at least
   this many collisions to test hypersensitivity to initial conditions.

5. For initial conditions that place you in the chaotic regime, explore the difference
   in behavior for a relatively slight ($\leq 10^{-3}$) variation in initial conditions.

6. Try initial conditions that differ at the machine precision level to gauge just how
   sensitive chaotic trajectories really are to initial conditions (be patient).

7. How much does the number of steps to reach chaos change for a 10^{-3} variation?

3.11 Lagrangian and Hamiltonian Dynamics

3.11.1 Hamilton's Principle

As illustrated in Figure 3.22, Hamilton's principle states that the actual trajectory
of a physical system described by a set of generalized coordinate $\mathbf{q} = (q_1, q_2, \cdots, q_N)$
between two specific states 1 and 2 is such that the action integral is stationary under
variation in $\mathbf{q}$:

$$S[\mathbf{q}] = \int_{t_1}^{t_2} L(\mathbf{q(t)}, \dot{\mathbf{q}}, t)\, dt, \qquad \frac{\delta S}{\delta \mathbf{q}(t)} = 0, \qquad (3.55)$$

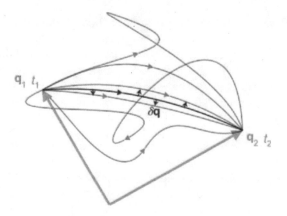

Figure 3.22. A physical system evolves such that out of the many possibilities, the actual path taken through configuration space is that which produces a stationary action, $\delta S = 0$ [Penrose(07)].)

where L is the Lagrangian of the system.

1. You are given the fact that a particle falls a known distance d in a known time $t = \sqrt{2D/g}$. Assume a quadratic dependence on distance and time,

$$d = \alpha t + \beta t^2. \tag{3.56}$$

   Show that the action $S = \int_0^{t_0} L\, dt$ for the particle's trajectory is an extremum only when $\alpha = 0$ and $\beta = g/2$.

2. Consider a mass m attached to a harmonic oscillator with characteristic frequency $\omega = 2\pi$ that undergoes one period $T = 1$ of an oscillation:

$$x(t) = 10\cos(\omega t). \tag{3.57}$$

   a. Propose a modification of the known analytic form that agrees with it at $t = 0$ and $t = T$, though differs for intermediate values of t. Make sure that your proposed form has an adjustable parameter that does not change the $t = 0$ or $t = T$ values.

   b. Compute the action for an entire range of values for the parameter in your proposed trajectory and thereby verify that only the known analytic form yields a minimum action.

3.11.2 Lagrangian & Hamiltonian Problems

1. A bead of mass m moves without friction on a circular hoop of wire of radius R. The hoop's axis is vertical in a uniform gravitational field and the hoop is rotating at a constant angular velocity ω.

a. Derive the Lagrange equations of motion for this system.

b. Determine the angle that the bead assumes at equilibrium, that is, an angle that does not change with time.

c. For small perturbations about this equilibrium configuration, what is the frequency ω_0 of oscillations?

d. Choose parameter values and solve Lagrange equations numerically.

e. Verify that the analytic expression you have derived for the equilibrium angle and for the frequency of oscillations about the equilibrium position agrees with the numerical results.

f. Examine some initial conditions that lead to nonequilibrium positions of the bead.

g. Plot the time dependence of the bead's position and its trajectory in phase space for a wide range of initial conditions.

2. Consider a 1-D harmonic oscillator with displacement q and momentum p. The energy

$$E(p, q) = \frac{p^2}{2m} + \frac{m\omega^2 q^2}{2} \tag{3.58}$$

is an integral of the motion, and the area of a periodic orbit is

$$A(E) = \oint p\, dq = 2 \int_{q_{min}}^{q_{max}} p\, dq. \tag{3.59}$$

a. Use the analytic, or numeric, solution for simple harmonic motion to compute the area $A(E)$.

b. Compute the derivative $T = dA(E)/dE$ via a central-difference approximation and compare to the analytic answer.

c. Now repeat this problem using a nonlinear oscillator for which there is only a numerical solution. (Oscillators of the form $V = kx^p$ with p even should work just fine.) You can determine the period from the time dependence of your solution, and then use your solution to compute $A(E)$ for different initial conditions.

3. Verify **Liouville's Theorem** for the realistic (nonlinear) pendulum without friction. In particular, verify that the flow of trajectories in phase space is similar to that of an incompressible fluid.

a. The equations of motion for this problem are

$$\frac{dy^{(0)}}{dt} = y^{(1)}(t), \quad \frac{dy^{(1)}}{dt} = -\sin(y^{(0)}(t)). \tag{3.60}$$

You should already have worked out solutions to these equations in your study of nonharmonic motion.

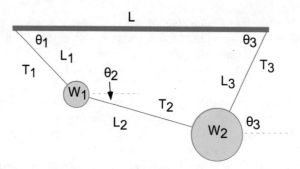

Figure 3.23. A free body diagram for one weight in equilibrium. Balancing the forces in the x and y directions for all weights leads to the equations of static equilibrium.

    b. Start out with a set of initial configurations for the pendulum that fall upon a circle in the (y_1, y_2) phase space below the separatrix, and then follow the paths taken by different points higher up on the circle.

    c. Verify that different points on the starting circle and at the center of the circle take different times to orbit through phase space.

    d. Examine the orbits for at least 5 times: $0, \tau/4, \tau/2, 3\tau/4, \tau$, where τ is the period for small oscillations.

    e. Verify that the domain of points that started on the original circle gets progressively more deformed as they proceeds through phase space. Make sure also to include the point at the center of the circle.

    f. Verify that the density of points in phase space remains constant in time.

    g. Follow the passage through phase space for an initial circle of points that starts off with its center along the separatrix.

    h. Note the effect on the phase space orbits of including a frictional torque.

3.12 Weights Connected by Strings (Hard)

Two weights $(W_1, W_2) = (10, 20)$ are hung from three pieces of string with lengths $(L_1, L_2, L_3) = (3, 4, 4)$ and a horizontal bar of length $L = 8$ (Figure 3.23). Determine the angles assumed by the strings and the tensions exerted by the strings. (An analytic solution exists, but it is painful, and in any case, the techniques used here are applicable to more complicated problems.)

    1. Write out the equations for the geometric constraints that the horizontal length of the structure is L and that the left and right ends of the string are at the same height.

    2. Write out the equations resulting from balancing the horizontal and vertical forces to zero for each mass.

3. Why can't we assume the equilibrium of torques?

4. You now have a set of simultaneous, nonlinear equations to solve. You may treat θ as a single unknown or you may treat $\sin\theta$ and $\cos\theta$ as separate unknowns with an additional identity relating them. In any case, you will need to search for a solution.

5. We recommend setting the problem up as a matrix problem with all of the unknowns as a vector, and using a Newton-Raphson algorithm to search in a multi–dimensional space for a solution. For example, our solution [LPB(15)] was of the form

$$\vec{f} + F' \, \vec{\Delta x} = 0, \quad \Rightarrow \quad F' \vec{\Delta x} = -\vec{f}, \tag{3.61}$$

$$\vec{\Delta x} = \begin{pmatrix} \Delta x_1 \\ \Delta x_2 \\ \cdot \cdot \\ \Delta x_9 \end{pmatrix}, \quad \vec{f} = \begin{pmatrix} f_1 \\ f_2 \\ \cdot \cdot \\ f_9 \end{pmatrix}, \quad F' = \begin{pmatrix} \partial f_1/\partial x_1 & \cdots & \partial f_1/\partial x_9 \\ \partial f_2/\partial x_1 & \cdots & \partial f_2/\partial x_9 \\ & \ddots & \\ \partial f_9/\partial x_1 & \cdots & \partial f_9/\partial x_9 \end{pmatrix}, \tag{3.62}$$

$$\Rightarrow \quad \vec{\Delta x} = -F'^{-1}\vec{f},$$

where the inverse must exist for a unique solution.

6. At some stage, derivatives are needed, and it is easiest to use a forward-difference approximation for them:

$$\frac{\partial f_i}{\partial x_j} \simeq \frac{f_i(x_j + \delta x_j) - f_i(x_j)}{\delta x_j}, \tag{3.63}$$

where δ_j is an arbitrary small value.

3.13   Code Listings

```
# ForcedOsc.py Driven Oscillator with Matplotlib

import numpy as np, matplotlib.pylab as plt
from rk4Algor import rk4Algor

F=1; m=1; mu=0.001; omegaF=2; k=1               # Constants
omega0 = np.sqrt(k/m)                       # Natural frequency
y = np.zeros((2))
tt = []; yPlot=[]                            # Empty list init

def f(t,y):                             # RHS force function
    freturn = np.zeros((2))                  # Set up 2D array
    freturn[0] = y[1]
    freturn[1] = 0.1*np.cos(omegaF*t)/m-mu*y[1]/m-omega0**2*y[0]
    return freturn
```

```
y[0] = 0.1                                      # Initial x
y[1] = 0.3                                      # Inital speed
f(0,y)                                          # Function at t=0
dt = 0.01;    i = 0
for t in np.arange(0,100,dt):
    tt.append(t)
    y = rk4Algor(t,dt,2, y, f)                          # Call rk4
    yPlot.append(y[0])                  # Use y[1] for velocity
    i = i+1
plt.figure()
plt.plot(tt,yPlot)
plt.title('$\omega_f$=2,k=1,m=1,$\omega_0$=1,$\lambda = 0.001$')
plt.xlabel('t')
plt.ylabel('x')
plt.show()
```

Listing 3.1. ForcedOscillate.py solves the ODE for forced oscillator using rk4 and plotting with Matplotlib.

```
# ODEsympy.py: symbolic soltn of HO ODE using sympy

from sympy import *

f, g = symbols('f g',cls=Function)          # Make f a function
t, kap, w0 = symbols('t kap w0')
f(t)
f(t).diff(t)
print( "\n ODE to be solved:" )
diffeq = Eq(f(t).diff(t,t) + kap*(f(t).diff(t)) + (w0*w0)*f(t))
print( diffeq)
print( "\n Solution of ODE:")
ff = dsolve(diffeq,f(t))                            # Solves ODE
F = ff.subs(t,0)
print( ff)
```

Listing 3.2. ODEsympy.py uses the symbolic manipulation package Sympy to evaluate the constants in the harmonic oscillator solution (3.2).

```
# SqBillardCM.py: Animated classical billiards on square table

from vpython import *

dt = 0.01;    Xo = -90.;   Yo =   -5.4;
v = vector(13.,13.1,0)
r0 = r= vector(Xo,Yo,0); eps = 0.1;  Tmax = 500; tp = 0
scene = canvas(width=500, height=500, range=120,
               background=color.white, foreground=color.black)
table = curve(pos=([(-100,-100,0),(100,-100,0),(100,100,0),
    (-100,100,0),(-100,-100,0)]))
ball = sphere(pos=vector(Xo,Yo,0), color=color.red, radius=0.1,
```

```
      make_trail=True)
for t in arange(0,Tmax,dt):
    rate(5000)
    tp = tp + dt
    r = r0 + v*tp
    if(r.x>= 100 or r.x<=-100):          # Right and left walls
        v = vector(-v.x,v.y,0)
        r0 = vector(r.x,r.y,0)
        tp = 0
    if(r.y>= 100 or r.y<=-100):          # Top and bottom walls
        v = vector(v.x,-v.y,0)
        r0 = vector(r.x,r.y,0)
        tp = 0
    ball.pos=r
```

Listing 3.3. SqBilliardCM.py Computes trajectories within a square billiard, and produces animated output using the Visual package.

```
# ProjectileAir.py: VPython, O(dt^2) projectile + drag

from vpython import *
from numpy import *

v0 = 22.;    angle = 34.;    g = 9.8;    kf = 0.8;    N = 5
v0x = v0*cos(angle*pi/180.);    v0y = v0*sin(angle*pi/180.)
T = 2.*v0y/g;    H = v0y*v0y/2./g;    R = 2.*v0x*v0y/g
print('No Drag T =', T,', H =', H,', R =', R)

graph1 = graph(title='Projectile +- Drag', xtitle='x',
   ytitle='y', xmax=R, xmin=-R/20., ymax=8, ymin=-6.0)
funct = gcurve(color=color.red)
funct1 = gcurve(color=color.green)

def plotNumeric(k):
 vx = v0*cos(angle*pi/180.)
 vy = v0*sin(angle*pi/180.)
 x = 0.0
 y = 0.0
 dt = vy/g/N/2.
 print("\n       With Friction   ")
 print("      x               y")
 for t in arange(0,0.7*T,0.1):
     #rate(30)
     dt=0.1
     vx = vx - k*vx*dt
     vy = vy - g*dt - k*vy*dt
     x = x + vx*dt
     y = y + vy*dt
     funct.plot(pos=(x,y))
     print(" %13.10f  %13.10f "%(x,y))

def plotAnalytic():
    v0x = v0*cos(angle*pi/180.)
    v0y = v0*sin(angle*pi/180.)
    dt = 2.*v0y/g/N
    print("\n        No Friction   ")
```

```
    print("           x              y")
    for t in arange(0,N,0.1):
        x = v0x*t
        y = v0y*t -g*t*t/2.
        funct1.plot(pos=(x,y))
        print(" %13.10f   %13.10f"%(x ,y))

plotNumeric(kf)
plotAnalytic()
```

Listing 3.4. **ProjectileAir.py** solves for a projectile's trajectory using a form of the velocity–
Verlet algorithm accurate to second order in time.

```
# UranusNeptune.py:  Orbits of Neptune & Uranus

from visual.graph import *
scene = display(width=600,height=600,
    title = 'White Neptune & Black Uranus', range=40)
sun = sphere(pos=(0,0,0), radius=2, color=color.yellow)
escenau = gdisplay(x=600,width=400,height=400,
    title='Pertubation of Uranus Angular Position')
graphu = gcurve(color=color.cyan)

escenan = gdisplay(x=800,y=400,width=400,height=400)
graphn = gcurve(color=color.white)
rfactor = 1.8e-9
G = 4*pi*pi          # in units T in years, R AU, Msun=1
mu = 4.366244e-5     # mass Uranus in solar masses
M = 1.0              # mass Sun
mn = 5.151389e-5     # Neptune mass in solar masses
du = 19.1914         # distance Uranus Sun in AU
dn = 30.0611         # distance Neptune sun in AU
Tur = 84.0110        # Uranus Orbital Period yr
Tnp = 164.7901       # Neptune Orbital Period yr
omeur = 2*pi/Tur     # Uranus angular velocity (2pi/T)
omennp = 2*pi/Tnp    # Neptune angular velocity
omreal = omeur
urvel = 2*pi*du/Tur    # Uranus orbital velocity UA/yr
npvel = 2*pi*dn/Tnp    # Neptune orbital velocity UA/yr
# 1 Uranus at lon 2 gr 16 min sep 1821
radur = (205.64)*pi/180.   # to radians in 1690 -wrt x-axis
urx = du*cos(radur)        # init x- pos ur. in 1690
ury = du*sin(radur)        # init y-pos ur in 1690
urvelx = urvel*sin(radur)
urvely = -urvel*cos(radur)
# 1690 Neptune at long.
radnp = (288.38)*pi/180. # 1690 rad Neptune wrt x-axis
Uranus = sphere(pos=(urx,ury,0), radius=0.5,color =(.88,1,1),
    make_trail=True)
urpert = sphere(pos=(urx,ury,0), radius=0.5,color =(.88,1,1),
    make_trail=True)
fnu = arrow(pos=Uranus.pos,color=color.orange,axis=vector(0,4,0))
npx = dn*cos(radnp)          #init coord x Neptune 1690
npy = dn*sin(radnp)          #              y
npvelx = npvel*sin(radnp)
npvely = -npvel*cos(radnp)
Neptune = sphere(pos=(npx,npy,0), radius=0.4,color=color.cyan,
```

```
   make_trail=True)
fun = arrow(pos=Neptune.pos,color=color.orange,axis=vector(0,-4,0))
nppert = sphere(pos=(npx,npy,0), radius=0.4, color=color.white,
   make_trail=True)
velour = vector(urvelx,urvely,0)      #initial vector velocity Uranus
velnp = vector(npvelx,npvely,0)       #initial vector velocity Neptune
dt = 0.5                   # time increment in terrestrial year
r = vector(urx,ury,0)           # initial position Uranus wrt Sun
rnp = vector(npx,npy,0)          # initial position Neptune wrt Sun
veltot = velour
veltotnp = velnp
rtot = r
rtotnp = rnp

def ftotal(r,rnp,i):    # i==1 Uranus   i==2 Neptune
    Fus = -G*M*mu*r/(du**3)  # Force sun over URANUS
    Fns = -G*M*mn*rnp/(dn**3)  # Force Sun over NEPTUNE
    dnu = mag(rnp-r)          # distance Neptune-Uranus
    Fnu = -G*mu*mn*(rnp-r)/(dnu**3)  # force N on U
    Fun = -Fnu               # force Uranus on Neptune
    Ftotur = Fus+Fnu            # total force on U (sun + N)
    Ftotnp = Fns+Fun            # On Neptune F sun +F urn
    if i==1: return Ftotur
    else:     return Ftotnp

def rkn(r,veltot,rnp,m,i):     # on Neptune
    k1v = ftotal(r,rnp,i)/m
    k1r = veltot
    k2v = ftotal(r,rnp+0.5*k1r*dt,i)/m
    k2r = veltot+0.5*k2v*dt
    k3v = ftotal(r,rnp+0.5*k2r*dt,i)/m
    k3r = veltot+0.5*k3v*dt
    k4v = ftotal(r,rnp+k3r*dt,i)/m
    k4r = veltot+k4v*dt
    veltot = veltot+(k1v+2*k2v+2*k3v+k4v)*dt/6.0
    rnp = rnp+(k1r+2*k2r+2*k3r+k4r)*dt/6.0
    return r,veltot

def rk(r,veltot,rnp,m,i):    # on Uranus
    k1v = ftotal(r,rnp,i)/m
    k1r = veltot
    k2v = ftotal(r+0.5*k1r*dt,rnp,i)/m
    k2r = veltot+0.5*k2v*dt
    k3v = ftotal(r+0.5*k2r*dt,rnp,i)/m
    k3r = veltot+0.5*k3v*dt
    k4v = ftotal(r+k3r*dt,rnp,i)/m
    k4r = veltot+k4v*dt
    veltot = veltot+(k1v+2*k2v+2*k3v+k4v)*dt/6.0
    r = r+(k1r+2*k2r+2*k3r+k4r)*dt/6.0
    return r,veltot

for i in arange(0,320):# estaba 1240
    rate(10)
    rnewu,velnewu = rk(r,velour,rnp,mu,1) # Uranus
    rnewn,velnewn = rkn(rnp,velnp,r,mn,2) # Neptune
    r = rnewu                # Uranus position
    velour = velnewu          # Uranus velocity
    du = mag(r)
    omeur = mag(velour)/du        # Angular velocity of Uranus
    degr = 205.64*pi/180- omeur*i*dt # Angular position Uranus
    rnp = rnewn               # Neptune pos
    velnp = velnewn           # Neptune pos
    dn = mag(rnp)
    omenp = mag(velnp)/dn
```

```
radnp = radnp - dt*omenp              # Radians Neptune
npx = dn*cos(radnp)
npy = dn*sin(radnp)
rnp = vector(npx,npy,0)               # Neptune position
deltaomgs = -omeur+omreal
graphu.plot(pos=(i,deltaomgs*180/pi*3600))
urpert.pos = r
fnu.pos = urpert.pos                  # position of arrow on Uranus
dnu = mag(rnp-r)                           # distance Neptune-Uranus
fnu.axis = 75*norm(rnp-r)/dnu         # axes  the arrow over Uranus
Neptune.pos = rnp                          # radiovector Neptune
fun.pos = Neptune.pos
fun.axis = -fnu.axis                   # arrow on Neptune
```

Listing 3.5. UranusNeptune.py solves for the orbits of Uranus and Neptune and their interaction.

Wave Equations & Fluid Dynamics

4.1 Chapter Overview

The chapter applies classical mechanics to the description of continuous media, with an emphasis on problems not usually seen in traditional texts. We start with waves on strings, making them realistic by including friction, variable tension and density, and ultimately nonlinear effects, and then go on to solve the equations with a "leapfrog" (time stepping) algorithm. We then extend the solution to study normal modes as an eigenvalue problem, masses on vibrating strings, and waves on membranes. The middle sections of the chapter cover shock waves and their extension to solitary waves, as described by the Korteweg-de Vries and the Sine-Gordon equations. These extensions of the wave equation produce fascinating results. We have found that the recent occurrences of tsunamis lead to high interest in these solitons, and we recommend their inclusion.

The last part of this chapter deals with hydrodynamics as described by the Navier-Stokes equation. This is not easy material, though having running simulations on hand makes the material more accessible. Study of the vorticity form of the equation is inherently interesting and provides an insightful extension of the role of a potential in E&M. We provide sample solutions of these equations, along with a discussion of the multitude of necessary boundary conditions.

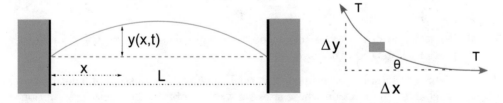

Figure 4.1. *Left:* A stretched string of length L tied down at both ends with a vertical disturbance $y(x, t)$. *Right:* An element of the string showing how the string's tension T produces a restoring force.

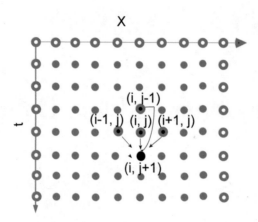

Figure 4.2. Schematic of the algorithm used to solve the wave equation. Four sites (black centers) are used to advance the solution a single time step ahead (black circle). The boundary and initial conditions are indicated by the white-centered dots.

4.2   String Waves

Consider a string of length L and density $\rho(x)$ per unit length, tied down at both ends, and under tension $T(x)$ (Figure 4.1 left). Assume that the relative displacement of the string from its rest position $y(x,t)/L$ is small and that the slope $\partial y/\partial x$ is also small.

1. Consider the infinitesimal section of the string shown on the right of Figure 4.1 and how the difference in the components of the tensions at x and $x + \Delta x$ results in a vertical restoring force. Show that application of Newton's laws to this section leads to the wave equation:

$$\frac{dT(x)}{dx}\frac{\partial y(x,t)}{\partial x} + T(x)\frac{\partial^2 y(x,t)}{\partial x^2} = \rho(x)\frac{\partial^2 y(x,t)}{\partial t^2}. \tag{4.1}$$

2. What conditions are necessary to obtain the familiar wave equation

$$\frac{\partial^2 y(x,t)}{\partial x^2} = \frac{1}{c^2}\frac{\partial^2 y(x,t)}{\partial t^2}, \qquad c = \sqrt{T/\rho} \ ? \tag{4.2}$$

3. What conditions must be met to obtain a unique solution to this second-order partial differential equation?

4. Figure 4.2 shows a spacetime lattice with time varying in steps of length Δt and space in steps of Δx on whose sites we want to obtain the numerical solution

$$y(x,t) = y(i\Delta x, j\Delta t) \stackrel{\text{def}}{=} y_{i,j} \tag{4.3}$$

to the wave equation upon these lattice sites.

a. Express the second derivatives in (4.2) in terms of finite differences, and show that this results in the difference form of the wave equation:

$$\frac{y_{i,j+1} + y_{i,j-1} - 2y_{i,j}}{c^2(\Delta t)^2} = \frac{y_{i+1,j} + y_{i-1,j} - 2y_{i,j}}{(\Delta x)^2}. \quad (4.4)$$

b. Show that a rearrangement of (4.4) leads to a "leapfrog" algorithm that predicts $y(x,t)$ at a future time in terms of the known values at present and past times and three nearby positions (Figure 4.2):

$$y_{i,j+1} = 2y_{i,j} - y_{i,j-1} + \frac{c^2}{c'^2}[y_{i+1,j} + y_{i-1,j} - 2y_{i,j}], \qquad c' \stackrel{\text{def}}{=} \Delta x/\Delta t. \quad (4.5)$$

Here c', the grid or lattice velocity, is the ratio of numerical parameters.

c. Where in Figure 4.2 do the initial conditions enter?

d. Where in Figure 4.2 do the boundary conditions enter?

e. (Optional) Show that in order for the solution to be stable, the step size must satisfy the *Courant condition* [Press et al.(94), Courant et al.(28)],

$$\frac{c}{c'} = \frac{c\Delta t}{\Delta x} \leq 1. \quad (4.6)$$

Equation (4.6) means that the solution gets better with smaller *time* steps, though gets worse for smaller space steps (unless you simultaneously make the time step smaller).

f. Write a program that implements the leapfrog algorithm (4.5) and plots up the motion of the string, or, better yet, produces an animation of the motion.

g. Examine a variety of initial conditions with the string at rest. The program EqStringMovMat.py in Listing 4.1 uses Matplotlib, while EqStringMov.py uses Visual, both with the initial conditions:

$$y(x,t=0) = \begin{cases} 1.25x/L, & x \leq 0.8L, \\ (5-5x/L), & x > 0.8L, \end{cases} \qquad \frac{\partial y}{\partial t}(x,t=0) = 0. \quad (4.7)$$

h. In order to start the algorithm, you will need to know the solution at a negative time ($j = 0$). Show that use of the central-difference approximation for the initial velocity leads to needed value, $y_{i,0} = y_{i,2}$.

i. Change the time and space steps used in *your* simulation so that sometimes they satisfy the Courant condition (4.6), and sometimes they don't. Describe what happens in each case.

j. Use the plotted time dependence to estimate the peak's propagation velocity c. Compare to (4.2).

k. When a string is plucked near its end, a pulse reflects off the ends and bounces back and forth. Change the initial conditions of the program to one corresponding to a string plucked exactly in its middle, and see if a traveling or a standing wave results.

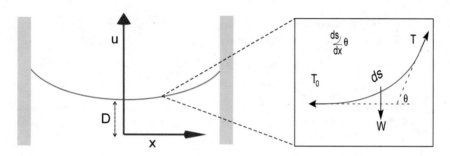

Figure 4.3. *Left:* A uniform string suspended from its ends in a gravitational field assumes a catenary shape. *Right:* A force diagram of a section of the catenary at its lowest point. Because the ends of the string must support the entire weight of the string, the tension varies along the string.

5. A standard procedure is to analyze waves as the sum of normal modes, with each mode having a distinct frequency:

$$y(x,t) = \sum_{n=0}^{\infty} B_n \sin k_n x \, \cos \omega_n t. \qquad (4.8)$$

Because the wave equation is linear, each individual product in (4.8) is an independent solution.

a. Determine the coefficients B_n in (4.8) for your chosen initial conditions.
b. Compare the numerical solution to the sum of normal modes, taking note of the number of terms kept in the sum.
c. Increase the number of terms included in your sum of normal modes until you notice random noise arising from round-off error.
d. Change your initial conditions so that only a single normal mode is excited. Does your simulation remain in this normal mode for all times?

6. Outline a procedure that would solve the wave equation for all times simultaneously, and estimate how much memory would be required.

7. Examine the possibilities of solving the wave equation via a relaxation technique like that used for Laplace's equation in §5.2.1.

4.2.1 Extended Wave Equations

With Friction

1. Consider a string vibrating in a viscous medium with a frictional force opposing motion, proportional to the string velocity, and proportional to the length of a

string element:

$$F_f \simeq -2\kappa \, \Delta x \, \frac{\partial y}{\partial t}. \tag{4.9}$$

Here κ is proportional to the viscosity of the medium.

a. Modify the wave equation so that it includes the frictional force (4.9).
b. Extend the algorithm used to solve the wave equation to now include friction.
c. Check that the results of your simulation show waves that damp in time.
d. As a check, reverse the sign of κ and see if the wave grows in time.

Variable Density & Tension

1. Realistic cables on bridges may be thicker near the ends in order to support the additional weight of the cable and other elements. Accordingly, the wave equation should incorporate variable density and correspondingly tension.

   a. Extend your wave equation algorithm so that it is now appropriate to (4.1) including a $T(x)$ and a $\rho(x)$.
   b. Assume that the string's tension and density vary as

   $$\rho(x) = \rho_0 e^{\alpha x}, \quad T(x) = T_0 e^{\alpha x}, \tag{4.10}$$

   and explore the effect of using (4.10) in your simulation.
   c. In which regions would you expect faster wave propagation? Is that what you find?
   d. Since the extended wave equation including (4.10) is still linear, normal-mode solutions that vary like $u(x,t) = A\cos(\omega t)\sin(\gamma x)$ should still exist. Explore this possibility.
   e. Examine standing wave solutions including (4.10) and verify that the string acts like a high-frequency filter, that is, that there is a frequency below which no waves occur.

2. Telephone wires and power lines tend to sag in the middle under their own weight (Figure 4.3). Their shape is a catenary, and their tension is nonuniform:

   $$y(x) = D\cosh\left(\frac{\rho g x}{T_0}\right), \quad T(x) = T_0\cosh\left(\frac{\rho g x}{T_0}\right), \tag{4.11}$$

   where T_0 is the tension in the absence of gravity [Becker(54)].

   a. Modify your simulation to compute waves on a catenary with $\alpha = 0.5$, $T_0 = 40$ N, and $\rho_0 = 0.01$ kg/m.
   b. Search for normal-mode solutions of this variable-tension wave equation, that is, solutions that vary as $u(x,t) = A\cos(\omega t)\sin(\gamma x)$.
   c. The wave speed should be higher in regions of high density. Develop an empirical measure of the propagation velocity and determine the percentage differences in speeds.

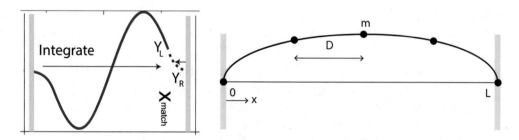

Figure 4.4. *Left:* Integrating a string wave $Y_L(x)$ out from the left to x_{match}, and checking if it matches the string wave $Y^R(x)$ obtained by integrating in from the right. *Right:* Identical masses of m and uniform spacing D on stretched string with fixed ends.

    d. Extend the simulation to produce waves on a catenary including friction. The program `CatFriction.py` in Listing 4.2 does this.

    e. Plot your results showing *both* the disturbance $u(x,t)$ about the catenary and the actual height $y(x,t)$ above the horizontal for a plucked string initial condition.

4.2.2 Computational Normal Modes

The normal modes of a system are defined as oscillations in which all parts of the system move with a definite phase relation and with a sinusoidally time dependence with one frequency. The subject is treated extensively in most classical mechanics texts. We repeat (4.1) describing a vibrating string with variable mass and tension:

$$\frac{dT(x)}{dx}\frac{\partial y(x,t)}{\partial x} + T(x)\frac{\partial^2 y(x,t)}{\partial x^2} = \rho(x)\frac{\partial^2 y(x,t)}{\partial t^2}. \tag{4.12}$$

If the system is excited in a normal mode with frequency ω_n, and is at rest at time $t = 0$, then the time dependence of the solution to (4.12) must separate as

$$y(x,t) = Y(x)\sin(\omega_n t). \tag{4.13}$$

1. Show that substitution of (4.13) into the PDE (4.12) leads to the ODE

$$\frac{dT(x)}{dx}\frac{dY(x)}{dx} + T(x)\frac{d^2Y(x)}{dx^2} = \rho(x) - \omega_n^2 Y(x). \tag{4.14}$$

    The approach used to search for normal modes of (4.14) is indicated on the left of Figure 4.4. It follows the same path as that discussed in §6.2.2, and indicated in Figure 6.1, for the bound state eigenvalues of the Schrödinger equation:

1. Guess a value for the frequency ω_n.

2. Impose the boundary conditions $Y(0) = Y(L) = 0$, where L is the length of the string (your choice).

3. Start at $x = 0$ and use an ODE solver to step $Y_L(x)$ to the right until you reach the *matching point* x_{match} (Figure 4.4 left). The exact value of this matching point is not important, and your final solution should be independent of it.

4. Start at $x = L$ and use an ODE solver to step $Y_R(x)$ to the left until you reach the *matching point* x_{match}.

5. In order for amplitude and transverse velocity to be continuous at $x = x_{match}$, Y and dY/dx must be continuous there. Requiring continuity of the logarithmic derivative Y'/Y combines both requirements into a single condition that is independent of Y's normalization. Incorporate expressions in your program for the logarithmic derivative of both the left and right waves.

6. It is unlikely the left and right waves Y_L and Y_R will match for an arbitrary ω_n guess, and so measure the mismatch in logarithmic derivatives as:

$$\Delta(\omega_n, x) = \left. \frac{Y_L'(x)/Y_L(x) - Y_R'(x)/Y_R(x)}{Y_L'(x)/Y_L(x) + Y_R'(x)/Y_R(x)} \right|_{x=x_{match}}, \qquad (4.15)$$

where the denominator is included in order to limit the magnitude of Δ.

7. Conduct a search for those ω_n's for which $\Delta = 0$. Continue each search until $\Delta \simeq 0$ within some set tolerance. The bisection algorithm can be used for the search.

8. Test your procedure for the standard wave equation whose eigenfreqencies $\omega_n = n\pi c/L$ are known.

9. Obtain the eigenfreqencies for several normal modes of the realistic wave equation and compare them to the frequencies obtained from the simple wave equation.

4.2.3 Masses on Vibrating String

Figure 4.4 right shows N identical masses m attached to a massless string under tension T, with distance D between the masses [Beu(13)]. We denote the vertical displacement of mass i from equilibrium by $y_i(t)$, and assume that the first and last masses are kept fixed, $y_1(t) = y_N(t) = 0$.

1. Show that for small $y_i(t)$, the equation of motion for mass i is:

$$(y_{i-1} - 2y_i + y_{i+1})\frac{T}{D} = m\frac{d^2 y_i}{dt^2}, \qquad i = 1, 2, \ldots, N-1. \qquad (4.16)$$

2. Assume the system is in a normal mode, in which case all masses vibrate at the same frequency ω_n and with related phases:

$$y_i(t) = A_i \sin(\omega_n t), \qquad (4.17)$$

where the amplitudes A_i are constants. What are the equations of motions now?

3. Show that the equations of motion now reduce to the set of N linear equations in the amplitudes A_i and the parameter $\lambda = Dm\omega_n^2/T$:

$$A_1 = 0,$$
$$2A_2 - A_3 = \lambda A_2,$$
$$-A_{i-1} + 2A_i - A_{i+1} = \lambda A_i, \qquad (i = 2, \ldots, N-1)$$
$$\ddots \qquad\qquad (4.18)$$
$$-A_{N-2} + 2A_{N-1} = \lambda A_{N-1},$$
$$A_N = 0.$$

4. Show that theses linear equations can be cast into the form of a matrix eigenvalue problem with a tridiagonal matrix:

$$
\begin{bmatrix}
1 & 0 & 0 & 0 & 0 & 0 & 0 \\
0 & 2 & -1 & 0 & 0 & 0 & 0 \\
0 & -1 & 2 & -1 & 0 & 0 & 0 \\
\ddots & \ddots & \ddots & \ddots & \ddots & \ddots & \ddots \\
0 & 0 & 0 & -1 & 2 & -1 & 0 \\
0 & 0 & 0 & 0 & -1 & 2 & 0 \\
0 & 0 & 0 & 0 & 0 & 0 & 1
\end{bmatrix}
\begin{bmatrix}
A_1 \\ A_2 \\ A_3 \\ \vdots \\ A_{N-2} \\ A_{N-1} \\ A_N
\end{bmatrix}
= \lambda
\begin{bmatrix}
A_1 \\ A_2 \\ A_3 \\ \vdots \\ A_{N-2} \\ A_{N-1} \\ A_N
\end{bmatrix}. \qquad (4.19)
$$

5. Assign values to the parameters. Good choices are $Dm/T = 1$ and $N = 150$.

6. Use a matrix library (§1.2.2) to solve for the five lowest normal modes.

7. Compare the eigenvalues obtained from the numerical solution to those of a string with the same mass density and tension.

8. Plot the eigenfunctions for the masses on a string along with those of a string with the same mass density and tension.

Single Mass on Vibrating String

1. [Fetter & Walecka(80)] discuss the normal modes for a stretched string with a point mass m attached to its center. The string has length L, mass density ρ, and fixed ends (Figure 4.4 right, with just the center mass).

a. Show that the asymmetric normal modes with a node at the center of the string are unperturbed by the mass and have eigenfrequencies

$$\omega_n = n\frac{c\pi}{L}, \qquad n = 2, 4, 5, \ldots, \tag{4.20}$$

where $c = \sqrt{T/\rho}$ is the propagation speed.

b. Show that the symmetric normal modes in which the mass moves have eigenfrequencies ω_n that are roots of the transcendental equation

$$\cot\frac{\omega_n L}{2c} = \frac{m}{2c\rho}\omega_n. \tag{4.21}$$

c. Find the lowest five roots of (4.21) for $c = 1$, $L = 100$, $\rho = 10$, and $m = 10$.

d. Compare the perturbed eigenvalues so obtained to those of a string with the same mass density and tension, though no mass at the center.

e. Are the eigenfrequencies for the symmetric and antisymmetric modes interlaced?

f. Plot the x dependence of the eigenfunctions for these five eigenvalues along with those of a string with the same mass density and tension, though no mass at the center.

2. Another approach to solving for the normal modes of a string with a mass m at its center is to view it as a string with a mass density $\rho(x)$ that has a delta function at $x = L$, and then to solve the wave equation (4.1). Although one cannot place a singularity into a computer program, one can introduce a $\rho(x)$ that is very large in a very small region near the center of the string.

a. Construct a $\rho(x)$ that approximates the parameters given in Problem (1), namely: $c = 1$, $L = 100$, $\rho = 10$, and $m = 10$.

b. You will want to make Δx very small so that the high density region is included in a number of steps.

c. Compute the normal modes of the system by following the procedures given in §4.2.2.

d. Compare to the results found in Problem (1). Are the symmetric modes affected more than the asymmetric modes?

4.2.4 Wave Equation for Large Amplitudes

The wave equations we have studied so far all follow from the assumption of small displacements, and are , accordingly, linear equations.

1. Show that an extension of the wave equation so that it contains the next order in displacements is [Taghipour et al.(14), Keller(59)]:

$$c^2\frac{\partial^2 y(x,t)}{\partial x^2} = \left[1 + \frac{\partial^2 y(x,t)}{\partial x^2}\right]^2 \frac{\partial^2 y(x,t)}{\partial t^2}. \tag{4.22}$$

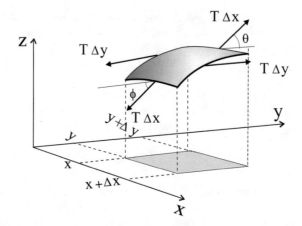

Figure 4.5. A small part of an oscillating membrane and the forces that act on it.

2. Extend the leapfrog algorithm to solve this nonlinear equation, making whatever assumptions about initial conditions are needed.

   a. Repeat some of the solutions to the wave equation studied previously, only now for large values of y/L, in which case nonlinear effects are important.
   b. Examine what were normal modes for the linear problem, but now for large amplitude oscillations.

4.3   Membrane Waves

Figure 4.5 shows a square section of the vertical displacement of a membrane, and the resulting components of the tension in the x and y directions.

1. We want to extend the simulation of waves on a string to now describe waves on a 2-D stretched membrane under tension per unit length T, and a mass per unit area of ρ [Kreyszig(98)].

   a. Show that the forces on the membrane in the z direction as a result of a change in just y, or in just x, are

   $$\sum F_z(x_{fixed}) \approx T\Delta x \frac{\partial^2 u}{\partial y^2}\Delta y, \qquad \sum F_z(y_{fixed}) \approx T\Delta y \frac{\partial^2 u}{\partial x^2}\Delta x. \quad (4.23)$$

   b. Show that applying Newton's second law to the square section leads to the wave equation

   $$\frac{1}{c^2}\frac{\partial^2 u}{\partial t^2} = \frac{\partial^2 u}{\partial x^2} + \frac{\partial^2 u}{\partial y^2}, \qquad c = \sqrt{T/\rho}. \quad (4.24)$$

c. Assume that the solution is separable into the product of individual function of x, y, and t, $u(x, y, t) = X(x)Y(y)T(t)$, and show that this leads to the solution:

$$X(x) = A \sin kx + B \cos kx, \quad Y(y) = C \sin qy + D \cos qy, \qquad (4.25)$$
$$T(t) = E \sin c\xi t + F \cos c\xi t. \qquad (4.26)$$

d. The boundary condition is that the membrane's ends are fixed to the top edges of a square box of sides π. The initial condition is that the membrane is initially at rest, with a vertical displacement from equilibrium $u(x, y, t = 0) = \sin 2x \sin y$. Write down the equations describing the initial and the boundary conditions.

e. Show that imposition of the initial and boundary conditions require

$$u(x, y, t) = \cos c\sqrt{5} \, \sin 2x \, \sin y. \qquad (4.27)$$

f. Use the central-difference expression for derivatives to express the wave equation (4.24) in terms of finite differences.

g. Discretize the variables so that $u(x = i\Delta, y = i\Delta y, t = k\Delta t) \equiv u_{i,j}^k$, and rearrange the difference equation to obtain the leapfrog algorithm predicting future values in terms of the present and past ones:

$$u_{i,j}^{k+1} = 2u_{i,j}^k - u_{i,j}^{k-1} \frac{c^2}{c'^2} \left[u_{i+1,j}^k + u_{i-1,j}^k - 4u_{i,j}^k + u_{i,j+1}^k + u_{i,j-1}^k \right], \quad (4.28)$$

where $c' \overset{\text{def}}{=} \Delta x / \Delta t$ is the grid velocity.

h. To initiate this algorithm, we need to know the solution at $t = -\Delta t$, that is, before the initial time. Show that applying the initial condition that the membrane is released from rest implies $u_{i,j}^{-1} = u_{i,j}^1$

i. Use the previous result to show that the algorithm for the first step is

$$u_{i,j}^1 = u_{i,j}^0 + \frac{c^2}{2c'^2} \left[u_{i+1,j}^0 + u_{i-1,j}^0 - 4u_{i,j}^0 + u_{i,j+1}^0 + u_{i,j-1}^0 \right]. \qquad (4.29)$$

j. Solve the 2-D wave equation numerically. The program Waves2D.py in Listing 4.4 solves the 2-D wave equation numerically, while Waves2Danal.py computes the analytic solution.

k. Compare the numerical solution to the analytic one (4.27).

l. Tune your numeric solution to obtain the best possible agreement with the analytic solution (which in this case is *not* a slowly converging infinite series).

2. A membrane with the shape of a sector of a circle has an opening angle α and radius R. Its edges are fixed [Fetter & Walecka(80)].

a. Show that the solution to the wave equation is separable in plane polar co-ordinates (r, ϕ) and leads to the eigenfunctions

$$u_{mn}(r, \phi) = C J_{m\pi/\alpha}\left(\frac{X_{mn}r}{R}\right) \sin\frac{m\pi\phi}{\alpha} \quad m = 1, 2, 3, \ldots, \infty, \quad (4.30)$$

where the eigenvalues X_{mn} are determined by the condition

$$J_{m\pi/\alpha}(X_{mn}) = 0 \quad n = 1, 2, 3, \ldots, \infty. \quad (4.31)$$

Here C is a constant and J is the Bessel function.
b. Determine the values of X_{mn} for the first three eigenfunctions.
c. Make plots of the normal modes and nodal lines for a several low-lying modes.
d. Find numerical values for the three lowest eigenfrequencies for waves with $\alpha = \pi/2, \pi$, and 2π.

3. Modify the program in Problem 1 for waves on a square drumhead to one that solves for waves on the sectors of Problem 12.

a. Choose sets of initial conditions that lead to the excitation of individual eigenfunctions of the membrane. (You may want to use the form of the analytic solution as a guide.)
b. Deduce the lowest three eigenvalues. We discuss one way to do this via matching in §4.2.2. Or you may want to use the form of the analytic solution as a guide.

4. Problem 8.1 in [Fetter & Walecka(80)] considers a stretched circular membrane with a point mass m attached to its center. They use perturbation theory to deduce the change in eigenfrequencies of the circularly symmetric modes.

a. Model the point mass as a very large increase in the density $\rho(x)$ near the center of the membrane.
b. Make the step size small enough in your wave equation solution that several steps are needed to cover the $\rho(x)$ variation.
c. Solve the wave equation for the perturbed normal modes and compare to the unperturbed case.

4.4 Shock Waves

4.4.1 Advective Transport

The continuity equation describes conservation of mass:

$$\frac{\partial \rho(\mathbf{x}, t)}{\partial t} + \vec{\nabla} \cdot \mathbf{j} = 0, \quad \mathbf{j} \overset{\text{def}}{=} \rho\mathbf{v}(\mathbf{x}, t). \quad (4.32)$$

Here $\rho(\mathbf{x}, t)$ is the mass density, $\mathbf{v}(\mathbf{x}, t)$ is the velocity of a fluid, and the product $\mathbf{j} = \rho\mathbf{v}$ is the mass current. For 1-D flow in the x direction, and for a fluid that is

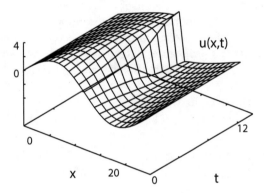

Figure 4.6. A visualization showing the wave height versus position for increasing times showing the formation of a shock wave (sharp edge) from an initial sine wave.

moving with a constant velocity $v = c$, the continuity equation (4.32) takes a simple form known as the *advection equation*:

$$\frac{\partial \rho}{\partial t} + c\frac{\partial \rho}{\partial x} = 0. \tag{4.33}$$

1. Use the substitution $u(x,t) = f(x-ct)$ to prove that any function with the form of a traveling wave is a solution of the advection equation.

2. Prove that wavefronts of constant phase move with along the fluid $c = dx/dt$.

3. Develop a leapfrog algorithm for the solution of the advection equation.

4. The simple leapfrog algorithm is known to lead to instabilities when applied to the advection equation [Press et al.(94)]. Soon we will examine a better algorithm, the Lax-Wendroff method.

5. Test your algorithm for a number of initial conditions corresponding to differing initial wave forms, and look for instabilities.

4.4.2 Burgers' Equation

A simple model for shock waves and turbulence is Burgers' equation [Burgers(74), Falkovich & Sreenivasan(06)]:

$$\frac{\partial u}{\partial t} + \epsilon u\,\frac{\partial u}{\partial x} = 0. \tag{4.34}$$

This nonlinear equation is an extension of the advection equation (4.33) in which the wave speed $c = \epsilon u$ is proportional to the amplitude of the wave [Tabor(89)]. Accordingly, the shapes of waveforms are not preserved in time since parts of the wave with large amplitudes propagate faster than those with small amplitudes, and in time this leads to *shock waves*, such as that shown in Figure 4.6.

1. Express the derivatives in Burgers' equation as central differences, and show that this leads to the leapfrog algorithm for time stepping:

$$u(x, t + \Delta t) = u(x, t - \Delta t) - \beta \left[\frac{u^2(x + \Delta x, t) - u^2(x - \Delta x, t)}{2} \right],$$

$$\Rightarrow \quad u_{i,j+1} = u_{i,j-1} - \beta \left[\frac{u^2_{i+1,j} - u^2_{i-1,j}}{2} \right], \qquad \beta = \frac{\epsilon}{\Delta x / \Delta t}. \qquad (4.35)$$

Here u^2 is the square of u and is not its second derivative, and β is a ratio of constants known as the *Courant-Friedrichs-Lewy* (CFL) *number*.

2. Shock wave with their large gradients are notoriously difficult to compute and require a higher order version of the leapfrog algorithm. The *Lax-Wendroff method* retains second-order differences for the time derivative and uses Burgers' equation itself to relate derivatives:

$$u_{i,j+1} = u_{i,j} - \frac{\beta}{4} \left(u^2_{i+1,j} - u^2_{i-1,j} \right) \qquad (4.36)$$

$$+ \frac{\beta^2}{8} \left[(u_{i+1,j} + u_{i,j}) \left(u^2_{i+1,j} - u^2_{i,j} \right) \quad -(u_{i,j} + u_{i-1,j}) \left(u^2_{i,j} - u^2_{i-1,j} \right) \right].$$

3. Write a program to solve Burgers' equation via the Lax-Wendroff method.

   a. Define arrays u0[100] and u[100] for the initial data and the solution.
   b. Take the initial wave to be sinusoidal, u0[i] = $3\sin(3.2x)$, with speed $c = 1$.
   c. Incorporate the boundary conditions u[0]=0 and u[100]=0.
   d. Keep the CFL number $\beta = \epsilon/(\Delta x/\Delta t) < 1$ for stability.
   e. Plot the initial wave and the solution for several time values on the same graph in order to see the formation of a shock wave (like Figure 4.6).
   f. Run the code for several increasingly large CFL numbers. Is the stability condition $\beta < 1$ correct for this nonlinear problem?

AdvecLax.py in Listing 4.5 presents our implementation of the Lax-Wendroff method.

4.5   Solitary Waves (Solitons)

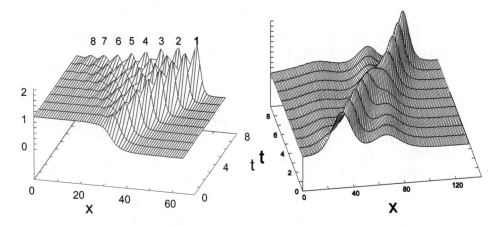

Figure 4.7. *Left:* A single two-level waveform progressively breaking up into eight solitons (labeled) as time increases, with the tallest soliton (1) becoming narrower and faster in time. *Right:* Two solitons crossing, with the taller soliton (left at $t = 0$) catching up and overtaking the shorter one at $t \simeq 5$. The waves resume their original shapes after the collision.

4.5.1   Including Dispersion, KdeV Solitons

1. *Dispersion* broadens a wavefront as it travels through a medium. Start with a plane wave traveling to the right,

$$u(x, t) = e^{\pm i(kx - \omega t)}. \tag{4.37}$$

a. Show that when (4.37) is substituted into the advection equation (4.33), the dispersion relation $\omega = ck$ results.

b. Evaluate the *group velocity* $v_g = \partial\omega/\partial k$ as a function of frequency ω for a wave obeying this dispersion relation.

c. Why is this called *dispersionless* propagation?

d. Consider a wave is propagating with a small amount of *dispersion*,

$$\omega \simeq ck - \beta k^3, \tag{4.38}$$

with β small. Evaluate the group velocity corresponding to this dispersion relation.

e. Show that if the plane-wave solution (4.37) arises from a wave equation, then the ω term of the dispersion relation (4.38) would arise from a first-order time derivative, the ck term from a first-order space derivative, and the k^3 term from a third-order space derivative:

$$\frac{\partial u(x, t)}{\partial t} + c\,\frac{\partial u(x, t)}{\partial x} + \beta\,\frac{\partial^3 u(x, t)}{\partial x^3} = 0. \tag{4.39}$$

2. The analytic description of *solitons*, unusual water waves that occur in shallow, narrow canals, was given by [Korteweg & deVries(1895)] based on the partial differential equation [Abarbanel et al.(93), Tabor(89)]:

$$\frac{\partial u(x,t)}{\partial t} + \varepsilon u(x,t)\frac{\partial u(x,t)}{\partial x} + \mu\frac{\partial^3 u(x,t)}{\partial x^3} = 0. \qquad (4.40)$$

The nonlinear term $\varepsilon u\,\partial u/\partial t$ sharpens the wave and causes a *shock* wave, the $\partial^3 u/\partial x^3$ term produces dispersive broadening, and the $\partial u/\partial t$ term produces traveling waves.

a. Show by substitution that the soliton waveform,

$$u(x,t) = -c\,\text{sech}^2[\sqrt{c}(x - ct - \xi_0)/2]/2, \qquad (4.41)$$

   is a solution of the KdeV equation (4.40) with c the wave speed.

b. The c term in the wave's amplitude leads to faster waves having larger amplitudes. Plot the sech^2 term and show that it produces a single lump-like wave.

3. Show that application of time and space derivatives given by central-difference approximations leads to the finite-difference form of the KdeV equation:

$$u_{i,j+1} \simeq u_{i,j-1} - \frac{\epsilon}{3}\frac{\Delta t}{\Delta x}\left[u_{i+1,j} + u_{i,j} + u_{i-1,j}\right]\left[u_{i+1,j} - u_{i-1,j}\right]$$
$$- \mu\frac{\Delta t}{(\Delta x)^3}\left[u_{i+2,j} + 2u_{i-1,j} - 2u_{i+1,j} - u_{i-2,j}\right]. \qquad (4.42)$$

4. Show that a good starting condition is:

$$u_{i,2} \simeq u_{i,1} - \frac{\epsilon}{6}\frac{\Delta t}{\Delta x}\left[u_{i+1,1} + u_{i,1} + u_{i-1,1}\right]\left[u_{i+1,1} - u_{i-1,1}\right]$$
$$- \frac{\mu}{2}\frac{\Delta t}{(\Delta x)^3}\left[u_{i+2,1} + 2u_{i-1,1} - 2u_{i+1,1} - u_{i-2,1}\right]. \qquad (4.43)$$

5. Modify or run the program `Soliton.py` that solves the KdeV equation (4.40) for the initial condition and parameters:

$$u(x,0) = 0.5[1 - \tanh(x/5 - 5)], \quad \epsilon = 0.2, \ \mu = 0.1, \ \Delta x = 0.4, \ \Delta t = 0.1.$$

These constants satisfy the stability condition with $|u| = 1$.

a. Plot the KdeV solutions as a 3-D graph of disturbance u versus position *and versus* time. The program `SolitonAnimate.py` produces an animated solution.

6. Observe the wave profile as a function of time and try to confirm the experimental observation that a taller soliton travels faster than a smaller one.

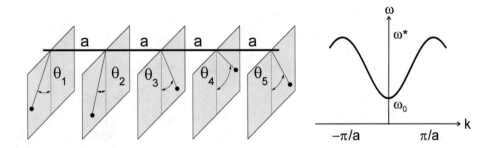

Figure 4.8. *Left:* A 1-D chain of pendula attached to a torsion bar. The pendula swing in planes perpendicular to the length of the bar. *Right:* The dispersion relation for a linearized chain of pendula.

7. Explore what happens when a tall soliton collides with a short one.

   a. Start with a tall soliton at $x = 12$ and a smaller one in front of it at $x = 26$:

$$u(x, t = 0) = 0.8[1 - \tanh^2(3x/12 - 3)] + 0.3[1 - \tanh^2(4.5x/26 - 4.5)]. \quad (4.44)$$

   b. Identify reflection, transparency, interference, and different propagation speeds (Figure 4.7 right).

8. Construct phase-space plots of $\dot{u}(t)$ versus $u(t)$ for various parameter values.

   a. Show that soliton solutions correspond to the *separatrix* solutions to the KdeV equation, analogous to the infinite period for a vertical pendulum.

9. A two-meter bore 5000 m offshore spawns a series of solitons. What force is exerted upon a solid breakwater when the soliton reaches land?

4.5.2   Pendulum Chain Solitons, Sine-Gordon Solitons

Consider the 1-D chain of coupled pendula shown on left of Figure 4.8. The pendula are identical, equally spaced by distance, and attached to a torsion bar that twists as the pendula swing.

1. Assume that each pendulum is acted upon by a gravitational torque as well as two torques arising from the twisting of the connecting rod. Prove that Newton's law for rotational motion leads to a set of coupled nonlinear equations:

$$\kappa(\theta_{j+1} - 2\theta_j + \theta_{j-1}) - mgL \sin \theta_j = I \frac{d^2\theta_j(t)}{dt^2}, \quad (4.45)$$

where I is the moment of inertia of each pendulum, L is the length of each pendulum, and κ is the torque constant of the bar.

2. How many equations are there?

3. Show that the linear version of (4.45) is

$$\frac{d^2\theta_j(t)}{dt^2} + \omega_0^2\theta_j(t) = \frac{\kappa}{I}(\theta_{j+1} - 2\theta_j + \theta_{j-1}), \qquad (4.46)$$

where $\omega_0 = \sqrt{mgL/I}$ is the natural frequency for any one pendulum.

a. Show that a traveling-wave $\theta_j(t) = A\exp[i(\omega t - k\dot{x}_j)]$ can propagate on the chain in the linear approximation if its frequency and wave number are related by the *dispersion relation* (Figure 4.8 right):

$$\omega^2 = \omega_0^2 - \frac{2\kappa}{I}(1 - \cos ka). \qquad (4.47)$$

b. Under what conditions might there be dispersionless propagation in which all frequencies propagate with the same velocity?

c. Prove that the dispersion relation (4.47) limits which frequencies that can propagate on the chain to the range

$$\omega_0 \le \omega \le \omega^* \qquad (\omega^*)^2 = \omega_0^2 + \frac{4\kappa}{I}. \qquad (4.48)$$

d. Prove that waves with $\omega > \omega^*$ are nonphysical because they correspond to wavelengths $\lambda < 2a$, that is, oscillations where there are no particles.

4. Show that if the wavelengths in a pulse are much longer than the pendulum-pendulum repeat distance a, that is, if $ka \ll 1$, then a can be replaced by the continuous variable x, and the system of coupled, nonlinear equations becomes a single nonlinear equation:

$$\frac{1}{c^2}\frac{\partial^2\theta}{\partial t^2} - \frac{\partial^2\theta}{\partial x^2} = \sin\theta, \qquad \text{(Sine-Gordon Equation)}, \qquad (4.49)$$

where time is measured in $\sqrt{I/mgL}$ units and distances in $\sqrt{\kappa a/(mgLb)}$ units. This equation is called the Sine-Gordon equation since it is similar to the Klein-Gordon equation studied in §6.2.4.

5. Prove that the Sine-Gordon equation supports the two soliton solutions

$$\theta(x - vt) = 4\tan^{-1}\left(\exp\left[+\frac{x - vt}{\sqrt{1 - v^2}}\right]\right), \qquad (4.50)$$

$$\theta(x - vt) = 4\tan^{-1}\left(\exp\left[-\frac{x - vt}{\sqrt{1 - v^2}}\right]\right) + \pi. \qquad (4.51)$$

6. Plot up these two solutions and thereby show that they correspond to a solitary *kink* traveling with velocity $v = -1$ that flips the pendulums around by 2π as it moves down the chain, and an *antikink* in which the initial $\theta = \pi$ values are flipped to final $\theta = -\pi$ values.

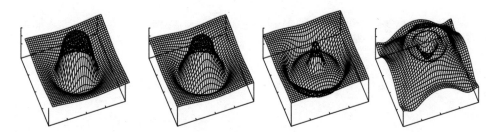

Figure 4.9. A circular ring, Sine-Gordon soliton at several times.

The 2-D generalization of the SGE equation (4.49) is

$$\frac{1}{c^2}\frac{\partial^2 u}{\partial t^2} - \frac{\partial^2 u}{\partial x^2} - \frac{\partial^2 u}{\partial y^2} = \sin u \qquad \text{(2-D SGE)}. \qquad (4.52)$$

Whereas the 1-D SGE describes wave propagation along a chain of connected pendulums, the 2-D form describes wave propagation in nonlinear elastic media, or elementary particles in some field theoretic models [Christiansen & Lomdahl(81), Christiansen & Olsen(78), Argyris(91)]. The equation can be solved numerically via the finite difference technique:

$$u_{m,l}^2 \simeq \frac{1}{2}\left(u_{m+1,l}^1 + u_{m-1,l}^1 + u_{m,l+1}^1 + u_{m,l-1}^1\right)$$
$$- \frac{\Delta t^2}{2}\sin\left[\frac{1}{4}\left(u_{m+1,l}^1 + u_{m-1,l}^1 + u_{m,l+1}^1 + u_{m,l-1}^1\right)\right], \qquad (4.53)$$

where for simplicity we have made the time and space steps proportional, $\Delta t = \Delta x/\sqrt{2}$ [LPB(15)]. Figure 4.9 shows the time evolution of a circular ring soliton. We note that the ring at first shrinks in size, then expands, and then shrinks back into another (though not identical) ring soliton. Our 2-D Sine-Gordon code `TwoDsol.py` is given in Listing 4.8, with a pseudocode below:

```
# TwoDsol.java: solves Sine-Gordon equation for 2D soliton
Import packages
Set parameter values
Declare arrays

Function initial(u) # Set initial conditions for x & y arrays
    u[i,j] = 4 * atan(3-sqrt(x^2 + y^2)

Function solution(ninit)  # Solves KGE
    Determine u along borders
    Assign otherwise undefined u values
    Iterate
    Assign past u values = present values
    Iterate again
```

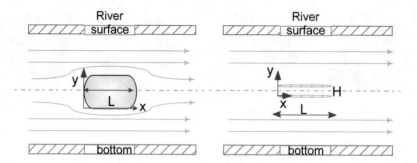

Figure 4.10. Side view of the flow of a stream around a submerged beam (*left*) and around two parallel plates (*right*). Both beam and plates have length L along the direction of flow. The flow is seen to be symmetric about the centerline and to be unaffected at the bottom and at the surface by the submerged object.

```
Call initial
assign x & y values
Set up plotting grid
Solve for u for 22 iterations
Plot sin(u(x,y)/2) as surface plot
```

4.6 Hydrodynamics

4.6.1 Navier-Stokes Equation

The basic equation of hydrodynamics is the continuity equation (4.32):

$$\frac{\partial \rho(\mathbf{x}, t)}{\partial t} + \vec{\nabla} \cdot \mathbf{j} = 0, \qquad \mathbf{j} \overset{\text{def}}{=} \rho \mathbf{v}(\mathbf{x}, t). \tag{4.54}$$

A more general description is provided by the *Navier-Stokes equation*,

$$\frac{D\mathbf{v}}{Dt} = \nu \nabla^2 \mathbf{v} - \frac{1}{\rho} \vec{\nabla} P(\rho, T, x), \tag{4.55}$$

which describes fluid flow including viscosity. Here ν is the kinematic viscosity and P is the pressure. The explicit functional dependence of the pressure on the fluid's density and temperature $P(\rho, T, x)$ is known as the *equation of state of the fluid*, and would have to be known before trying to solve the Navier-Stokes equation. The new operator here is the *hydrodynamic* derivative,

$$\frac{D\mathbf{v}}{Dt} \overset{\text{def}}{=} (\mathbf{v} \cdot \vec{\nabla})\mathbf{v} + \frac{\partial \mathbf{v}}{\partial t}, \tag{4.56}$$

that gives the rate of change, as viewed from a stationary frame, of the velocity of material in *an element of flowing fluid*. This derivative incorporates changes as a result of the motion of the fluid (first, nonlinear term) as well as any explicit time dependence of the velocity. The explicit form of the Navier-Stokes equation is:

$$\frac{\partial v_x}{\partial t} + \sum_{j=x}^{z} v_j \frac{\partial v_x}{\partial x_j} = \nu \sum_{j=x}^{z} \frac{\partial^2 v_x}{\partial x_j^2} - \frac{1}{\rho} \frac{\partial P}{\partial x},$$

$$\frac{\partial v_y}{\partial t} + \sum_{j=x}^{z} v_j \frac{\partial v_y}{\partial x_j} = \nu \sum_{j=x}^{z} \frac{\partial^2 v_y}{\partial x_j^2} - \frac{1}{\rho} \frac{\partial P}{\partial y}, \qquad (4.57)$$

$$\frac{\partial v_z}{\partial t} + \sum_{j=x}^{z} v_j \frac{\partial v_z}{\partial x_j} = \nu \sum_{j=x}^{z} \frac{\partial^2 v_z}{\partial x_j^2} - \frac{1}{\rho} \frac{\partial P}{\partial z}.$$

For simplicity, we assume that the pressure is independent of density and temperature, and that there is steady-state fluid flow (velocity independent of time). Under these conditions, the Navier-Stokes equation becomes

$$\sum_{i} \frac{\partial v_i}{\partial x_i} = 0, \qquad (\mathbf{v} \cdot \vec{\nabla})\mathbf{v} = \nu \nabla^2 \mathbf{v} - \frac{1}{\rho} \vec{\nabla} P. \qquad (4.58)$$

The first equation expresses the equality of inflow and outflow and is known as the *condition of incompressibility*. If the channel in which the fluid is flowing is wide (in z), we can ignore the z dependence of the velocity, and so have three PDEs:

$$\frac{\partial v_x}{\partial x} + \frac{\partial v_y}{\partial y} = 0, \qquad (4.59)$$

$$\nu \left(\frac{\partial^2 v_x}{\partial x^2} + \frac{\partial^2 v_x}{\partial y^2} \right) = v_x \frac{\partial v_x}{\partial x} + v_y \frac{\partial v_x}{\partial y} + \frac{1}{\rho} \frac{\partial P}{\partial x}, \qquad (4.60)$$

$$\nu \left(\frac{\partial^2 v_y}{\partial x^2} + \frac{\partial^2 v_y}{\partial y^2} \right) = v_x \frac{\partial v_y}{\partial x} + v_y \frac{\partial v_y}{\partial y} + \frac{1}{\rho} \frac{\partial P}{\partial y}. \qquad (4.61)$$

The algorithm for solution of the Navier-Stokes and continuity PDEs uses successive overrelaxation (discussed in §5.2.1 for Poisson's equation). We divide space into a rectangular grid with the spacing h in both the x and y directions:

$$x = ih, \quad i = 0, \ldots, N_x; \qquad y = jh, \quad j = 0, \ldots, N_y, \qquad (4.62)$$

and assume $\nu = 1\,\mathrm{m^2/s}$ and $\rho = 1\,\mathrm{kg/m^3}$. For problems in which the velocity varies only along the direction of flow x, we obtain a relaxation algorithm for v^x, which we

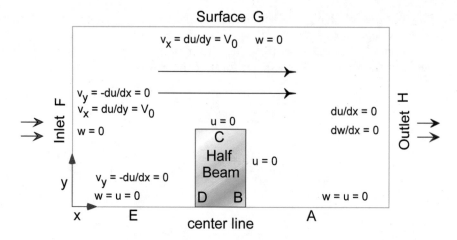

Figure 4.11. Boundary conditions for flow around a submerged beam. The flow is symmetric about the centerline, and the beam has length L in the x direction (along flow).

express as the new value of v^x as given as the old value plus a correction (residual):

$$v_{i,j}^x = v_{i,j}^x + \omega r_{i,j}, \quad r \stackrel{\text{def}}{=} v_{i,j}^{x(\text{new})} - v_{i,j}^{x(\text{old})} \tag{4.63}$$

$$r = \frac{1}{4} \left\{ v_{i+1,j}^x + v_{i-1,j}^x + v_{i,j+1}^x + v_{i,j-1}^x - \frac{h}{2} v_{i,j}^x \left[v_{i+1,j}^x - v_{i-1,j}^x \right] \right.$$

$$\left. - \frac{h}{2} v_{i,j}^y \left[v_{i,j+1}^x - v_{i,j-1}^x \right] - \frac{h}{2} \left[P_{i+1,j} - P_{i-1,j} \right] \right\} - v_{i,j}^x \tag{4.64}$$

Successive iterations sweep the interior of the grid, continuously adding in the residual (4.63) until the change becomes smaller than some set level of tolerance. In this *successive overrelaxation* form, convergence is accelerated via an amplifying factor $\omega \geq 1$.

4.6.2 Flow over Submerged Beam

Boundary conditions for hydrodynamic flow can be challenging. For a submerged beam (Figure 4.11), those used in our simulation are:

$$
\begin{array}{llll}
u & = 0; & w & = 0 & \text{Centerline EA} \\
u & = 0, & w_{i,j} & = -2(u_{i+1,j} - u_{i,j})/h^2 & \text{Beam back B} \\
u & = 0, & w_{i,j} & = -2(u_{i,j+1} - u_{i,j})/h^2 & \text{Beam top C} \\
u & = 0, & w_{i,j} & = -2(u_{i-1,j} - u_{i,j})/h^2 & \text{Beam front D} \quad (4.65) \\
\partial u/\partial x & = 0, & w & = 0 & \text{Inlet F} \\
\partial u/\partial y & = V_0, & w & = 0 & \text{Surface G} \\
\partial u/\partial x & = 0, & \partial w/\partial x & = 0 \quad \text{Outlet H} &
\end{array}
$$

1. Modify the program `Beam.py`, or write your own, to solve the Navier-Stokes equation for the velocity of a fluid in 2-D flow. Represent the x and y components of the velocity by the arrays `vx[Nx,Ny]` and `vy[Nx,Ny]`.

2. Specialize your solution to the rectangular domain and boundary conditions indicated in Figure 4.11.

3. Using the following parameter values,

$$\nu = 1\,\mathrm{m}^2/\mathrm{s}, \quad \rho = 10^3\,\mathrm{kg/m}^3 \quad \text{(flow parameters)},$$
$$N_x = 400, \quad N_y = 40, \quad h = 1 \quad \text{(grid parameters)},$$

leads to the analytic solution

$$\frac{\partial P}{\partial x} = -12, \quad \frac{\partial P}{\partial y} = 0, \quad v^x = \frac{3j}{20}\left(1 - \frac{j}{40}\right), \quad v^y = 0. \tag{4.66}$$

4. For the relaxation method, output the iteration number and the computed v^x.

5. Repeat the calculation and see if overrelaxation speeds up convergence.

4.6.3   Vorticity Form of Navier-Stokes Equation

Solution of the hydrodynamic equations is made easier by the introduction of two potentials from which the velocity is obtained by differentiation. The first is the *stream function* $\mathbf{u}(\mathbf{x})$ whose curl gives the velocity:

$$\mathbf{v} \stackrel{\text{def}}{=} \vec{\nabla} \times \mathbf{u}(\mathbf{x}) = \hat{\epsilon}_x\left(\frac{\partial u_z}{\partial y} - \frac{\partial u_y}{\partial z}\right) + \hat{\epsilon}_y\left(\frac{\partial u_x}{\partial z} - \frac{\partial u_z}{\partial x}\right), \tag{4.67}$$

where we assumed no flow in the z direction. Because $\vec{\nabla}\cdot(\vec{\nabla}\times\mathbf{u}) \equiv 0$, any $\mathbf{v}$ that can be written as the curl of $\mathbf{u}$ automatically satisfies the continuity equation $\vec{\nabla}\cdot\mathbf{v} = 0$. Furthermore, because $\mathbf{v}$ has only x and y components, the stream function $\mathbf{u}(\mathbf{x})$ needs have only a z component:

$$u_z \equiv u \quad \Rightarrow \quad v_x = \frac{\partial u}{\partial y}, \quad v_y = -\frac{\partial u}{\partial x}. \tag{4.68}$$

For 2-D flows the contour lines $u =$ constant are called *streamlines*.

The second potential function is the *vorticity* field $\mathbf{w}(\mathbf{x})$, and is related to the angular velocity ω of the fluid:

$$\mathbf{w} \stackrel{\text{def}}{=} \vec{\nabla} \times \mathbf{v}(\mathbf{x}), \quad w_z = \left(\frac{\partial v_y}{\partial x} - \frac{\partial v_x}{\partial y}\right), \tag{4.69}$$

where the second form is for flow in the z direction. We see from (4.67) that the vorticity $\mathbf{w}$ is related to the stream function via:

$$\mathbf{w} = \vec{\nabla} \times \mathbf{v} = \vec{\nabla} \times (\vec{\nabla} \times \mathbf{u}) = \vec{\nabla}(\vec{\nabla} \cdot \mathbf{u}) - \nabla^2 \mathbf{u}, \tag{4.70}$$

$$\Rightarrow \quad \vec{\nabla}^2 \mathbf{u} = -\mathbf{w}, \tag{4.71}$$

where the second form follows for flow in the z direction. Equation (4.71) is analogous to Poisson's equation of electrostatics, $\nabla^2 \phi = -4\pi\rho$, only now each component of vorticity $\mathbf{w}$ is a source for the corresponding component of the stream function $\mathbf{u}$. The vorticity form of the Navier-Stokes equation is obtained by taking the curl of the velocity form:

$$\nu \nabla^2 \mathbf{w} = [(\vec{\nabla} \times \mathbf{u}) \cdot \vec{\nabla}]\mathbf{w}. \tag{4.72}$$

Equations (4.71) and (4.72) yield the two simultaneous PDEs that we need to solve. In 2-D, with $\mathbf{u}$ and $\mathbf{w}$ having only z components for no flow in the z direction:

$$\frac{\partial^2 u}{\partial x^2} + \frac{\partial^2 u}{\partial y^2} = -w, \tag{4.73}$$

$$\nu \left(\frac{\partial^2 w}{\partial x^2} + \frac{\partial^2 w}{\partial y^2} \right) = \frac{\partial u}{\partial y}\frac{\partial w}{\partial x} - \frac{\partial u}{\partial x}\frac{\partial w}{\partial y}. \tag{4.74}$$

After expressing the derivatives in terms of finite differences on the previous lattice, we obtain the algorithm:

$$u_{i,j} = u_{i,j} + \omega\, r_{i,j}^{(1)}, \qquad w_{i,j} = w_{i,j} + \omega\, r_{i,j}^{(2)} \quad \text{(SOR)}, \tag{4.75}$$

$$r_{i,j}^{(1)} = \frac{1}{4}\left(u_{i+1,j} + u_{i-1,j} + u_{i,j+1} + u_{i,j-1} + w_{i,j} \right) - u_{i,j}, \tag{4.76}$$

$$r_{i,j}^{(2)} = \frac{1}{4}\Big(w_{i+1,j} + w_{i-1,j} + w_{i,j+1} + w_{i,j-1} - \frac{R}{4}\left\{ [u_{i,j+1} - u_{i,j-1}] \right.$$
$$\left. \times [w_{i+1,j} - w_{i-1,j}] - [u_{i+1,j} - u_{i-1,j}][w_{i,j+1} - w_{i,j-1}] \right\} \Big) - w_{i,j}.$$

Here ω is the overrelaxation parameter used to accelerate convergence. It should lie in the range $0 < \omega < 2$ for stability. The residuals are just the changes in a single step, $r^{(1)} = u^{\text{new}} - u^{\text{old}}$ and $r^{(2)} = w^{\text{new}} - w^{\text{old}}$. The parameter R is a *Reynolds number*. When we solve the problem in natural units, we measure distances in units of grid spacing h, velocities in units of initial velocity V_0, stream functions in units of $V_0 h$, and vorticity in units of V_0/h. This R is known as the *grid Reynolds number* and differs from the physical R, which has a pipe diameter in place of the grid spacing h.

Beam.py in Listing 4.9 is our program for the solution of the vorticity form of the Navier-Stokes equation. You will notice that although the relaxation algorithm, which is applied to the stream- and vorticity functions separately, is rather simple, implementing the boundary conditions is not.

1. Use Beam.py as a basis for your solution for the stream function u and the vorticity w using the finite-differences algorithm (4.75).

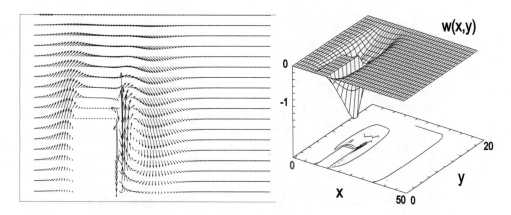

Figure 4.12. *Left:* The velocity field around the beam as represented by vectors. *Right:* The vorticity as a function of x and y for the same velocity field. Fluid rotation is seen to be largest behind the beam.

2. A good place to start your simulation is with a beam of size $L = 8h$, $H = h$, Reynolds number $R = 0.1$, and intake velocity $V_0 = 1$. Keep your grid small during debugging, say, $N_x = 24$ and $N_y = 70$.

3. Explore the convergence of the algorithm.

   a. Print out the iteration number and u values upstream from, above, and downstream from the beam.

   b. Determine the number of iterations necessary to obtain three-place convergence for successive relaxation ($\omega = 0$).

   c. Determine the number of iterations necessary to obtain three-place convergence for successive *over*relaxation ($\omega \simeq 0.3$). This should be a significant speedup, and so use this number for future calculations.

   d. Change the beam's horizontal placement so that you can see the undisturbed current entering on the left and then developing into a standing wave. Note that you may need to increase the size of your simulation volume to see the effect of all the boundary conditions.

   e. Make surface plots including contours of the stream function u and the vorticity w. Explain the behavior seen.

   f. The results of the simulation (Figure 4.12 right) are for the one-component stream function u. Make several visualizations showing the fluid velocity throughout the simulation region. Note that the velocity is a vector with two components (or a magnitude and direction), and both degrees of freedom are interesting to visualize. A plot of vectors would work well here.

   g. Explore how increasing the Reynolds number R changes the flow pattern. Start at $R = 0$ and gradually increase R while watching for numeric instabili-

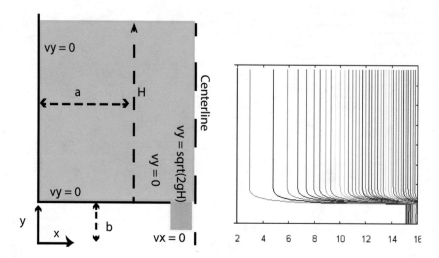

Figure 4.13. *Left:* A tank with a fluid of depth H and with the fluid flowing out from an orifice on the bottom. *Right:* The contour lines of constant u for the half tank.

ties. To overcome the instabilities, reduce the size of the relaxation parameter ω and continue to larger R values.

   h. Verify that the flow around the beam is smooth for small R values, but that it separates from the back edge for large R, at which point a small vortex develops.

4. Determine the flow behind a circular rock in the stream.

5. The boundary condition at an outlet far downstream should not have much effect on the simulation. Explore the use of other boundary conditions there.

6. Determine the pressure variation around the beam.

4.6.4 Torricelli's Law, Orifice Flow

Torricelli's law states that the speed v of a fluid flowing out of a hole at the bottom of a tank filled to height H is the same as the speed $v = \sqrt{2gH}$ that a body would acquire in falling freely from a height H. Figure 4.13 shows a tank filled with fluid flowing out an orifice on the bottom. The fluid's upper surface is moving downwards with a velocity $V_0 = 8 \times 10^{-4}$, the viscosity of the fluid $\nu = 0.5$ poise, and the Reynolds number $R = V_0 h/\nu$ [Hoffmann & Chiang(00)].

As before, it is challenging to specify all of the boundary conditions. Here are some reasonable choices, where h is the step size:

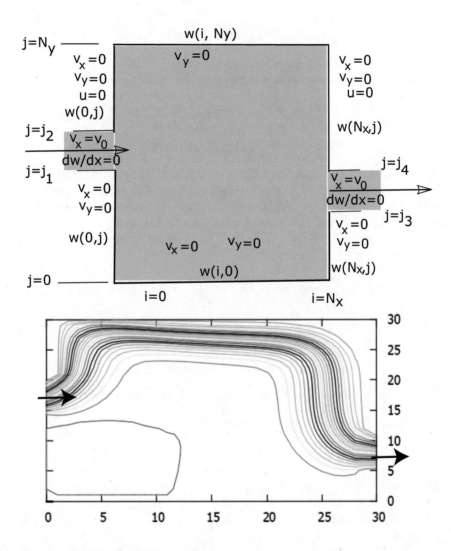

Figure 4.14. *Left:* A tank filled with fluid entering on the left and leaving on the right. Various boundary conditions are indicated. The cavity has inflow at the left between J_1 and J_2 and outflow at the right surface between J_3 and J_4. *Right:* The contour lines of constant u for the inflow and outflow tank.

The fluid tends to stick to the side of the tank, and so:

$$v_y = -\partial u/\partial x = 0, \qquad w(0,y) = -2[u(0,y) - u(h,y)]/h^2. \qquad (4.77)$$

Except at hole, the fluid cannot move down along the bottom of the tank $(y = b)$, and so

$$v_y = -\partial u/\partial y = 0, \qquad w(x,b) = -2[u(x,b) - u(x,b+h)]/h^2. \qquad (4.78)$$

By symmetry, the velocity v_x along the centerline (right side of $x = a$) has no x component, while the y component is given by Torricelli's law $v_y = -\sqrt{2gH}$:

$$v_x = \frac{\partial u}{\partial y} = 0, \qquad v_y = -\frac{\partial u}{\partial x} \approx -\frac{u(a,y) - u([a-h,y])}{h} = -\sqrt{2gh}, \qquad (4.79)$$

$$w(a,y) = -2[u(a,y) - u(a,y-h)]/h^2. \qquad (4.80)$$

At the very edge of the hole, the fluid is still in contact with the tank and so $v_x = v_y = 0$. Below the hole the fluid flows smoothly out with constant vorticity w, and so at the hole $v_x = 0$, $v_y = \sqrt{2gH}$, and $\partial w/\partial y = 0$. Since the hole is very narrow, choose the y components of velocity as $v_y = 0$ at the beginning of the hole, $v_y = 0$, at the right surface $v_y = \sqrt{2g(H-y)}$ with H height from bottom ($y = 0$) to the upper surface of the tank:

$$v_x = \frac{\partial u}{\partial y} = 0, \qquad v_y = \sqrt{2g(H-y)} = -\frac{\partial u}{\partial x}, \qquad w(x,y=0) = w(x,y=h).$$

1. Solve the vorticity form of the Navier-Stokes equation for the stream lines (constant u) and the vorticity lines (constant w) using the algorithm (4.75).

2. Construct a velocity vector plot of the flow within the tank.

3. Use $\omega = 0.1$ as the overrelaxation parameter and $h = 0.4$ as the space step size.

4. Confirm that on the lattice the boundary conditions are:

On top ($j = N_y$):

$$u[i, Ny] = u[i, Ny], \quad 1 \le i < N_x, \ w[i, Ny] = 0.$$

On the left wall:

$$u[0, j] = u[1, j], \ N_{down} \le j < N_y, \ w[0, j] = -2(u[0, j] - u[1, j])/h^2.$$

At bottom, left of hole:

$$u[i, N_{down}] = u[i, N_{down}-1], \ 1 \le i \le N_b, \ w[i, N_{down}] = -2(u[i, 0] - u[i, 1])/h^2.$$

Along the centerline:

$$v_y = -\sqrt{2gh(Ny-j)}, \quad 1 \le j \le N_y, \quad u[Nx, j] = u[Nx-1, j] + v_y h,$$
$$u[Nx, j] = u[Nx, j-1], \quad w[Nx, j] = -2(u[Nx, j] - u[Nx, j-1])/h^2.$$

At hole and below:

$$u[i,0] = u[i-1,1], \ N_b+1 \le i \le N_x; \ w[i-1,0] = w[i-1,1](\text{bottom}),$$

$$v_y = 0, \ i = N_b, \ 0 \le j \le N_{down}; \ v_y = -\sqrt{2gh(Ny+Nb-j)}, \ i = N_x,;$$

$$v_y = -\sqrt{2gh(Ny+Nb-j)}/2. \ i = Nx-1; \ u[i,j] = u[i-1,j] - v_y h.$$

Some results from our program `Torricelli.py` are show on the right of Figure 4.13. The contours were constructed with gnuplot after writing out the data to files. Here's a pseudocode of it:

```
Torricelli.py:  Flow through a container
Define constants
Declare  u, ua, w arrays
Open files
Boundary Condition Functions
    Borders calls: BelowHole, Borderight, BottomBefore, Top, Left
Relax function
    solve u & w
    relax
    compute, print out r1, r2 residuals
Iterate <= Niter times
    Relax
Write u & w to file Torri.dat
```

4.6.5   Inflow and Outflow from Square Box

This is a variation on the Torrecelli problem above. Figure 4.14 shows a square tank with dimensions $30 \ cm \times 30 \ cm$, filled with an incompressible fluid into which fluid flows in from the left with velocity $V_0 = 18 \ cm/sec$, and exits at a lower level to the right also with velocity V_0. The kinematic viscosity of the fluid is $25 \ cm^2/sec$ [Hoffmann & Chiang(00)]. The boundary conditions, assuming a step size $\Delta x = \Delta y = h = 1 \ cm$, are indicated in Figure 4.14, and are:

Left surface $(i = 0, \ 0 \le j < J1)$ has constant stream function u, and so:

$$\frac{\partial^2 u}{\partial y^2} = 0 \quad \Rightarrow \quad -w = \frac{\partial^2 u}{\partial x^2} + \frac{\partial^2 u}{\partial y^2} = \frac{\partial^2 u}{\partial x^2},$$

$$v_y = -\frac{\partial u}{\partial x} = 0, \quad \Rightarrow \quad \frac{h^2}{2}\frac{\partial^2 u}{\partial x}\Big|_{0j} = 0,$$

$$\Rightarrow \quad \frac{\partial^2 u}{\partial x^2}\Big|_{0j} = \frac{2[u(h,j)-u(0,j)]}{h^2}, \quad \Rightarrow \quad w(0,j) = \frac{2(u(0,j)-u(h,j))}{h^2}$$

Left surface $(i = 0, \ J1 \le j \le J2)$:

$$\frac{\partial w}{\partial x} = 0, \quad v_x = v_0.$$

For j > j2:

$$w(0,j) = \frac{2[u(0,j) - u(h,j)]}{h^2}, \quad \frac{\partial u}{\partial y} = \frac{\partial u}{\partial x} = 0.$$

Right surface: same as left surface, though with $i = N_x$.

Upper surface: moves to the right with a velocity u_0:

$$v_y = -\frac{\partial u}{\partial x} = 0 \;\Rightarrow\; -w = \frac{\partial^2 u}{\partial x^2} + \frac{\partial^2 u}{\partial y^2} = \frac{\partial^2 u}{\partial y^2},$$

$$\Rightarrow\; u_{i,N_y-1} = u_{i,N_y} - \frac{\partial u}{\partial y}\Big|_{i,N_y} h + \frac{\partial^2 u}{\partial y^2}\Big|_{i,N_y} \cdots = u_{i,N_y} - \frac{\partial u}{\partial y}\Big|_{i,N_y} h$$

$$\Rightarrow\; w_{i,N_y} = \frac{2(u_{i,N_y} - u_{i,N_y-1})}{h^2} - \frac{2u_0}{h}.$$

Bottom surface:

$$v_x = 0., \quad v_y = 0, \quad w[i,0] = \frac{2(u[i,0] - u[i,1])}{h^2}.$$

1. Confirm the validity of the boundary conditions given above.

2. Solve the vorticity form of the Navier-Stokes equation for the stream lines (constant u) and the vorticity lines (constant w) in the fluid using the algorithm (4.75).

3. Construct a velocity vector plot of the flow within the tank.

   The program `CavityFlow.py` in Listing 4.11 employs overrelaxation to compute the constant u streamlines and outputs to the disk file `Cavity.dat`. The contours shown on the right of Figure 4.14 were constructed with gnuplot. The pseudocode is essentially the same as for `Torricelli.py` above.

1. Solve the vorticity form of the Navier-Stokes equation for the stream lines (constant u) and the vorticity lines (constant w) in the fluid using the algorithm (4.75).

2. Construct a velocity vector plot of the flow within the tank.

4.6.6   Chaotic Convective Flow

In 1961 Edward Lorenz used a simplified version of the hydrodynamic equations including convection, viscosity, and gravity to predict weather patterns. The computations were so sensitive to parameter values that at first he thought he had a numerical problem, though eventually realized that the system was chaotic [Peitgen et al.(94),Motter & Campbell(13)]. Lorenz's equations with simplified variables are

$$\dot{x} = \sigma(y - x), \quad \dot{y} = \rho\, x - y - x\, z, \quad \dot{z} = -\beta\, z + x\, y, \tag{4.81}$$

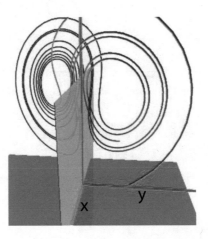

Figure 4.15. A 3-D plot of a Lorenz attractor.

where $x(t)$ is a measure of fluid velocity as a function of time t, $y(t)$ and $z(t)$ are measures of the y and z temperature distributions, and σ, ρ, and β are parameters. The xz and xy terms make these equations nonlinear.

1. Modify your ODE solver to handle these three, simultaneous Lorenz equations.

2. To start, use parameter values $\sigma = 10$, $\beta = 8/3$, and $\rho = 28$.

3. Make sure to use a small enough step size so that good precision is obtained. You must have confidence that you are seeing chaos and not numerical error.

4. Makes plots of x versus t, y versus t, and z versus t.

5. The initial behaviors in these plots are called "transients" and are not considered dynamically interesting. Leave off these transients in the plots to follow.

6. Make a "phase space" plot of $z(t)$ versus $x(t)$. The distorted, number eight-like figures you obtain (Figure 4.15) are called Lorenz attractors, "attractors" because even chaotic solutions tend to be attracted to them.

7. Make phase space plots of $y(t)$ versus $x(t)$ and $z(t)$ versus $y(t)$.

8. Make a 3-D plot of $x(t)$ versus $y(t)$ versus $z(t)$.

9. The parameters given to you should lead to chaotic solutions. Check this claim by finding the smallest change you can make in a parameter that still produce different answers.

4.7 Code Listings

```
# EqStrigMovMat.py:   Matplotlib, animated wave eqn for string

from numpy import *
import numpy as np, matplotlib.pyplot as plt
import matplotlib.animation as animation

rho = 0.01;    ten = 40.; c = sqrt(ten/rho)       # Density, tension
c1 = c;         ratio =  c*c/(c1*c1)              # CFL criterion = 1
xi = np.zeros((101,3), float)                     # Declaration
k = range(0,101)

def Initialize():                                 # Initial conditions
    for i in range(0, 81):     xi[i, 0] = 0.00125*i
    for i in range (81, 101): xi[i, 0] = 0.1 - 0.005*(i - 80)

def animate(num):
    for i in range(1, 100):
        xi[i,2] = 2.*xi[i,1] - xi[i,0] + ratio*(xi[i+1,1]
                + xi[i-1,1] - 2*xi[i,1])
    line.set_data(k,xi[k,2])                      # Data to plot,x,y
    for m in range (0,101):
        xi[m, 0] = xi[m, 1]                       # Recycle array
        xi[m, 1] = xi[m, 2]
    return line

Initialize()                                      # Plot initial string
fig = plt.figure()
ax = fig.add_subplot(111, autoscale_on=False, xlim=(0, 101),
  ylim=(-0.15, 0.15))
ax.grid()                                         # Plot   grid
plt.title("Vibrating String")
line, = ax.plot(k, xi[k,0], lw=2)
for i in range(1,100):
    xi[i,1] = xi[i,0] + 0.5*ratio*(xi[i+1,0] + xi[i-1,0]
            - 2*xi[i,0])
ani = animation.FuncAnimation(fig, animate,1)     # Dummy 1
plt.show()
print("finished")
```

Listing 4.1. **EqStringMovMat.py** uses a leapfrog algorithm to solve the wave equation.

```
# CatFriction.py: Solve for wave on catenary with friction

from numpy import *

dt = 0.0001; dx = 0.01; T = 1.; rho = 0.1
maxtime = 100; kappa = 30; D = T/(rho*9.8)

x = zeros((512,3),float)
q = open('CatFriction.dat','w');   rr = open('CatFunct.dat','w+t')

for i in range (0,101): x[i][0] = -0.08*sin(pi*i*dx)   # IC
```

```
for i in range(1,100):  # First step
    x[i][1] = (  dt*(T/rho)*(( x[i+1][0]-x[i][0] )
            /dx*( exp((i-50)*dx/D)
            -exp(-(i-50)*dx/D))/D +(exp((i-50)*dx/D)
            +exp(-(i-50)*dx/D))* ( x[i+1][0]+x[i-1][0]
            -2.0*x[i][0] )/(pow(dx,2)) )
            -2*kappa*x[i][0]+2*x[i][0]/dt   )/(2*kappa+(2./dt))  ↩

for k in range (0,300):  # Other steps
    for i in range(1,100):
        x[i][2] = (dt*(T/rho)*((x[i+1][1]-x[i][1])
            /dx*(exp((i-50)*dx/D) - exp(-(i-50)*dx/D))/D
            + (exp((i-50)*dx/D) + exp(-(i-50)*dx/D)) *
            (x[i+1][1]+x[i-1][1] -2.0*x[i][1])/(pow(dx,2)))
            -2*kappa*x[i][1]
            -(-2*x[i][1]+x[i][0])/dt)/(2*kappa+(1./dt))
    for i in range(1,101):
        x[i][0] = x[i][1]
        x[i][1] = x[i][2]
    if (k%4==0 or k==0):
        for i in range(0,100):
            a1=exp((i-50.)*dx/D)
            a2=exp(-(i-50.)*dx/D)
            rr.write("%7.3f"%(D*(a1+a2)))
            rr.write("\n")
            q.write('%7.3f'%(x[i,2]))
            q.write("\n")
        q.write("\n");
        rr.write("\n");
rr.closed
q.closed
print("Data stored in CatFrict.dat and CatFunct.dat")
```

Listing 4.2. **CatFriction.py** solves for waves on a catenary with friction.

```
# Waves2Danal.py: analytical solution Helmholtz eqn rectangular ↩
    membrane

import numpy as np; import matplotlib.pyplot as plt
import mpl_toolkits.mplot3d.axes3d

t = 0; c = np.sqrt(180/390.)   # speed, tension N/m2, density kg/m2
s5 = np.sqrt(5); N = 32

def membrane(t,X,Y):
    return np.cos(c*s5*t) * np.sin(2*X)*np.sin(Y)

plt.ion(); fig=plt.figure()              # Interactive on
ax = fig.add_subplot(111, projection='3d')
xs = np.linspace(0,np.pi,32); ys = np.linspace(0,np.pi,32)
X, Y = np.meshgrid(xs,ys);   Z = membrane(0, X, Y)  # x,y grid
wframe = None
ax.set_xlabel('x'); ax.set_ylabel('y')
ax.set_title('Vibrating Membrane')

for t in np.linspace(0,10,40):              # Total time 10/40
    oldcol = wframe
    Z = membrane(t,X,Y)                     # Membrane at t !=0
```

```
        wframe = ax.plot_wireframe(X,Y,Z)              # Plot wireframe
        plt.pause(0.01)
        if oldcol is not None:                          # Remove old frame
            ax.collections.remove(oldcol)
        plt.draw()                                       # Plot new frame
plt.show()
```

Listing 4.3. **Waves2Danal.py** computes the analytic solution for 2-D waves

```
# Waves2D.py: Helmholtz equation, rectangular vibrating membrane

""" Initial: u(x,y,t=0)=0 at borders, du/dt(x,y,t=0)=0
    Tension = 180 N/m^2, density = 390.0 kg/m^2 (rubber)"""

import matplotlib.pylab as p; from numpy import *
from mpl_toolkits.mplot3d import Axes3D

tim = 30;      N = 71    # use several values for tim (time)
c = sqrt(180./390)         # Speed = sqrt(ten[]/den[kg/m2;])
u = zeros((N,N,3),float);      v = zeros((N,N),float)
incrx = pi/N;               incry = pi/N
cprime = c;
covercp = c/cprime;    ratio = 0.5*covercp*covercp

def initial():
    y = 0.0
    for j in range(0,N):                              # Initial position
        x = 0.0
        for i in range(0,N):
            u[i][j][0] = 3*sin(2.0*x)*sin(y)          # Initial shape
            x += incrx
        y += incry
initial()

def vibration(tim):
 for k in range(1,tim):                               # Later time steps
  for j in range(1,N-1):
   for i in range(1,N-1):
      u[i][j][2] = 2*u[i][j][1] -u[i][j][0] +ratio*(u[i+1][j][1]
          + u[i-1][j][1] +u[i][j+1][1]+u[i][j-1][1] - 4*u[i][j][1])
   for j in range(0,N):
       for i in range(0,N):
           u[i][j][0]=u[i][j][1]
           u[i][j][1]=u[i][j][2]
   for j in range(0,N):
      for i in range(0,N):
          v[i][j] = u[i][j][2]      # Convert to 2D for matplotlib
 return v

v = vibration(tim)
x1 = range(0, N)
y1 = range(0, N)
X, Y = p.meshgrid(x1,y1)

def functz(v):
    z = v[X,Y];  return z

Z = functz(v)
```

```
fig = p.figure()
ax = Axes3D(fig)
ax.plot_wireframe(X, Y, Z, color = 'r')
ax.set_xlabel('x')
ax.set_ylabel('y')
ax.set_zlabel('u(x,y)')
p.show()
```

Listing 4.4. **Waves2D.py** solves the wave equation numerically for a vibrating membrane.

```
# AdvecLax.py:        Solve advection eqnt via Lax-Wendroff scheme
# du/dt+ c*d(u**2/2)/dx=0;    u(x,t=0)=exp(-300(x-0.12)**2)
# initial condition: u(x,0)=exp(-300(x-0.12)**2) gaussian shape

from numpy import *
import numpy as np, matplotlib.pyplot as plt

N = 100;   c = 1.;       dx = 1./N;       beta = 0.8   # beta = c*dt/dx
dt = beta*dx/c;     T_final = 0.5;   n = int(T_final/dt)
u0 = [];      uf = [];      xx = [];      u=np.zeros((N+1),float)
plt.figure(0)

def plotIniExac():          # Plot initial and exact solution
    for i in range(0, N):
        x = i*dx
        xx.append(x)
        u0.append(exp(-300.* (x-0.12)**2))   # Gaussian initial data
        # initfn.plot(pos = (0.01*i, u0[i]) )  # plot initial ←
            function
        uf.append(exp(- 300.*(x-0.12-c*T_final)**2))   # Exact in cyan
    plt.plot(xx,u0,'b')   # in blue
    plt.plot(xx,uf,'g')   # in orange
plotIniExac()

def numerical():                 # Lax Wendroff solution
    for j in range(0, n+1):         # loop for time
        for i in range(0, N - 2):   #  loop for x
            u[i+1] = (1.-beta*beta)*u0[i+1] - (0.5*beta)*(1.-beta) ←
                *u0[i+2] + 0.5*beta*(1. + beta)*u0[i]
            u[0] = 0.
            u[N-1] = 0.
            u0[i] = u[i]
    plt.plot(xx,u[:-1],'r')   # in red

numerical()
plt.title('Initial in blue, final in green, numerical in red')
plt.xlabel('x')
plt.show()
```

Listing 4.5. **AdvecLax.py** solves the advection equation via the Lax-Wendroff scheme.

```
# Soliton.py:       Korteweg de Vries equation for a soliton
```

```
import matplotlib.pylab as p;
from mpl_toolkits.mplot3d import Axes3D ;
from numpy import *

ds = 0.4;        dt = 0.1;       max = 2000
mu = 0.1;   eps = 0.2;      mx = 131
u   = zeros( (mx, 3), float); spl = zeros( (mx, 21), float); m = 1

for  i in range(0, 131):                        # Initial wave
     u[i, 0] = 0.5*(1  -((math.exp(2*(0.2*ds*i -5.))) -1)/
         (math.exp(2*(0.2*ds*i -5.))) +1)))
u[0,1] = 1.; u[0,2] = 1.; u[130,1] = 0.; u[130,2] = 0.   # Ends

for i in range (0, 131, 2): spl[i, 0] = u[i, 0]
fac = mu*dt/(ds**3)
print("Working. Please hold breath and wait while I count to 20")
for  i in range (1, mx-1):                          # First time step
     a1 = eps*dt*(u[i + 1, 0] + u[i, 0] + u[i - 1, 0])/(ds*6.)
     if i > 1 and  i < 129:
          a2 = u[i+2,0]+2.*u[i-1,0]-2.*u[i+1,0]-u[i-2,0]
     else:    a2 = u[i-1, 0] - u[i+1, 0]
     a3 = u[i+1, 0] - u[i-1, 0]
     u[i, 1] = u[i, 0] - a1*a3 - fac*a2/3.
for j in range (1, max+1):                          # Next time steps
     for i in range(1, mx-2):
          a1 = eps*dt*(u[i+1,1] + u[i, 1] + u[i-1,1])/(3.*ds)
          if i > 1 and i < mx-2:
               a2 = u[i+2,1] + 2.*u[i-1,1]- 2.*u[i+1,1]-u[i-2,1]
          else:    a2 = u[i-1, 1] - u[i+1, 1]
          a3       = u[i+1, 1] - u[i-1, 1]
          u[i, 2] = u[i,0] - a1*a3 - 2.*fac*a2/3.
     if j%100 == 0:                      # Plot every 100 time steps
          for i in range (1, mx - 2): spl[i, m] = u[i, 2]
          print(m)
          m = m + 1
     for k in range(0, mx):            # Recycle array saves memory
          u[k, 0] = u[k, 1]
          u[k, 1] = u[k, 2]

x = list(range(0, mx, 2))                    # Plot every other point
y = list(range(0, 21))        # Plot 21 lines every 100 t steps
X, Y = p.meshgrid(x, y)

def functz(spl):
     z = spl[X, Y]
     return z

fig  = p.figure()                                 # Create figure
ax = Axes3D(fig)                                  # Plot axes
ax.plot_wireframe(X, Y, spl[X, Y], color = 'r') # Red wireframe
ax.set_xlabel('Positon')                      # Label axes
ax.set_ylabel('Time')
ax.set_zlabel('Disturbance')
p.show()                                          # Show figure
print("That's all folks!")
```

Listing 4.6. **Soliton.py** solves the KdeV equation for 1-D solitons corresponding to a "bore" initial conditions.

```python
# SolitonAnimate.py:        Korteweg de Vries soliton equation

from numpy import *
import numpy as np, matplotlib.pyplot as plt
import matplotlib.animation as animation

ds = 0.4
dt = 0.1
mu = 0.1
eps = 0.2
mx = 101
fac = mu*dt/ds**3
u = np.zeros( (mx, 3), float)              # Soliton amplitude

def init():
    for i in range(0, mx):                 # Initial wave
        u[i, 0] = 0.5*(1-( (math.exp(2*(0.2*ds*i-5))-1)
             /(math.exp(2*(0.2*ds*i-5) )+ 1))))
    u[0, 1] = 1.
    u[0, 2] = 1.
    u[mx-1, 1] = 0.
    u[mx-1, 2] = 0.                         # End points
init()

k = range(0,mx)
fig = plt.figure()
# select axis; 111: only one plot, x,y, scales given
ax = fig.add_subplot(111, autoscale_on=False, xlim=(0,mx),
    ylim=(0,3))
ax.grid()                                  # plot a grid
plt.ylabel("Height")                       # temperature of each point
plt.title("Soliton (runs very slowly)")
line, = ax.plot(k, u[k,0],"b", lw=2)

for i in range (1, mx - 1 ):               # First time step
    a1 = eps*dt*(u[i + 1, 0] + u[i, 0] + u[i - 1, 0])/(ds*6.)
    if i>1 and  i < mx-2:
        a2 = u[i+2,0]   +   2*u[i-1,0] -2*u[i+1,0] - u[i-2,0]
    else:
        a2 = u[i - 1, 0] - u[i + 1, 0]
    a3 = u[i + 1, 0] - u[i - 1, 0]
    u[i, 1] = u[i, 0] - a1*a3 - fac*a2/3.

# later time steps

def animate(num):                          # Following next time steps
    for i in range(1, mx - 2):
        a1 = eps*dt*(u[i + 1, 1] + u[i, 1] + u[i - 1, 1])/(3.*ds)
        if i>1 and  i < mx - 2:
            a2 = u[i+2,1] + 2*u[i-1, 1] - 2*u[i+1,1]   - u[i-2,1]
        else:
            a2 = u[i - 1, 1] - u[i + 1, 1]
        a3 = u[i + 1, 1] - u[i - 1, 1]
        u[i, 2] = u[i, 0] - a1*a3 - 2.*fac*a2/3.
    line.set_data(k,u[k,2])        # Plot (position,height)
    u[k, 0] = u[k, 1]     # Recycle array
    u[k, 1] = u[k, 2]
    return line,

ani = animation.FuncAnimation(fig, animate,1)
plt.show()
print("finished")
```

Listing 4.7. SolitonAnimate.py solves the KdeV equation for 1-D solitons and animates them.

```python
# TwoDsol.py: solves Sine-Gordon equation for 2D soliton

from numpy import *; from mpl_toolkits.mplot3d import Axes3D
import numpy as np, matplotlib.pylab as plt, math

D = 201; dx = 14./200.; dy = dx; dt = dx/sqrt(2.);
dts = (dt/dx)*(dt/dx)

u = zeros((D,D,3),float);   psi = zeros((D,D),float)

def initial( u):  # initial conditions
    yy = -7.
    for i in range(0,D ):
        xx = -7.
        for j in range(0,D ):
            tmp = 3.- sqrt(xx*xx + yy*yy)
            u[i][j][0] = 4.*(math.atan(tmp))
            xx = xx + dx
        yy = yy + dy

def solution(nint):
    time = 0.
    for m in range(1,D-1):
        for l in range(1,D-1):
            a2 = u[m+1][l][0]+u[m-1][l][0]+u[m][l+1][0]+u[m][l-1][0]
            tmp = .25*a2
            u[m][l][1] = 0.5*(dts*a2-dt*dt*sin(tmp))
    for mm in range(1, D-1):              # Borders in second iteration
        u[mm][0][1]   = u[mm][1][1]
        u[mm][D-1][1] = u[mm][D-2][1]
        u[0][mm][1]   = u[1][mm][1]
        u[D-1][mm][1] = u[D-2][mm][1]
    u[0][0][1]     = u[1][0][1]           # Still undefined terms
    u[D-1][0][1]   = u[D-2][0][1]
    u[0][D-1][1]   = u[1][D-1][1]
    u[D-1][D-1][1] = u[D-2][D-1][1]
    tmp = 0.
    for k in range(0, nint+1):            # Following iterations
        print(k, "out of ",nint)
        for m in range(1,D-1):
            for l in range(1, D-1):
                a1 = u[m+ ←
                    1][l][1]+u[m-1][l][1]+u[m][l+1][1]+u[m][l-1][1]
                tmp = .25*a1
                u[m][l][2] = -u[m][l][0]  + dts*a1-dt*dt*sin(tmp)
                u[m][0][2]   = u[m][1][2]
                u[m][D-1][2] = u[m][D-2][2]
        for mm in range(1,D-1):
            u[mm][0][2]   = u[mm][1][2]
            u[mm][D-1][2] = u[mm][D-2][2]
            u[0][mm][2]   = u[1][mm][2]
            u[D-1][mm][2] = u[D-2][mm][2]
        u[0][0][2] = u[1][0][2]
```

```
        u[D−1][0][2] = u[D−2][0][2]
        u[0][D−1][2] = u[1][D−1][2]
        u[D−1][D−1][2] = u[D−2][D−1][2]
        for  l  in  range(0, D):  # New iterations now old
          for  m  in  range(0,D):
              u[l][m][0] = u[l][m][1]
              u[l][m][1] = u[l][m][2]
        if (k == nint):
          for  i  in  range(0,D, 5):
            for j  in  range(0,D, 5):  psi[i][j]=sin(u[i][j][2]/2)
        time = time + dt

def funcz(u):
    z = psi[X,Y]
    return z

initial(u)
time=0.
xx = arange(0,D,5)
yy =arange(0,D,5)
fig = plt.figure()
ax = Axes3D(fig)
X, Y = plt.meshgrid(xx, yy)
solution(22)        # Number of time iterations
Z = funcz(psi)
fig = plt.figure()
ax = Axes3D(fig)
ax.plot_wireframe(X, Y, Z, color = 'g')
ax.set_xlabel('X')
ax.set_ylabel('Y')
ax.set_zlabel('Z')
plt.show()
print('Done')
```

Listing 4.8. TwoDsol.py solves the Sine-Gordon equation for a 2-D soliton.

```
# Beam.py: solves Navier−Stokes equation for flow around beam

import matplotlib.pylab as p;
from mpl_toolkits.mplot3d import Axes3D;
from numpy import *;

print("Working, wait for the figure after 100 iterations")

Nxmax = 70;    Nymax = 20;    IL = 10;    H = 8;    T = 8;    h = 1.
u = zeros((Nxmax+1, Nymax+1), float)              # Stream
w = zeros((Nxmax+1, Nymax+1), float)              # Vorticity
V0 = 1.0;    omega = 0.1;    nu = 1.;    iter = 0;  R = V0 * h/nu

def borders():
    for i in range(0, Nxmax+1):              # Init stream
        for j in range(0, Nymax+1):          # Init vorticity
            w[i, j] = 0.
            u[i, j] = j * V0
    for i in range(0, Nxmax+1 ):             # Fluid surface
        u[i, Nymax] = u[i, Nymax−1] + V0*h
        w[i, Nymax−1] = 0.
```

```
        for j in range(0, Nymax+1):
            u[1, j] = u[0, j]
            w[0, j] = 0.                                    # Inlet
        for i in range(0, Nxmax+1):                         # Centerline
            if i <= IL and i>= IL+T:
                u[i, 0] = 0.
                w[i, 0] = 0.
        for j in range(1, Nymax ):                          # Outlet
            w[Nxmax, j] = w[Nxmax−1, j]
            u[Nxmax, j] = u[Nxmax−1, j]

def beam():                                                 # BC for beam
    for j in range (0, H+1):                                # Sides
        w[IL, j] = − 2 * u[IL−1, j]/(h*h)                   # Front
        w[IL+T, j] = − 2 * w[IL+T+1, j]/(h*h)               # Back
    for i in range(IL, IL+T + 1):
        w[i, H − 1] = − 2 * u[i, H]/(h*h);
    for i in range(IL, IL+T+1):
        for j in range(0, H+1):
            u[IL, j] = 0.                                   # Front
            u[IL+T, j] = 0.                                 # Back
            u[i, H] = 0;                                    # Top

def relax():                                                # Relax stream
    beam()                                                  # Reset
    for i in range(1, Nxmax):                               # Relax stream
        for j in range (1, Nymax):
            r1 = omega*((u[i+1,j]+u[i−1,j]+u[i,j+1]+u[i,j−1]
                + h*h*w[i,j])/4−u[i,j])
            u[i, j] += r1
    for i in range(1, Nxmax):                               # Relax vorticity
        for j in range(1, Nymax):
            a1 = w[i+1, j]  +  w[i−1,j]  +  w[i,j+1] + w[i,j−1]
            a2 = (u[i,j+1] − u[i,j−1])*(w[i+1,j] − w[i − 1, j])
            a3 = (u[i+1,j] − u[i−1,j])*(w[i,j+1] − w[i, j − 1])
            r2 = omega *( (a1 − (R/4.)*(a2 − a3) )/4. − w[i,j])
            w[i, j] += r2

borders()
while (iter <= 100):
    iter += 1
    if iter%10 == 0: print (iter)
    relax()
for i in range (0, Nxmax+1):
    for j in range(0, Nymax+ 1):  u[i,j] = u[i,j]/V0/h
x = range(0, Nxmax−1);          y = range(0, Nymax−1)
X, Y = p.meshgrid(x, y)

def functz(u):                                             # Stream flow
    z = u[X, Y]
    return z

Z = functz(u)
fig = p.figure()
ax = Axes3D(fig)
ax.plot_wireframe(X, Y, Z, color = 'r')
ax.set_xlabel('X')
ax.set_ylabel('Y')
ax.set_zlabel('Stream Function')
p.show()
```

Listing 4.9. **Beam.py** solves Navier-Stokes equation for flow over a submerged beam.

```
# Torricelli.py: solves Navier-Stokes equation for orifice flow

from numpy import *           # Need for zeros

Niter = 700;  Ndown = 20;  Nx  =  17;  N2x = 2*Nx;  Ny  = 156
Nb = 15;   h = 0.4;   h2 = h*h;  g = 980.;   nu = 0.5;   iter = 0;
Vtop = 8.0e-4;      omega = 0.1;   R = Vtop*h/nu

u = zeros((Nx+1, Ny+1), float);  ua =zeros((N2x,Ny), float)
w = zeros((Nx+1, Ny+1), float)

Torri = open('Torri.dat','w');   uall = open('uall.dat','w')

def BelowHole():
    for i in range(Nb+1,Nx+1):    # Below orifice
        u[i,0] = u[i-1,1]       # du/dy =vx=0
        w[i-1,0] = w[i-1,1]   # Water is at floor
        for j in range (0,Ndown+1):
            if i==Nb:     vy = 0
            if i==Nx:     vy = -sqrt(2.0*g*h*(Ny+Nb-j))
            if i==Nx-1:   vy = -sqrt(2.0*g*h*(Ny+Nb-j))/2.
            u[i,j] = u[i-1,j]-vy*h      # du/dx=-vy

def BorderRight():
    for j in range (1,Ny+1):     # Center orifice very sensitive
        vy = -sqrt(2.0*g*h*(Ny-j))
        u[Nx,j] = u[Nx-1,j]+vy*h
        u[Nx,j] = u[Nx,j-1]
        w[Nx,j] = -2*(u[Nx,j]-u[Nx,j-1])/h**2

def BottomBefore():
    for i in range (1,Nb+1):        # Bottom,  before  the  hole
        u[i,Ndown] = u[i,Ndown-1]
        w[i,Ndown] = -2*(u[i,0]-u[i,1])

def Top():
    for i in range(1,Nx):         # Top
        u[i,Ny] = u[i,Ny-1]
        w[i,Ny] = 0

def Left():
    for j in range (Ndown,Ny):      # Left wall
        w[0,j] =  -2*(u[0,j]-u[1,j])/h**2
        u[0,j] = u[1,j]   # du/dx=0

def Borders(iter):        # Method borders: init & B.C.
    BelowHole()
    BorderRight()                   #right (center of hole)
    BottomBefore()                  # Bottom before the hole
    Top()
    Left()

def Relax(iter):
    Borders(iter)
    for i in range(1, Nx):
        for j in range (1, Ny):
            if j<=Ndown:
                if i>Nb:
                    r1 = omega*((u[i+1,j]+u[i-1,j]+u[i,j+1]+u[i,j-1]
```

```
                                    +h*h*w[i,  j])*0.25-u[i,j])
                     u[i,j]+= r1
             if j>Ndown:
                     r1 = omega*((u[i+1,j]+u[i-1,j]+u[i,j+1]+u[i,j-1]
                                    +h*h*w[i,  j])*0.25-u[i,j])
                     u[i,j]+= r1
     if iter%50==0:
         print("Residual r1 ", r1)
     Borders(iter)
     for  i  in range(1, Nx):                    # Relax stream function
       for  j  in range  (1, Ny):
             if j<=Ndown:
                 if i>=Nb:
                     a1 = w[i+1, j]+ w[i-1,j]+w[i,j+1]+ w[i,j-1]
                     a2 = (u[i,j+1]-u[i,j-1])*(w[i+1,j]-w[i-1,j])
                     a3 = (u[i+1,j]-u[i-1,j])*(w[i,j+1]-w[i,j-1])
                     r2 = omega*( (a1 + (R/4.)*(a3 - a2) )/4.0-w[i,j])
                     w[i,j]+=r2
             if j>Ndown:
                     a1 = w[i+1, j]+ w[i-1,j]+w[i,j+1]+ w[i,j-1]
                     a2 = (u[i,j+1]-u[i,j-1])*(w[i+1,j]-w[i-1,j])
                     a3 = (u[i+1,j]-u[i-1,j])*(w[i,j+1]-w[i,j-1])
                     r2 = omega*( (a1 + (R/4.)*(a3 - a2) )/4.0-w[i,j])
                     w[i,j]+=r2

while (iter <=  Niter):
    if iter %100 == 0:
        print ("Iteration", iter)  # iterations counted
    Relax(iter)
    iter  += 1      # counter of iterations
for j in range(0,Ny): # Send w to disk in gnuplot format
    for i in range(0,Nx):
        Torri.write("%8.3e \n"%(w[i,j]))
    Torri.write("\n")
Torri.close()
for j in range(0,Ny):  # Send symmetric tank data to disk
    for i in range(0,N2x):
        if i <= Nx:
            ua[i,j] = u[i,j]
            uall.write("%8.3e \n"%(ua[i,j]))
        if i > Nx:
            ua[i,j] = u[N2x-i,j]
            uall.write("%8.3e \n"%(ua[i,j]))
    uall.write("\n")
uall.close()
utorr = open('Torri.dat','w')    # Send u data to disk
for j in range(0,Ny):
    utorr.write("\n")
    for i in range(0,Nx):
        utorr.write("%10.3e  \n"%(u[i,j]))
utorr.close()
```

Listing 4.10. **Torricelli.py** solves for flow out of an orifice at bottom of tank.

```
# CavityFlow.py: solves Navier-Stokes equation for cavity flow

from numpy import *          # needed for zeros
```

```
Niter = 200;    Nx = 30;   h = 1;   Ny = 30;    J1 = 15;    J2 = 20
J3 = 6;    J4 = 11;  V0 = 18.;  omega = 0.1;     h2 = h*h;   nu = 25
iter = 0;      u0 = 3.;   R = V0*h/nu

u = zeros((Nx+1, Ny+1), float);   w = zeros((Nx+1, Ny+1), float)

def top():
    for i in range(1,Nx+1):
        u[i-1,Ny] = u[i,Ny]                    # vy=-du/dx=0
        w[i,Ny] = 2*(u[i,Ny]-u[i,Ny-1])/h**2 -2*u0/h
def bottom():
    for i in range (1,Nx+1):           # Bottom
        w[i,0] = 2*(u[i,0]-u[i,1])/h**2
        u[i,1] = u[i,0]                        # du/dy=0
        u[i-1,0] = u[i,0]                      # du/dx=0
def borderleft():
    for j in range (1,Ny+1):                   # Right
        if j < J1:
            w[0,j] =  2*(u[0,j]-u[1,j])/h**2    # Below hole
            u[0,j] = u[1,j]
            u[0,j] = u[0,j-1]
        if j >= J1 and j <= J2:
            w[1,j] = w[0,j]
            u[0,j] = u[0,j-1]+V0*h
        if j > J2:
            u[0,j] = u[0,j-1]                   # du/dy=0
            u[1,j] = u[0,j]                     # du/dx=0
            w[0,j] =  2*(u[0,j]-u[1,j])/h**2
def borderight():
    for j in range (1,Ny+1):        # Right
        if j< J3:
            w[Nx-1,j] =  2*(u[Nx-1,j]-u[Nx,j])/h**2   # Below hole
            u[Nx-1,j] = u[Nx,j]
            u[Nx,j] = u[Nx,j-1]
        if j >= J3 and j <= J4:
            u[Nx,j] = u[Nx-1,j]
            u[Nx,j] = u[Nx,j-1]+V0*h
            w[Nx-1,j] = w[Nx,j]
        if j > J4:
            u[Nx,j] = u[Nx,j-1]
            u[Nx-1,j] = u[Nx,j]
            w[Nx-1,j] =  2*(u[Nx-1,j]-u[Nx,j])/h**2
def Borders(iter):         # Method borders: init & B.C.
    top()
    bottom()
    borderight()
    borderleft()
def Relax(iter):
    Borders(iter)
    for  i in range(1, Nx):
        for  j in range (1, Ny):
            r1 = omega*((u[i+1,j]+u[i-1,j]+u[i,j+1]
                +u[i,j-1]+h*h*w[i, j])*0.25-u[i,j])
            u[i,j] += r1
    if iter%100 == 0:  print( "Residual r1 ", r1)
    Borders(iter)
    for i in range(1, Nx):                 # Relax stream function
        for  j in range (1, Ny):
            a1 = w[i+1, j] + w[i-1,j] + w[i,j+1] + w[i,j-1]
            a2 = (u[i,j+1] - u[i,j-1])*(w[i+1,j] - w[i-1,j])
            a3 = (u[i+1,j] - u[i-1,j])*(w[i,j+1] - w[i,j-1])
            r2 = omega*((a1 + (R/4)*(a3 - a2) )/4.- w[i,j])
            w[i,j] += r2
```

```
while (iter <= Niter):
    if iter %100 ==0:  print ("Iteration Number", iter)
    Relax(iter)
    iter    += 1
utorr = open('Cavity.dat','w')      # Send data to disk  of u
for j in range(0, Ny+1):
    utorr.write("\n")
    for i in range(0,Nx+1): utorr.write("%10.3e  \n"%(u[i,j]))
utorr.close()
print("data in Cavity,dat in disk")
```

Listing 4.11. **CavityFlow.py** solves for flow through a cavity.

5

Electricity & Magnetism

5.1 Chapter Overview

Thanks to the richness of Maxwell's equations there are many problems in this chapter.[1] We start by expressing Laplace's and Poisson's equations as finite difference equations, and then solving them using relaxation techniques. These techniques are elegant in their simplicity and powerful in their wide applicability. (The finite element approach is computationally more efficient, though requires significantly more development.) After electrostatics, we examine the solutions of the Maxwell's equations as electromagnetic vector waves. The algorithm, finite difference time domain, is a variation of the leapfrog algorithm also used in Chapters 4 and 6 to solve wave equations. A latter part of the chapter continues with a number of problems requiring the calculations of electric fields by direct integration, and of trajectories of charged particles in magnetic fields. The chapter ends with relativity, first the study of the Lorentz transformations of E&M fields between frames, and then the transformation of the resulting motion in the two frames. To conclude, there is the difficult problem of two interacting relativistic particles.

We have tried to restate all problems here in SI units in which:

$$F = \frac{q_1 q_2}{4\pi\epsilon_0 r^2}, \qquad \mathbf{B}(\mathbf{r}) = \frac{\mu_0}{4\pi} \int \frac{I d\mathbf{l} \times \hat{\mathbf{r}}'}{|\mathbf{r}'|^2}, \qquad \nabla^2 \phi = -\frac{\rho}{\epsilon}. \qquad (5.1)$$

$$\nabla \cdot \mathbf{E} = \frac{1}{\epsilon_0}\rho, \qquad \nabla \times \mathbf{B} = \mu_0 \mathbf{J} + \mu_0 \epsilon_0 \frac{\partial \mathbf{E}}{\partial t}, \qquad (5.2)$$

$$\nabla \times \mathbf{E} + \frac{\partial \mathbf{B}}{\partial t} = 0, \qquad \nabla \cdot \mathbf{B} = 0. \qquad (5.3)$$

---

[1]We thank Viktor Podolski for his contributions to this chapter.

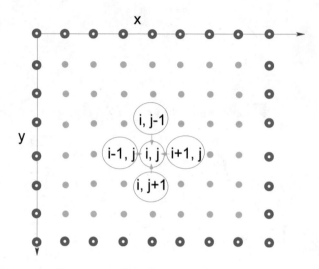

Figure 5.1. The algorithm for Laplace's equation in which the potential at the point $(x, y) = (i, j)\Delta$ equals the average of the potential values at the four nearest-neighbor points. The nodes with white centers correspond to fixed values of the potential along the boundaries.

5.2   Electric Potentials via Laplace's & Poisson's Equations

The electric potential $U(\mathbf{x})$ arising from a charge density $\rho(\mathbf{x})$ satisfies Poisson's partial differential equation (PDE):

$$\nabla^2 U(\mathbf{x}) = -\frac{1}{\epsilon_0}\rho(\mathbf{x}), \qquad (5.4)$$

where $\rho(\mathbf{x})$ is the charge density. In a charge-free region of space, $\rho(\mathbf{x}) = 0$, and so the potential there satisfies *Laplace's equation*:

$$\nabla^2 U(\mathbf{x}) = 0. \qquad (5.5)$$

5.2.1   Solutions via Finite Differences

Even for problems with cylindrical symmetry, the inherent simplicity of setting up rectangular grids and of expressing derivatives in terms of rectangular coordinates leads us to solve these equations in rectangular coordinates:

$$\frac{\partial^2 U(x,y,z)}{\partial x^2} + \frac{\partial^2 U(x,y,z)}{\partial y^2} + \frac{\partial^2 U(x,y,z)}{\partial z^2} = 0, \qquad \text{Laplace,} \qquad (5.6)$$

$$\frac{\partial^2 U(x,y,z)}{\partial x^2} + \frac{\partial^2 U(x,y,z)}{\partial y^2} + \frac{\partial^2 U(x,y,z)}{\partial z^2} = -\frac{1}{\epsilon_0}\rho(\mathbf{x}), \qquad \text{Poisson.} \qquad (5.7)$$

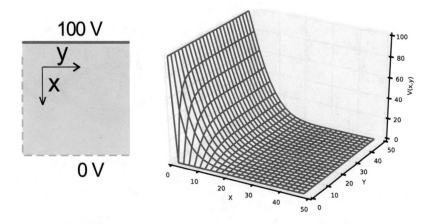

Figure 5.2. *Left:* A region of space in which a wire at the top is kept at 100 V with the sides and bottom grounded. *Right:* The computed electric potential as a function of x and y. The projections onto the shaded xy plane are equipotential (contour) lines.

Electric charge is the source of electrostatic fields, be it directly in the charge density, or indirectly through the imposition of boundary conditions.

Because the U's in (5.6) and (5.7) depend simultaneously on x, y, and z, the equations are partial differential equations (PDEs). And because we want to obtain a unique solution, we must ensure that it satisfies some specified boundary conditions. In contrast to solving ordinary differential equations where we can use a single algorithm (usually rk4) for nearly all equations, each type of PDE including its particular set of boundary conditions tends to demand a custom-built algorithm. Fortunately, we can use simple and powerful *finite difference* techniques for Laplace's and Poisson's equations.

We assume a 2-D problem and divide space up into a lattice (Figure 5.1), and solve for $U(x, y)$ only at each lattice site. Because we will express derivatives in terms of the finite differences in the values of U at the different lattice sites, the approach is called a *finite-difference* method. Specifically, we start with the Taylor expansions:

$$U(x + \Delta x, y) = U(x, y) + \frac{\partial U}{\partial x}\Delta x + \frac{1}{2}\frac{\partial^2 U}{\partial x^2}(\Delta x)^2 + \cdots, \qquad (5.8)$$

$$U(x - \Delta x, y) = U(x, y) - \frac{\partial U}{\partial x}\Delta x + \frac{1}{2}\frac{\partial^2 U}{\partial x^2}(\Delta x)^2 - \cdots, \qquad (5.9)$$

$$U(x, y + \Delta y) = U(x, y) + \frac{\partial U}{\partial y}\Delta y + \frac{1}{2}\frac{\partial^2 U}{\partial y^2}(\Delta y)^2 + \cdots, \qquad (5.10)$$

$$U(x, y - \Delta y) = U(x, y) - \frac{\partial U}{\partial y}\Delta y + \frac{1}{2}\frac{\partial^2 U}{\partial y^2}(\Delta y)^2 - \cdots. \qquad (5.11)$$

Rather cleverly, the additions $U(x+\Delta x, y)+U(x-\Delta x, y)$ and $U(x, y+\Delta y)+U(x, y-$

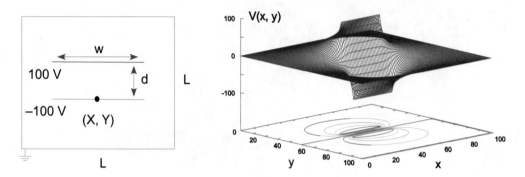

Figure 5.3. *Left:* A simple model of a parallel-plate capacitor within a box. A realistic model would have the plates close together, in order to condense the field, and the enclosing grounded box so large that it has no effect on the field near the capacitor. *Right:* A numerical solution for the electric potential for this geometry. The projection on the xy plane gives the equipotential lines.

Δy) improves precision by canceling off the $\mathcal{O}(\Delta)$ terms. With the central-difference algorithm for derivatives, we obtain approximations good to order Δ^4:

$$\frac{\partial^2 U(x,y)}{\partial x^2} \simeq \frac{U(x+\Delta x, y) + U(x-\Delta x, y) - 2U(x,y)}{(\Delta x)^2}, \tag{5.12}$$

$$\frac{\partial^2 U(x,y)}{\partial y^2} \simeq \frac{U(x, y+\Delta y) + U(x, y-\Delta y) - 2U(x,y)}{(\Delta y)^2}. \tag{5.13}$$

Substitution of these approximations in Poisson's equation (5.7) produces the finite-difference equation:

$$\frac{U(x+\Delta x, y) + U(x-\Delta x, y) - 2U(x,y)}{(\Delta x)^2}$$
$$+ \frac{U(x, y+\Delta y) + U(x, y-\Delta y) - 2U(x,y)}{(\Delta y)^2} = -\frac{\rho}{\epsilon_0}. \tag{5.14}$$

For simplicity, we take the x and y grids in Figure 5.1 to be of equal spacings Δ, and replace the x and y variables by the discrete labels i and j:

$$x = x_0 + i\Delta, \quad y = y_0 + j\Delta, \quad i, j = 0, \ldots, N_{max-1}. \tag{5.15}$$

This leads to a difference equation, which we rearrange into an algorithm:

$$U_{i+1,j} + U_{i-1,j} + U_{i,j+1} + U_{i,j-1} - 4U_{i,j} = -\frac{\rho_{i,j}}{\epsilon_0}, \tag{5.16}$$

$$\Rightarrow \quad U_{i,j} = \pi\rho_{i,j} + \tfrac{1}{4}\left[U_{i+1,j} + U_{i-1,j} + U_{i,j+1} + U_{i,j-1}\right]. \tag{5.17}$$

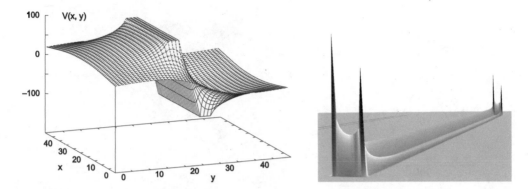

Figure 5.4. *Left:* A visualization of the computed electric potential for a capacitor with finite width plates. *Right:* A visualization of the charge distribution along the top plate determined by evaluating $\nabla^2 U(x, y)$ (courtesy of J. Wetzel). The bottom plate shows a negative charge distribution.

Equation (5.16) expresses a relation among the values of $U(x, y)$ at five points in space. Rather than solve these equations as a (big) matrix problem, we take advantage of the simple idea expressed by (5.17) that when we have a solution, it will be the average of the potential at the four nearest neighbors in Figure 5.1, plus a contribution from the local charge density. As an algorithm, (5.17) does not provide a direct solution to Poisson's equation, but rather is a condition that must be satisfied by a solution. We start with an initial guess for the potential and improve it by repeatedly sweeping through all space as we take the average over nearest neighbors at each lattice site; the process continues until the solution no longer changes below some level of precision, or fails to converge. When converged, the initial guess is said to have *relaxed* into the solution.

There are a number of ways in which the algorithm can applied. The *Jacobi method* updates the potential values only after (5.17) has been applied at each lattice site. This maintains the symmetry of the initial guess and boundary conditions. The *Gauss-Seidel method* utilizes the latest values for the potential as soon as they have been computed. This usually speeds up convergence, which in turn leads to less round-off error, though does break the symmetry of the boundary conditions.

5.2.2 Laplace & Poisson Problems

1. Find the electric potential for all points *inside* the charge-free square shown on the left of Figure 5.2. The bottom and sides of the square are grounded, while the wire at the top is kept at 100 V. Listing 5.2 is our code `LaplaceLine.py` that solves Laplace's equation within the square of Figure 5.2.

   a. Create a surface plot of the potential $V(x, y)$.
   b. Run 10-1000 iterations, and note when convergence occurs.

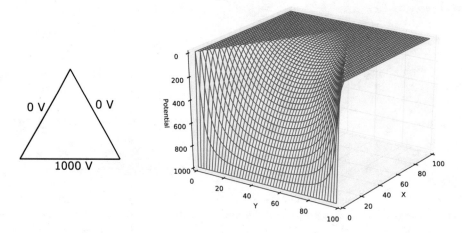

Figure 5.5. *Left:* An equilateral triangle has one side at 1000V and the other two sides at 0 V. *Right:* The corresponding solution of Laplace's equation.

    c. Modify the code so that it quits iterating once the sum of the diagonal elements $\sum |V_{i,i}|$ converges to some measure of precision such as 10^{-4}.

    d. Investigate the effect of varying the step size Δ. Draw conclusions regarding the stability and accuracy of the solution for various Δ's.

    e. Investigate the effect of using Gauss-Seidel versus Jacobi relaxation. Which converges faster? Do the answers agree?

2. A **Realistic Capacitor** has an electric field that is no longer uniform near the edges (edge effects) and extends beyond the edges of the capacitor as well (fringe fields). Figure 5.3 left shows a model of a realistic capacitor as two finite plates (wires) within a grounded box. (The box should be much bigger if it is not to have a significant effect on the fields.)

    a. Assume that the plates are very thin sheets of conductors, with a battery that maintains the top plate at $100\,\text{V}$ and the bottom at $-100\,\text{V}$. Write or modify LaplaceLine.py to solve this problem and create a surface plot of your results.

    b. Make your model more realistic by making the plates longer, the separation smaller, and the box bigger. How does this change the results?

    c. Next, assume that the plates are composed of a dielectric material with uniform charge densities ρ on the top and $-\rho$ on the bottom, though with no batteries. Solve Poisson's equation (5.4) in the region including the plates, and Laplace's equation elsewhere. The potential we found is shown in Figure 5.4.

    d. Investigate the distribution of charges on a capacitor with thick ($> 2\Delta y$) conducting plates. Because the plates are conductors, they remain at 100

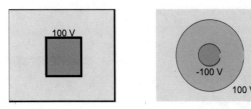

Figure 5.6. *Left:* The geometry of a capacitor formed by placing two long, square cylinders within each other. *Right:* The geometry of a capacitor formed by placing two long, circular cylinders within each other. The cylinders are cracked on the side so that wires connected to a battery can enter the region.

and -100 V. One approach here is to solve Laplace's equation as before to determine $U(x, y)$. Then substitute $U(x, y)$ into Poisson's equation (5.4) to determine $\rho(x, y)$. Figure 5.4 shows some of the results we found.

3. **Capacitance of Realistic Capacitor** Now that you have computed the charge distribution on the realistic capacitor, go one step further and compute the total charge on one plate.

   a. Now that you know the charge on the capacitor and the voltage of the capacitor, compute the capacitance of the realistic capacitor and compare that to the capacitance of the ideal capacitor that does not include edge effects.

   b. Run a series of simulations that vary the plate separation and determine which plate separations agree best with the formula for the ideal capacitor.

4. Figure 5.5 shows an **Equilateral Triangle** formed from conducting wires separated by insulators, with one side at 1000V and the other two sides at 0 V. Find the potential within the triangle.

5. For the preceding triangle, find the potential outside of the triangle. Assume that the potential vanishes at infinity (*Hint:* think big box.).

6. Figure 5.6 left shows a capacitor formed by a box within a box. Determine the electric potential between the boxes.

7. Figure 5.6 right shows a cracked cylindrical capacitor. Determine the electric potential between the conductors.

8. The PDE algorithm can be applied to arbitrary boundary conditions. Two boundary conditions to explore for the two cyclinders are triangular and sinusoidal:

$$U(x) = \begin{cases} 200x/w, & x \leq w/2, \\ 100(1 - x/w), & x \geq w/2, \end{cases} \quad \text{or} \quad U(x) = 100\sin\left(\frac{2\pi x}{w}\right). \quad (5.18)$$

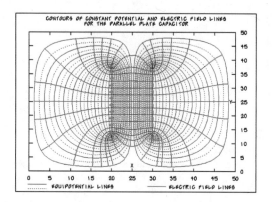 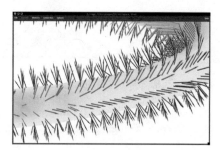

Figure 5.7. *Left:* Computed equipotential surfaces and electric field lines for a realistic capacitor. *Right:* Equipotential surfaces and electric field lines mapped onto the surface for a 3-D capacitor constructed from two tori.

5.2.3 Fourier Series vs. Finite Differences

Consider again the simple problem illustrated in Figure 5.2 of a wire in a box. In this case there exists what might be called an analytic solution of Laplace's equation (5.6) in the form of an infinite series [LPB(15)]:

$$U(x,y) = \sum_{n=1,3,5,\dots}^{\infty} \frac{400}{n\pi} \sin\left(\frac{n\pi x}{L}\right) \frac{\sinh(n\pi y/L)}{\sinh(n\pi)}. \qquad (5.19)$$

There are problems in using (5.19) as an algorithm. First, we must terminate the sum at some point. Nevertheless the convergence of the series may be so painfully slow that many terms are needed, and so the method is slow and round-off error may be large. In addition, the sinh functions tend to overflow for large n, and so one would need to take a large n limit:

$$\frac{\sinh(n\pi y/L)}{\sinh(n\pi)} = \frac{e^{n\pi(y/L-1)} - e^{-n\pi(y/L+1)}}{1 - e^{-2n\pi}} \underset{n\to\infty}{\to} e^{n\pi(y/L-1)}. \qquad (5.20)$$

Yet this means that the terms in the series get to be large and oscillatory, which may lead to unacceptable levels of subtractive cancellation. And finally, a Fourier series converges only in the *mean square*, that is, to the *average* of the left- and right-hand limits in the regions where the solution is discontinuous [Kreyszig(98)]. Explicitly, note in Figure 5.8 the large oscillations that tend to overshoot the function at corners. This is called **Gibbs overshoot** and occurs when a Fourier series with a finite number of terms is used to represent a discontinuous function.

1. Explicitly sum a Fourier series (either the one here or some other one with a finite conductor).

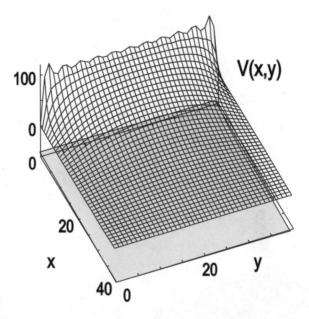

Figure 5.8. The analytic (Fourier series) solution of Laplace's equation summing 21 terms. Gibbs-overshoot leads to the oscillations near $x = 0$, and persist even if a large number of terms is summed.

    a. Stop summing the series when it has converged, for instance, when the ratio $|(\text{Last Term})/\text{Sum}|$ is $\leq 10^{-6}$.

    b. Note the effect of not including the large n limit (5.20).

2. Compare the series summation to the solution obtained with the relaxation algorithm.

3. Explore the Gibb's overshoot that occurs at the edge of the conductor in your solution by varying the number of terms used in your summation.

As seen in Figure 5.9, a point charge q is placed at point (ρ', ϕ', z') inside a cylindrical box of length L and radius a, with its sides and end caps grounded. [Jackson(88)] solves for the potential as the infinite series

$$U(\mathbf{x}, \mathbf{x}') = \frac{q}{\pi \epsilon_0 a} \sum_{m=-\infty}^{\infty} \sum_{n=1}^{\infty} \frac{e^{im(\phi-\phi')} J_m(x_{mn}\rho/a) J_m(x_{mn}\rho'/a)}{x_{mn} J_{m+1}^2(x_{mn}) \sinh(x_{mn}L/a)} \tag{5.21}$$

$$\times \sinh\left(\frac{x_{mn}z_<}{a}\right) \sinh\left(\frac{x_{mn}(L - z_>)}{a}\right), \tag{5.22}$$

where $z_<$ is the smaller of z and z', $z_>$ the larger of z and z', and x_{mn} is the nth zero of the Bessel function of order m.

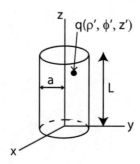

Figure 5.9. A cylindrical box of length L and radius a.

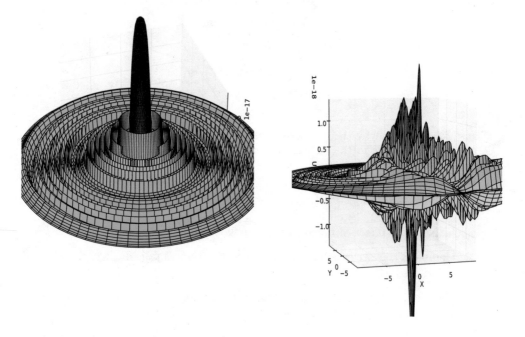

Figure 5.10. $U(x, y, z = 18)$ for a charge within a cylinder at *Left:* $(\rho' = 0, \phi' = 0, z' = 15)$, *Right:* $(\rho' = 12, \phi' = 7\pi/4, z' = 15)$.

1. Write a program to perform the summations in (5.21).

    a. As before, avoid overflows from the sinh function.
    b. Utilize a software library, such as `scipy` in Python, that has a method for computing the zeros of a Bessel function.
    c. Use visualization software that will permit you to visualize a cylindrical function of this sort. For example, we used `mplot3d` in Matplotlib.

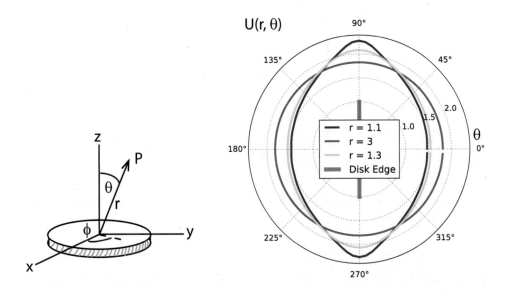

Figure 5.11. *Left:* A charged disc in the xy plane. The potential $U(r, \theta, \phi)$ is to be computed at point P. *Right:* A polar plot in θ of the potential $U(r, \theta, \phi)$ for three values of r and fixed ϕ, created by `LaplaceDisk.py` using Matplotlib.

    d. Experiment to find the maximum values of m and n that can be used in the sums. You should see convergence, and beyond that random fluctuations as round-off error accumulates.

    e. Comment on the effectiveness of the sums as an algorithm.

2. Evaluate the potential numerically using our previously-described relaxation method and the same parameters as you used for the summation. You may want to still use a rectangular grid.

3. Make a comparison of the results using the series to that using the relaxation method. Which appears more accurate?

4. Our program `LaplaceCyl.py` in Listing 5.4 evaluates these sums. The outputs for two positions of the charge are shown in Figure 5.10 for a cylinder of length $L = 20$. Verify that when the charge is at $\rho = 0$, we obtain rotational symmetry, though not for the case of $\rho = 12$.

5. Do you think the oscillations on the right in Figure 5.10 right may be due to breakdown of the algorithm?

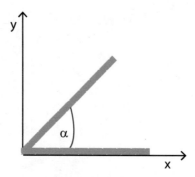

Figure 5.12. A wedge constructed from two infinite, conducting planes intersecting at an angle α with both sides kept at a potential V.

5.2.4 Disk in Space, Polar Plots

Consider the flat, conducting, circular disk of radius R shown on the left of Figure 5.11 [Jackson(88)]. The disk is kept at a constant potential V, and has a charge density $\rho \propto (R^2 - \rho^2)^{-1/2}$, where ρ is the radial distance from the center of the disk.

1. Show that for $r > R$ the potential is:

$$
U(r, \theta, \phi) = \frac{2V}{\pi} \frac{R}{r} \sum_{\ell=0}^{\infty} \frac{(-1)^\ell}{2\ell+1} \left(\frac{R}{r}\right)^{2\ell} P_{2\ell}(\cos\theta). \tag{5.23}
$$

2. Write a program that creates a polar plot of the potential at fixed values of r and ϕ as a function of the polar angle θ. Make the plot for several values of r. (Note: in §6.5.1 we give a program for computing Legendre polynomials.)

3. Now create some surface plots of the potential $U(r, \theta, \phi)$. Take note that the potential is symmetric with respect to ϕ, so in a sense you will be rotating the polar plots created above.

4. Compare the analytic solution given above to a numeric one that solves Laplace's or Poisson's equation using finite differences and relaxation. How do the two compare?

Our program `LaplaceDisk.py` created the polar plot shown on the right of Figure 5.11 using Matplotlib.

5.2.5 Potential within Grounded Wedge

Consider the wedge shown in Figure 5.12 that is formed by the interaction of two infinite, conducting planes at an angle α that are kept at a potential V.

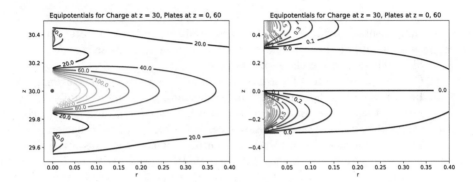

Figure 5.13. Some equipotential surfaces of a point charge at $z = 30$ between two parallel, grounded conducting planes at $z = 0$ and $z = 30$. The plot on the right shows a magnified view near the $z = 0$ plane.

1. Compute the potential within the wedge by solving Laplace's equation using a relaxation technique.

2. Plot with care the solution near the apex where the error will be larger.

3. Adjust the step size as a way of reducing the size of the error near the vertex.

4. Show that the potential near the apex drops off with a power law dependence, $\phi \sim r^{\pi/\alpha}$ (make a semilog plot of $\log \phi$ versus r, and determine the slope).

5. Use Poisson's equation to determine the surface charge density on the planes, and verify that charge tends to accumulate at edges.

6. Show that the amount of charge at the vertex and the electric field there decreases rapidly as the wedge angle α is made smaller.

5.2.6   Charge between Parallel Planes

A point charge q is placed at $z = 30$ between two infinite, parallel grounded conducting planes placed at $z_0 = 0$ and $z = L$. Problem 3.20 in [Jackson(88)] asks for you to show that the resulting potential can be written as

$$\Phi(z, r) = \frac{q}{\pi \epsilon_0 L} \sum_{n=1}^{\infty} \sin\left(\frac{n\pi z_0}{L}\right) \sin\left(\frac{n\pi z}{L}\right) K_0\left(\frac{n\pi r}{L}\right), \qquad (5.24)$$

where K_0 is the modified Bessel function of order 0.

1. Derive (5.24) and use it as a basis for its visualization (ours is in Figure 5.13).

    a. Use a subroutine library for the computation of the modified Bessel function.

b. One way of visualizing the field is to construct 2-D contour plots of the equipotential surfaces. The surfaces would be formed by rotations of these contours about the z axis. As we see in Figure 5.13, these contours look like distorted versions of those of a point charge.

c. Create the complete 3-D surface using a more powerful visualization package.

d. Investigate how the answer changes depending upon the number of terms used in the sum. In particular, increase the maximum number to such large values that noise appears due to excessive subtractive cancellation.

e. Make a separate plot of the equipotential surfaces near one of the grounded planes. These surfaces should have smooth approaches to zero, as seen on the right of Figure 5.13.

2. Use Poisson's equation to compute the induced surface density of charge on the upper and lower plates, and verify that they have an exponential falloff at large radial distances from the plates.

3. Solve this same problem directly using a relaxation technique applied to Poisson's equation. Clearly, you cannot represent the point charge numerically as a delta function, and so some approximation is necessary. We suggest two ways to do this, with the first being more interesting and the second more reliable. Here is the first way:

a. Instead of a point charge, use a large value for a charge density spread over a finite volume with integrated density equal to that of the point charge.

b. Make sure that your lattice spacing is small enough to have a number of lattice points within the density region.

c. Reduce the charged volume and increase the charge density until your solution becomes unstable. You then can move back to a stable solution.

d. Compare your best numerical solution to the analytic one, (5.24).

e. Use Poisson's equation to compute the induced surface density of charge on the upper and lower plates, and verify that they have an exponential falloff at large radial distances from the plates.

4. The second way to treat a point charge with a relaxation technique is to replace the charge by the potential it produces in a small circle around the charge, and include this potential as part of the boundary condition for Laplace's equation. Presumably, regardless of the plates, the potential very near to the charge should look like that of a point charge.

a. Ensure that your lattice spacing is small enough to have a number of lattice points included as part of the circle.

b. Try circles of various sizes, and compare the answers obtained.

c. Compare your best numerical solution to the analytic one, (5.24).

d. Use Poisson's equation to compute the induced surface density of charge on the upper and lower plates, and verify that they have an exponential falloff at large radial distances from the plates.

5.3 E&M Waves via FDTD

The basic technique used to solve for electromagnetic waves is essentially the same as that used for strings in §4.2 and quantum waves in §6.7.3: set up a grid in space and time, and step the initial solution forward in time one step at a time. When used for *E&M* simulations, this technique is known as the *finite difference time domain* (FDTD) method. What is new for E&M waves is that there are now two orthogonal vector fields with propagation in a third direction.

5.3.1 In Free Space

Maxwell's equations for the free space propagation of EM waves in the z direction reduces to four coupled PDEs:

$$\vec{\nabla} \cdot \mathbf{E} = 0 \quad \Rightarrow \quad \frac{\partial E_x(z,t)}{\partial x} = 0, \tag{5.25}$$

$$\vec{\nabla} \cdot \mathbf{H} = 0 \quad \Rightarrow \quad \frac{\partial H_y(z,t)}{\partial y} = 0, \tag{5.26}$$

$$\frac{\partial \mathbf{E}}{\partial t} = +\frac{1}{\epsilon_0 \mu_0} \vec{\nabla} \times \mathbf{H} \quad \Rightarrow \quad \frac{\partial E_x}{\partial t} = -\frac{1}{\epsilon_0 \mu_0} \frac{\partial H_y(z,t)}{\partial z}, \tag{5.27}$$

$$\frac{\partial \mathbf{H}}{\partial t} = -\vec{\nabla} \times \mathbf{E} \quad \Rightarrow \quad \frac{\partial H_y}{\partial t} = -\frac{\partial E_x(z,t)}{\partial z}. \tag{5.28}$$

Here we have chosen the electric field $\mathbf{E}(z,t)$ to be polarized in the x direction and the magnetic field $\mathbf{H}(z,t)$ to be polarized in the y direction (Figure 5.14 right), and accordingly the power flows in the direction of the bold arrow.

We express the space and time derivatives in central-difference approximations:

$$\frac{\partial E(z,t)}{\partial t} \simeq \frac{E(z, t + \frac{\Delta t}{2}) - E(z, t - \frac{\Delta t}{2})}{\Delta t}, \tag{5.29}$$

$$\frac{\partial E(z,t)}{\partial z} \simeq \frac{E(z + \frac{\Delta z}{2}, t) - E(z - \frac{\Delta z}{2}, t)}{\Delta z}. \tag{5.30}$$

We next look for solutions at only the discrete sites on the space time lattice of Figure 5.14 left. We increase the precision and robustness of the algorithm by solving for the E and H fields on separate lattices displaced from each other by half a time step and half a space step. For the present case this means that we use half-integer time steps as well as half-integer space steps, with H determined at integer time sites and half-integer space sites (open circles in Figure 5.14 left), and E determined at half-integer time sites and integer space sites (filled circles).

Because the fields already have subscripts indicating their directions, we indicate the lattice position as superscripts, for example,

$$E_x(z,t) \to E_x(k\Delta z, n\Delta t) \to E_x^{k,n}. \tag{5.31}$$

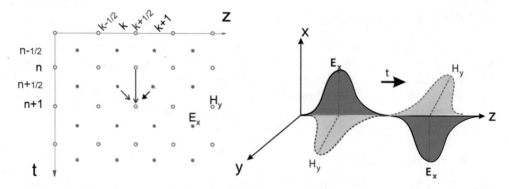

Figure 5.14. *Left:* The lattice used for converting the known values of E_x and H_y at three earlier times and three different space positions to obtain the solution at the present time. Note that the values of E_x are determined on the lattice of filled circles, corresponding to integer space indices and half-integer time indices. In contrast, the values of H_y are determined on the lattice of open circles, corresponding to half-integer space indices and integer time indices. *Right:* A single electromagnetic pulse traveling along the z axis. The coupled E and H pulses are indicated by solid and dashed curves, respectively, and the pulses at different z values correspond to different times.

After substituting the finite-difference approximations into Maxwell's equations (5.27) and (5.28), we obtain the discrete equations

$$\frac{E_x^{k,n+1/2} - E_x^{k,n-1/2}}{\Delta t} = -\frac{H_y^{k+1/2,n} - H_y^{k-1/2,n}}{\epsilon_0 \mu_0 \Delta z}, \tag{5.32}$$

$$\frac{H_y^{k+1/2,n+1} - H_y^{k+1/2,n}}{\Delta t} = -\frac{E_x^{k+1,n+1/2} - E_x^{k,n+1/2}}{\Delta z}. \tag{5.33}$$

We rearrange the equations into the form of an algorithm that advances the solution step by step through time, solving for E_x at time $n + \frac{1}{2}$, and H_y at time n:

$$E_x^{k,n+1/2} = E_x^{k,n-1/2} - \frac{\Delta t}{\epsilon_0 \mu_0 \, \Delta z}\left(H_y^{k+1/2,n} - H_y^{k-1/2,n}\right), \tag{5.34}$$

$$H_y^{k+1/2,n+1} = H_y^{k+1/2,n} - \frac{\Delta t}{\Delta z}\left(E_x^{k+1,n+1/2} - E_x^{k,n+1/2}\right). \tag{5.35}$$

Note that the algorithms must be applied simultaneously because the space variation of H_y determines the time derivative of E_x, while the space variation of E_x determines the time derivative of H_y (Figure 5.14). Furthermore, even though there are half step displacements of the lattices used for E and H, time always advances by a single time step, and successive space sites always differ by one space step. As an alternative viewpoint, we can double the index values and refer to even and odd times:

$$E_x^{k,n} = E_x^{k,n-2} - \frac{\Delta t}{\epsilon_0 \mu_0 \Delta z} \left(H_y^{k+1,n-1} - H_y^{k-1,n-1} \right), \qquad k \text{ even, odd,} \qquad (5.36)$$

$$H_y^{k,n} = H_y^{k,n-2} - \frac{\Delta t}{\Delta z} \left(E_x^{k+1,n-1} - E_x^{k-1,n-1} \right), \qquad k \text{ odd, even.} \qquad (5.37)$$

This makes it clear that E is determined for even space indices and odd times, while H is determined for odd space indices and even times. We simplify the algorithm and make its stability analysis simpler by renormalizing the electric fields to have the same dimensions as the magnetic fields, which leads to

$$\tilde{E}_x^{k,n+1/2} = \tilde{E}_x^{k,n-1/2} + \beta \left(H_y^{k-1/2,n} - H_y^{k+1/2,n} \right), \qquad (5.38)$$

$$H_y^{k+1/2,n+1} = H_y^{k+1/2,n} + \beta \left(\tilde{E}_x^{k,n+1/2} - \tilde{E}_x^{k+1,n+1/2} \right), \qquad (5.39)$$

$$\beta = \frac{c}{\Delta z/\Delta t}, \quad c = \frac{1}{\sqrt{\epsilon_0 \mu_0}}. \qquad (5.40)$$

Here c is the speed of light in a vacuum and β is the ratio of the speed of light to grid velocity $\Delta z/\Delta t$.

The space step Δz and the time step Δt must be chosen so that the algorithm is stable. The scales of the space and time dimensions are set by the wavelength and frequency, respectively, of the propagating wave. As a minimum, we want at least 10 grid points to fall within a wavelength:

$$\Delta z \leq \frac{\lambda}{10}. \qquad (5.41)$$

The time step is then determined by the Courant stability condition [Sullivan(00)]:

$$\beta = \frac{c}{\Delta z/\Delta t} \leq \tfrac{1}{2}. \qquad (5.42)$$

Equation (5.42) implies that making the time step smaller improves precision and maintains stability, but making the space step smaller must be accompanied by a simultaneous decrease in the time step in order to maintain stability.

In Listing 5.9 we present our implementation FDTD.py of the FDTD algorithm using the Visual package to create a 3-D animation of a propagating EM wave. Its pseudocode is:

```
Import packages
Set parameters
Declare Ex, Hy arrays (all space, 2 times)
Set up 3-D plots
Define PlotFields function
Initialize fields (all x)
```

```
while True (continues plotting over time until window closed)
    Calculate Ex(all x, future) from Ex(all x, present), Hy(all x, present)
    Calculate Hy(all x, future) from Ex(all x, present), Hy(all x, present)
    Impose boundary conditions on Ex & Hy at x = 0 & x = Xm-1
    PlotFields
    Ex(all x, present) = Ex(all x, future)          # Future becomes present
    Hy(all x, present) = Hy(all x, future)
```

1. In equations, not code, what are the initial conditions and boundary conditions on the fields used in FDTD.py?

2. Compare the solutions obtained with boundary conditions such that all fields vanish on the boundaries to those without explicit conditions for times less than and greater than those at which the pulses hit the walls.

3. Test the Courant condition (5.42) by determining the stability of the solution for different values of Δz and Δt.

4. The pulse propagates in the $\mathbf{E} \times \mathbf{H}$ direction. Verify that with no initial $\mathbf{H}$ field, we obtain pulses propagating to both the right and the left.

5. Modify the program so that there is an initial $\mathbf{H}$ pulse as well as an initial $\mathbf{E}$ pulse, both with a Gaussian times a sinusoidal shape.

6. Determine the direction of propagation if the $\mathbf{E}$ and $\mathbf{H}$ fields have relative phases of 0 and π.

7. Investigate the resonator modes of a wave guide by picking the initial conditions corresponding to plane waves with nodes at the boundaries.

8. For the resonator modes, investigate standing waves with wavelengths longer than the size of the integration region.

5.3.2   In Dielectrics

1. Extend the algorithm to include the effect of entering, propagating through, and exiting a dielectric material placed within the z integration region.

2. Ensure that you see both transmission and reflection at the boundaries.

3. Investigate the effect of varying the dielectric's index of refraction.

4. Place a medium with a periodic permittivity in the integration volume and verify that this act as a frequency-dependent filter that does not propagate certain frequencies.

5. Modify the program so that it plots the magnetic field in addition to the electric field.

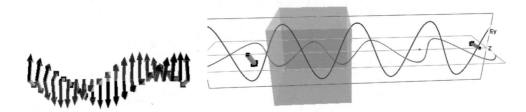

Figure 5.15. *Left: E* and *H* fields at $t = 100$ for a circularly polarized wave in free space. *Right:* One frame from the program `QuarterPlate.py` showing a linearly polarized electromagnetic wave entering a quarter wave plate from the left and leaving as a circularly polarized wave on the right.

The program `Dielect.py` in Listing 5.10 performs a FDTD simulation of a wave entering a dielectric, and produces a 2-D animation using Matplotlib. Likewise, the program `DielectVis.py` performs the same simulation and produces a 3-D animation using the Visual package.

5.3.3 Circularly Polarized Waves

To describe a circularly polarized wave propagating in the z direction we extend (5.27) and (5.28) to include:

$$\frac{\partial H_x}{\partial t} = \frac{\partial E_y}{\partial z}, \qquad \frac{\partial E_y}{\partial t} = \frac{1}{\epsilon_0 \mu_0} \frac{\partial H_x}{\partial z}. \tag{5.43}$$

When discretized in the same way as (5.34) and (5.35), we obtain

$$H_x^{k+1/2,n+1} = H_x^{k+1/2,n} + \frac{\Delta t}{\Delta z}\big(E_y^{k+1,n+1/2} - E_y^{k,n+1/2}\big), \tag{5.44}$$

$$E_y^{k,n+1/2} = E_y^{k,n-1/2} + \frac{\Delta t}{\epsilon_0 \mu_0 \Delta z}\big(H_y^{k+1/2,n} - H_y^{k-1/2,n}\big). \tag{5.45}$$

To produce a circularly polarized wave, we take E and H as having the same dimensions and set the initial conditions to:

$$E_x = \cos\left(t - \frac{z}{c} + \phi_y\right), \qquad H_x = \cos\left(t - \frac{z}{c} + \phi_y\right), \tag{5.46}$$

$$E_y = \cos\left(t - \frac{z}{c} + \phi_x\right), \qquad H_y = \cos\left(t - \frac{z}{c} + \phi_x + \pi\right), \tag{5.47}$$

where we have set the frequency $\omega = 1$. We take the phases to be $\phi_x = \pi/2$ and $\phi_y = 0$, so that their difference $\phi_x - \phi_y = \pi/2$, the requirement for circular polarization.

Listing 5.11 gives our implementation `CircPolarztn.py` for waves with transverse two-component **E** and **H** fields. Some results of the simulation are shown in Figure 5.15, where you will note the difference in phase between **E** and **H**.

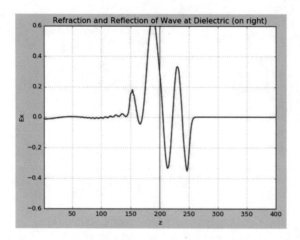

Figure 5.16. Three pulses enter and are then reflected from a dielectric medium on the right.

5.3.4 Wave Plates

A quarter-wave plate is an optical device that changes the polarization of light traveling through it by a quarter of a wavelength. We simulate a wave plate by starting with a linear polarized wave propagating along the z direction with both E_x and E_y components. The wave enters the plate and emerges from it still traveling in the z direction, though now with the relative phase of these fields shifted. The corresponding Maxwell equations are:

$$\frac{\partial H_x}{\partial t} = +\frac{\partial E_y}{\partial z}, \qquad \frac{\partial H_y}{\partial t} = -\frac{\partial E_x}{\partial z}, \tag{5.48}$$

$$\frac{\partial E_x}{\partial t} = -\frac{1}{\epsilon_0 \mu_0}\frac{\partial H_y}{\partial z}, \qquad \frac{\partial E_y}{\partial t} = +\frac{1}{\epsilon_0 \mu_0}\frac{\partial H_x}{\partial z}. \tag{5.49}$$

We take as initial conditions a wave incident from the left along the z axis with its E field at 45°, and with corresponding H components:

$$E_x(t=0) = 0.1\cos\frac{2\pi x}{\lambda}, \qquad E_y(t=0) = 0.1\cos\frac{2\pi y}{\lambda}, \tag{5.50}$$

$$H_x(t=0) = 0.1\cos\frac{2\pi x}{\lambda}, \qquad H_y(t=0) = 0.1\cos\frac{2\pi y}{\lambda}. \tag{5.51}$$

For a quarter-wave plate, the E_x and H_y components have their phases changed by $\lambda/4$ when they leave the plate, and then propagate through free space with no further changes in relative phase.

The algorithm again follows by discretizing time and space, and using a forward-difference approximation to express the derivatives. Because there are more coupled equations than before, it takes some more manipulations to solve for the fields at a

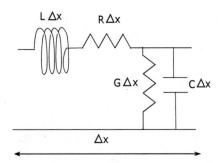 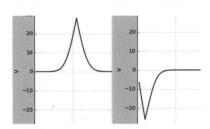

Figure 5.17. *Left:* A transmission line repeats indefinitely every Δx. *Right:* Two frames of an animation produced by `TeleMat.py` using Matplotlib, showing a power wave transmitted along a telegraph line and being reflected from an end.

future time in terms of the past fields. In this case two past values are required, though we do end up with very symmetric equations:

$$E_x^{k,n+1} = E_x^{k,n} + \beta \left(H_y^{k+1,n} - H_y^{k,n} \right), \qquad E_y^{k,n+1} = E_y^{k,n} + \beta \left(H_x^{k+1,n} - H_x^{k,n} \right),$$
(5.52)

$$H_x^{k,n+1} = H_x^{k,n} + \beta \left(E_y^{k+1,n} - E_y^{k,n} \right), \qquad H_y^{k,n+1} = H_y^{k,n} + \beta \left(E_x^{k+1,n} - E_x^{k,n} \right).$$
(5.53)

1. Modify the FDTD program of Listing 5.9 so that it solves the algorithm (5.52)–(5.53). Use $\beta = 0.01$.

2. After each time step, impose a gradual increment of the phase so that the total phase change will be one quarter of a wavelength.

3. Verify that the plate converts an initially linearly polarized wave into a circularly polarized one.

4. Verify that the plate converts an initially circularly polarized wave into a linearly polarized one.

5. What happens if you put two plates together? Three? Four? (Verify!)

5.3.5 Telegraph Line Waves

A model of a twin-lead transmission line consists of two parallel wires on which alternating current or pulses propagate [Sullivan(00),Inan & Marshall(11)]. The equivalent circuit for a segment of length Δx of a transmission line is shown on the left of Figure 5.17. There is an inductance $L\Delta x$, a resistance $R\Delta x$, a capacitance $C\Delta x$, and a conductance (inverse resistance of the dielectric material connecting the wires) $G\Delta x$. The telegrapher's equations describe the voltage and current:

$$\frac{\partial V(x,t)}{\partial x} = -RI - L\frac{\partial I(x,t)}{\partial t}, \tag{5.54}$$

$$\frac{\partial I(x,t)}{\partial x} = -GV - C\frac{\partial V(x,t)}{\partial t}. \tag{5.55}$$

For lossless transmission lines, that is those with $R = G = 0$, the equations become

$$\frac{\partial V(x,t)}{\partial x} = -L\frac{\partial I(x,t)}{\partial t}, \qquad \frac{\partial I(x,t)}{\partial x} = -C\frac{\partial V(x,t)}{\partial t}. \tag{5.56}$$

Differentiation of these equations and substitution into one another leads to a 1-D wave equation:

$$\frac{\partial^2 V(x,t)}{c^2 \partial t^2} - \frac{\partial^2 V(x,t)}{\partial x^2} = 0, \qquad c = \frac{1}{\sqrt{LC}}. \tag{5.57}$$

Wave equations are easy to solve using the *leapfrog* method, as we do with string waves in Chapter 4, the Schrödinger equation in Chapter 6, and Maxwell's equations in §5.3. One starts by using a central-difference approximation to express the derivatives in terms of finite differences, for instance,

$$\frac{\partial V(x,t)}{\partial x} \simeq \frac{V(x+\Delta x,t) - V(x-\Delta x,t)}{2\Delta}, \tag{5.58}$$

with a similar expression for the time derivative. Then, as we indicate in Figure 5.18, one rearranges the finite difference equations into a form that expresses the value of $V(x,t)$ at a future time in terms of its values at previous times. Finally, as also shown in Figure 5.18, one limits the solution to values on a space time lattice. The known initial conditions and boundary values are incorporated into the algorithm in order to propagate the initial values of $V(x,t)$ through all of space and time.

Algorithms often involve a balance between precision and robustness. If you arbitrarily keep decreasing the step size with the aim of achieving greater accuracy, you may find that your solution becomes unstable and eventually blows up exponentially. And when you have a time-dependent PDE, there are both time and space steps, and they must all be varied in a coordinated manner to avoid instabilities. For this type of leapfrog algorithm to be stable, the Courant condition requires

$$c\frac{\Delta t}{\Delta x} \leq 1. \tag{5.59}$$

In order to solve the two, coupled first-order PDEs (5.54), or the second-order PDE (5.57), we need to be given sufficient initial conditions and boundary conditions. Typically, one condition might be the initial voltage along the wire $V(x, t=0)$, and the other might be that the initial voltage on the line is a constant, $\partial V(x, t=0)/\partial t = 0$.

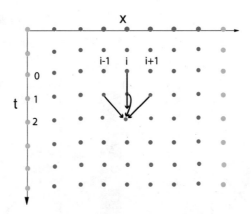

Figure 5.18. A space time lattice upon which the telegrapher's equations are solved using a leapfrog algorithm that uses values of V at three present times and one past time to compute a single value in the future.

We convert the velocity condition into something we can use in the time stepping algorithm by using the forward-difference form of the derivative:

$$\frac{\partial V(x,t)}{\partial t} \simeq \frac{V(x,\Delta t) - V(x,0)}{\Delta x} = 0, \qquad \Rightarrow \quad V(x,\Delta t) = V(x,0) = 0. \qquad (5.60)$$

1. Write down and program the leapfrog algorithm appropriate to the telegrapher's equations (5.56).

   a. The boundary conditions are $V(0,t) = V(L,t) = 0$, where L is the length of the transmission line.

   b. Use as initial conditions that a pulse was placed on a line that had a constant voltage:

   $$V(x, t=0) = 10\, e^{-x^2/0.1}, \qquad \frac{\partial V(x,t)}{\partial t} = 0. \qquad (5.61)$$

   c. Good values to try are $L = 0.1$, $C = 2.5$, $\Delta t = 0.025$, and $\Delta x = 0.05$.

2. Experiment with different values for Δx and Δt in order to obtain better precision (noisy solutions are usually not precise), or to speed up the computation.

3. Explore a selection of different time and space step sizes and see for which choices your solution becomes unstable.

4. Compare results from your experimental investigation of stability with the prediction of the Courant stability condition (5.59).

5. Investigate the effect of a nonzero value for the conductance G and the resistance R. Do your results agree with what you might expect?

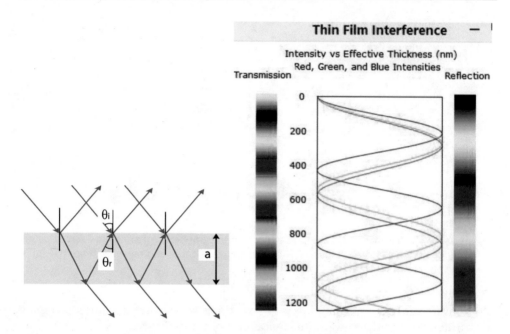

Figure 5.19. *Left:* A schematic of light rays reflected and refracted from the inner and outer surfaces of a thin film of thickness t. The film width is t. *Right:* Output from `ThinFilm.py` showing the intensity distributions and the spectra of light reflected from and transmitted through a thin film. The colors arise from the interference of the reflected and transmitted waves.

6. For nonzero R and G, investigate the distortion that occurs when a rectangular pulse is sent down the transmission line. At what point would you say the pulse shape becomes unrecognizable?

In Listing 5.13 we give our program `TeleMat.py` for solution of the telegrapher's equations with Matplotlib, and in Listing 5.14 we give the Visual version `TeleVis.py`.

5.4   Thin Film Interference of Light

As shown in Figure 5.19 left, a plane wave is incident at an angle of θ_i on a thin film of thickness a. The wave has wavelength λ in air and the film has an index of refraction n. Part of the wave reflects from top surface of the film; the remaining part enters the film, where part of it is transmitted through the film while the other part is reflected. The reflected rays from the top surface and the reflected rays from the bottom surface interfere, as does the wave directly transmitted with the transmitted wave that has been internally reflected. The phase difference, modulo 2π, between the reflected rays from the top and bottom surfaces is [Atkins & Elliot(10)]:

$$\delta = \frac{2\pi an \cos\theta_i}{\lambda} + \pi \qquad \text{(Reflected).} \qquad (5.62)$$

Here the π (an inversion) arises from the reflection from the top surface occurring from a medium with an index of refraction n larger than that of air. Likewise the phase difference between the directly transmitted wave and the transmitted wave that has undergone internal reflection is:

$$\delta = \frac{2\pi na \cos\theta_i}{\lambda} \qquad \text{(Transmitted).} \qquad (5.63)$$

No additional π occurs here because the internal reflection is from a medium (air) that has a smaller index of refraction than that of the film $(n > 1)$. In both cases, we see that the "effective thickness" of the film is an.

The amplitude of the reflected or transmitted wave is proportional to $\cos(\delta/2)$, and since intensity is proportional to the square of the amplitude, we have

$$I \propto \cos^2(\delta/2). \qquad (5.64)$$

For a particular value of the wavelength λ, complete destructive interference occurs when the phase shift is an odd multiple of π:

$$\delta = (2n+1)\pi, \quad n = 0, 1, \ldots \quad \text{(Destructive).} \qquad (5.65)$$

Accordingly, for normal incidence and reflection ($90°$), complete destructive interference occurs when

$$na = \frac{1}{2}n\lambda, \quad m = 0, 1 \ldots \quad \text{(Destructive).} \qquad (5.66)$$

1. Compute the intensity of the reflected wave as a function of the effective thickness na for the three colors: red (572 nm), green (540 nm), and blue (430 nm). Take the incident wave to be normal to the film.

2. Compute the intensity of the transmitted wave as a function of the effective thickness na for the three colors: red (572 nm), green (540 nm), and blue (430 nm). Take the incident wave to be normal to the film.

3. Convert the intensity profiles into a color panel (a spectrum) by having the intensity of each of the three colors vary according to (5.64).

4. A soap film in a bottle cap is tilted at an angle so that its thickness varies linearly with height. What are the expected intensity profiles and color panels?

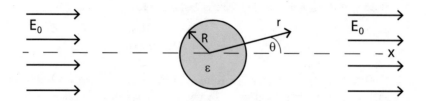

Figure 5.20. A dielectric sphere is placed in an initially uniform electric field.

5.5   Electric Fields

5.5.1   Vector Field Calculations & Visualizations

The preceding solutions of Laplace's equation for potentials can be extended to yield
the electric field vector, either by taking the derivatives of the potentials, or by solving
for the field directly. A traditional exercise involved drawing E-field curves by hand
orthogonal to the equipotential lines, beginning and ending on the boundaries (where
the charges lie). The regions of high line density are regions of high electric fields. A
more modern approach would be to extend finite-difference calculations of potentials
to compute electric fields as the negative gradient of the potential:

$$\mathbf{E} = -\nabla U(x,y) = -\frac{\partial U(x,y)}{\partial x}\,\hat{\epsilon}_x - \frac{\partial U(x,y)}{\partial y}\,\hat{\epsilon}_y. \tag{5.67}$$

If we use a central-difference approximation for the derivative, we obtain

$$E_x \simeq \frac{U(x+\Delta,y) - U(x-\Delta,y)}{2\Delta} = \frac{U_{i+1,j} - U_{i-1,j}}{2\Delta}. \tag{5.68}$$

We see that we have already calculated all that is needed, and only need to make
some subtractions. To represent a vector field, one may use software to plot arrows of
varying lengths and directions, or with just lines intersecting the equipotential surface
(Figure 5.7).

5.5.2   Fields in Dielectrics

Computing electric fields in dielectrics is much the same as computing them for conductors, but with the need to incorporate the appropriate boundary conditions. At
the interface of the dielectric with free space (or with a different dielectric) we must
require continuous tangential $\mathbf{E}$ and normal $\mathbf{D}$:

$$\mathbf{E}_\parallel\big|_1 = \mathbf{E}_\parallel\big|_2, \qquad \epsilon_1 \mathbf{E}_\perp\big|_1 = \epsilon_2 \mathbf{E}_\perp\big|_2. \tag{5.69}$$

Even though the boundary conditions are on the vector fields, and not the scalar potential, we can still solve Laplace's equation, though now with the boundary conditions
on the derivatives of the potential.

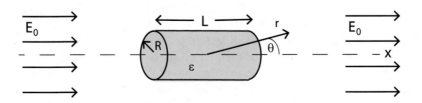

Figure 5.21. A dielectric cylinder of length L is placed in an initially uniform electric field.

The standard dielectric problem is the one illustrated in Figure 5.20, where there is a dielectric sphere of radius R and permittivity ϵ placed in an initially uniform electric field in the positive x direction [Jackson(88)]. As before, we employ Cartesian coordinates and express the boundary conditions in terms of them. We start off the simulation with the same initial field as in the problem,

$$V(x, y) = -E_0 x, \tag{5.70}$$

which is imposed over all of space, and which, presumably, will remain as the field at infinity. The boundary conditions at the dielectric-free space interface are simplest to express in polar coordinates:

$$\left.\frac{\partial V_{in}}{\partial \theta}\right|_{r=R} = \left.\frac{\partial V_{out}}{\partial \theta}\right|_{r=R}, \qquad \left.\epsilon \frac{\partial V_{in}}{\partial r}\right|_{r=R} = \left.\epsilon_0 \frac{\partial V_{out}}{\partial r}\right|_{r=R}. \tag{5.71}$$

We use the chain rule to express the derivatives in polar coordinates in terms of Cartesian derivatives:

$$\frac{\partial V}{\partial \theta} = \frac{\partial V}{\partial x}\frac{\partial x}{\partial \theta} + \frac{\partial V}{\partial y}\frac{\partial y}{\partial \theta} = -\frac{\partial V}{\partial x}\sin\theta + \frac{\partial V}{\partial y}\cos\theta, \tag{5.72}$$

$$\frac{\partial V}{\partial r} = \frac{\partial V}{\partial x}\frac{\partial x}{\partial r} + \frac{\partial V}{\partial y}\frac{\partial y}{\partial r} = +\frac{\partial V}{\partial x}\cos\theta + \frac{\partial V}{\partial y}\sin\theta. \tag{5.73}$$

The Cartesian derivatives are expressed simply in terms of finite differences of the potential, and, of course, $\sin\theta = y_i/(x_i^2 + y_i^2)$.

In the case of a dielectric sphere, the expansion of the field as an infinite series of Legendre polynomials reduces to just one term [Jackson(88)]:

$$V_{in} = -\frac{3}{\epsilon/\epsilon_0 + 2}E_0 r \cos\theta, \qquad V_{out} = -E_0 r \cos\theta + \frac{\epsilon/\epsilon_0 - 1}{\epsilon/\epsilon_0 + 2}E_0\frac{R^3}{r^2}\cos\theta. \tag{5.74}$$

1. Take the simulation used for the potential in the presence of conductors, and modify it to describe a dielectric sphere placed in an initially uniform electric field.

a. Make your lattice spacing fine enough to incorporate the curvature of the sphere.
b. Start with the potential on all lattice points given by (5.70).
c. Use a central-difference approximation to deduce expressions for the derivatives $\partial V/\partial\theta$ and $\partial V\partial r$ in terms of values of $V_{i,j}$ on the Cartesian grid. (Equations (5.72) and (5.72) will be useful for this.)
d. Use a relaxation technique to solve this problem by iteration.
e. Adjust your lattice spacing and possibly the number of iterations used so that you obtain good agreement with the analytic results (5.74).
f. Check that the potential at infinity remains unchanged.

2. Take your solution for the electric potential and compute the electric field within and without the sphere.

a. Visualize the electric field and confirm that there is a constant electric field within the sphere.

3. Modify your simulation to one that computes the electric field throughout all of space when a empty spherical cavity is placed within an infinite dielectric medium in which there was an initially uniform electric field in the x direction.

4. As shown in Figure 5.21, a dielectric cylinder is placed in an initially uniform electric field. Compute the potential and the electric field both within and without the cylinder. The field should increase with radius within the cylinder, but decrease outside.

5. A dielectric sphere, like that in Figure 5.20, has a uniform charge density of ρ.

a. Use Gauss's law to calculate the electric field and potential around the sphere.
b. Use a relation technique to solve Poisson's equation for the potential around the sphere, and from that deduce the electric field.
c. Compare the analytic and numeric results.

5.5.3 Electric Fields via Integration

Gauss's law relates the surface integral of the electric field to the charge enclosed by that surface. Often those integrals cannot be evaluated analytically, while sometimes they can be expressed in terms of the elliptic integral of the first kind,

$$K(m) = \int_0^{\pi/2} \frac{d\theta}{\sqrt{1 - m\sin^2\theta}}, \quad 0 \le m \le 1. \tag{5.75}$$

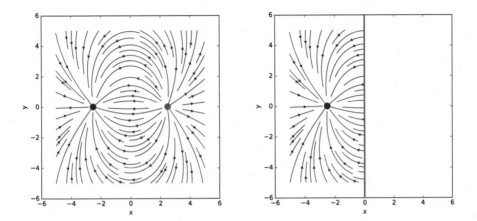

Figure 5.22. *Left:* The electric field due to two charges, the one on the right being an image charge. *Right:* The electric field due to a charge to the left of a conducting plane. The field vanishes within the conductor. Produced with Matplotlib from `ImagePlaneMat.py`.

We shall evaluate elliptic integrals numerically, and use as a check a comparison with the polynomial approximation [Abramowitz & Stegun(72)]:

$$K(m) \simeq a_0 + a_1 m_1 + a_2 m_1^2 - [b_0 + b_1 m_1 + b_2 m_1^2] \ln m_1 + \epsilon(m),$$

$$\begin{aligned} a_0 &= 1.38629\ 44 & a_1 &= 0.11197\ 23 & a_2 &= 0.07252\ 96 \\ b_0 &= 0.5 & b_1 &= 0.12134\ 78 & b_2 &= 0.02887\ 29 \end{aligned} ,$$

$$m_1 = 1 - m, \quad |\epsilon(m)| \leq 3 \times 10^{-5}.$$

1. Compute $K(m)$ by evaluating the integral in (5.75) numerically. Tune your integral evaluation until you obtain agreement at the $\leq 3 \times 10^{-5}$ level with the polynomial approximation.

2. Consider the problem of an infinite, grounded, thin, plane sheet of conducting material with a hole of radius a cut in it. The hole contains a conducting disc of slightly smaller radius kept at potential V and separated from the sheet by a thin ring of insulating material. The potential at a perpendicular distance z above the *edge* of the disk can be expressed in terms of an elliptic integral [Jackson(88)]:

$$\Phi(z) = \frac{V}{2}\left(1 - \frac{kz}{\pi a}\int_0^{\pi/2}\frac{d\phi}{\sqrt{1 - k^2 \sin^2 \phi}}\right), \quad k = 2a/(z^2 + 4a^2)^{1/2}. \quad (5.76)$$

Compute and plot the potential for $V = 1$, $a = 1$, and for values of z in the interval $(0.05, 10)$. Compare to a $1/r$ falloff.

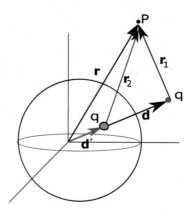

Figure 5.23. The geometry used to compute the position of an image charge that helps create the field due to a charge within a conducting sphere.

5.5.4   Electric Fields via Images

Given a problem with point charges and appropriate boundary conditions, the uniqueness theorem tells us that if we can insert an imaginary "image" charge external to the boundary that satisfies the boundary conditions, then we have solved the problem. As an example, if the boundary conditions correspond to a conductor being present, then the field within the conductor would still be zero, even though we have placed an image charge there. Likewise, as a reflection of the symmetry of images, if the image charge is considered to be a real charge, and the real charge is considered to be an image, then we can solve for the electric field *outside* of the boundary.

Some intuition is necessary to guess where to place the image charges, and that is helped along by the employment of visualization tools. Once the image charge is in place, solving for the field due to two charges is straightforward, and one visualizes the E_x and E_y components. In `ImagePlaneMat.py` in Listing 5.5 we use Matplotlib's built-in `streamplot` function to produce streamlines of the 2-D vector field (Figure 5.22 left). When using VPython, as we do in `ImagePlaneVP.py` in Listing 5.6, we plot a unit vector in the direction of the field at each lattice point (Figure 5.22 right).

1. Visualize the electric field due to a point charge.

2. Visualize the electric field due to two point charges.

3. Visualize the electric field due to an electric dipole.

4. Use the method of images to determine and visualize the electric field due to a point charge above a grounded, infinite conducting plane (Figure 5.22).

   a. Modify the VPython-based program `ImagePlaneVP.py` so that the vector **E** at each lattice point is proportional to the strength of the field.

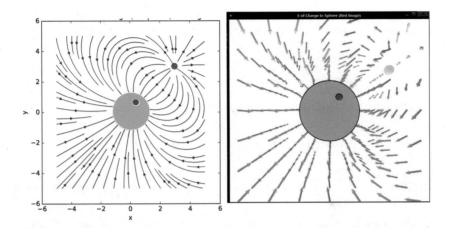

Figure 5.24. The electric field due to a charge outside of a spherical conductor, with the image charge within the conductor also shown. Because the image charge does not physically exist, the field within the conductor vanishes. Conversely, the field due to a physical charge within the conductor would vanish outside of the sphere. *Left:* Produced with Matplotlib from `ImageSphereMat.py`. *Right:* Produced with Visual from `ImageSphereVis.py`.

    b. Modify the Matplotlib-based program `ImagePlaneMat.py` that uses stream-lines to use other visualization methods such as `pcolormesh with levels` and `contourf with levels`.

5. Use the method of images to determine and visualize the electric field due to a point charge within a grounded conducting sphere. As seen in Figure 5.23, the conducting sphere of radius a is centered at the origin, the point charge q is located at $\mathbf{d}$, and the image charge q' is located at $\mathbf{d'}$. With $d' = a^2/d$, the net potential on the surface of the sphere is zero.

6. Use the method of images to determine and visualize the electric field due to a point charge outside of a grounded, conducting sphere.

    a. Modify the Visual-based program `ImageSphereVis.py` so that the vector $\mathbf{E}$ at each lattice point is proportional to the strength of the field.

    b. Modify the Matplotlib-based program `ImageSphereMat.py` to use other visualization methods such as `pcolormesh with levels` and `contourf with levels`.

5.6   Magnetic Fields via Direct Integration

The basic experimental law relating the flow of a steady current I to the magnetic induction $\mathbf{B}$ is

$$d\mathbf{B} = \frac{\mu_0}{2\pi} I \frac{d\mathbf{l} \times \hat{\mathbf{r}}}{r^2}, \tag{5.77}$$

where $\mathbf{r}$ is the observation point and $\mathbf{d\ell}$ is a differential length pointing the direction of current flow. The Biot-Savart law evaluates the basic law as a line integral,

$$d\mathbf{B} = \frac{\mu_0}{4\pi} I \int \frac{\mathbf{dl} \times \hat{\mathbf{r}}}{r^2},$$ (5.78)

or more generally as integration over a current density,

$$d\mathbf{B} = \frac{\mu_0}{4\pi} \int \mathbf{J}(\mathbf{r}') \times \frac{\widehat{\mathbf{r} - \mathbf{r}'}}{|\mathbf{r} - \mathbf{r}'|^2} d^3 r'.$$ (5.79)

1. Evaluate the integral in (5.78) numerically for the straight wire illustrated on the left of Figure 5.25.

   a. Compute the magnetic field and confirm that its magnitude is $B = \mu_0 I / 2\pi\rho$, where $\rho = r\sin\theta$ is radial distance to the wire in Figure 5.25 left. Do this by evaluating the cross product in (5.78) and computing the three components of the magnetic field.

   b. Show numerically that the lines of magnetic field are circles that lie in a plane perpendicular to the wire and centered on it.

2. Consider a coil of wire carrying a current I and wound uniformly upon a sphere, as shown on the right of Figure 5.25.

   a. Compute the magnetic field inside the sphere and confirm that it is uniform and in the direction shown in the figure.

   b. Estimate the precision of your computed magnetic field by evaluating the ratio of the "small" components of B, that is, the ones which should vanish, to the "large" components.

   c. Compute the magnetic field on the exterior of the sphere and confirm that it looks like that of a dipole.

5.6.1   Magnetic Field of Current Loop

The analytic expression for the field due to a magnetic dipole is a basic element in much of physics. The expression is usually derived under the assumption that the observation point is a long distance from the dipole. Here we wish to compute those fields without any assumptions as to distance. Figure 5.26 shows a current loop of radius a carrying a current I. The point P is a distance $\mathbf{r}$ from the center of the loop with spherical coordinates (r, θ, ϕ).

1. Rather than solving for the magnetic field directly, it is often simpler to solve for the vector potential $\mathbf{A}$, and then calculate the magnetic field as the derivative of $\mathbf{A}$. In terms of elliptic integrals, the ϕ component of the vector potential at

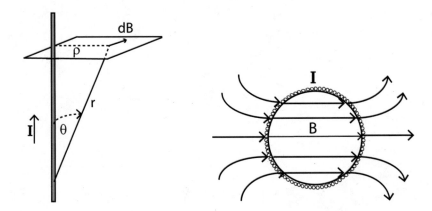

Figure 5.25. *Left:* A straight wire carrying a current I in the vertical direction. A magnetic field component in a plane perpendicular to the wire is shown. *Right:* A coil of wire carrying a current I wrapped uniformly around a sphere. The interior B field is uniform, while the exterior one is similar to a dipole.

point P is

$$A_\phi(r,\theta) = \frac{\mu_0}{4\pi} \frac{4Ia}{\sqrt{a^2 + r^2 + 2ar\sin\theta}} \left[\frac{(2-k^2)K(k) - 2E(k)}{k^2} \right], \qquad (5.80)$$

$$E(k) = \int_0^{\pi/2} \sqrt{1 - k^2 \sin^2\phi}\, d\phi, \qquad k^2 = \frac{4ar\sin\theta}{a^2 + r^2 + 2ar\sin\theta}. \qquad (5.81)$$

Here $K(k)$ is a complete elliptic integral of the first kind (5.75), and $E(k)$ is a complete elliptic integral of the second kind. Compute and plot the vector potential for $a = 1$, $I = 3$, and $\mu_0/4\pi = 1$, specifically:

a. $A_\phi(r = 1.1, \theta)$ versus θ.
b. $A_\phi(r, \theta = \pi/3)$ versus r.

2. Evaluate and plot the magnetic field by computing

$$B_r = \frac{1}{r\sin\theta}\frac{\partial}{\partial\theta}(\sin\theta\, A_\phi), \qquad B_\theta = -\frac{1}{r}\frac{\partial}{\partial r}(rA_\phi), \qquad B_\phi = 0. \qquad (5.82)$$

a. You may either evaluate the derivative of the expression for $A_\phi(r,\theta)$ in terms of elliptic integrals and then evaluate the integrals, or evaluate a numerical derivative of your numerical values for $A_\phi(r,\theta)$.

3. The magnetic dipole aspect of the current loop emerges when $A_\phi(r,\theta)$ and then **B** are evaluated for the far fields when $r \gg a$:

$$B_r \simeq \frac{\mu_0 m}{2\pi}\frac{\cos\theta}{r^3}, \qquad B_\theta \simeq \frac{\mu_0 m}{4\pi}\frac{\sin\theta}{r^3}, \qquad (5.83)$$

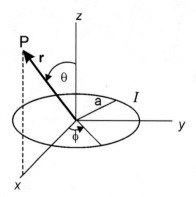

Figure 5.26. A ring of radius a carries a current I.

where $m = \pi I a^2$ is the magnetic dipole moment of the loop.

   a. Compare your numerical result for the far field with (5.83) for r values in the far field limit.

   b. Compare your numerical result for the near field with the approximate result for the far field (5.83) for small values of r.

4. Determine the magnetic field due to the current loop of Figure 5.26 by direct application of the Biot-Savart law.

   a. Construct a field map in a plane passing through the center of the loop and perpendicular to the plane of the loop. This should look like the field of a dipole.

   b. Compare your direct result to that obtained by first computing the vector potential $\mathbf{A}$, (5.82).

   c. Compare your result to the analytic, large r, result (5.83).

5.7   Motion of Charges in Magnetic Fields

5.7.1   Mass Spectrometer

Consider the 180° spectrometer sketched in Figure 5.27. It is used to focus and then detect particles of a fixed momentum as they enter the device at different angles.

1. Calculate the orbits of charged particles entering the spectrometer at different angles.

   a. Choose values for the distance between source and detector and for the width of the detector.

   b. Choose a value for the uniform magnetic field appropriate for electrons with an energy of 500 KeV.

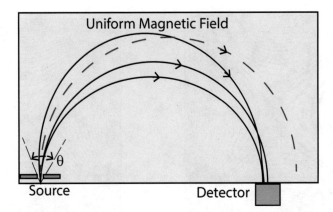

Figure 5.27. A uniform magnetic field perpendicular to the plane of the figure is used as a spectrometer to count particles of a single momentum originating from point source. The counter accepts particles entering at different angles, though not with too high a momentum (dashed curve).

    c. Run a series of simulations from which you can determine the acceptance angle θ for this device.

    d. How would you change the dimensions of the device so that it could measure electrons of 250 KeV? Verify numerically.

    e. What would happen if a dielectron with twice the mass of a single electron and twice the charge entered the spectrometer?

5.7.2 Quadruple Focusing

On the left of Figure 5.28 we show the magnetic field of a vertical focusing, magnetic quadrupole lens. On the right we show the magnetic field of a horizontal focusing lens. You can create fields like these by placing the poles of four bar-like magnets into an **X**, with the N and S poles alternating. When a positively charged particle passes perpendicular to the plane of the figure for the configuration on the left, those particles above the center will be focused down towards the center, while those below will be focused up towards the center (vertical focusing). Particles passing to the right or left of the center will be defocused away from the center. Just the reverse occurs for the configuration on the right, which focuses particles horizontally. Furthermore, because the fields get stronger as you move away from the center, specifically,

$$B_y = K\,x, \qquad B_x = K\,y, \tag{5.84}$$

those particles further from the center feel a greater push back to the center than those closer to the center. Accordingly, by having a particle beam pass through a series of

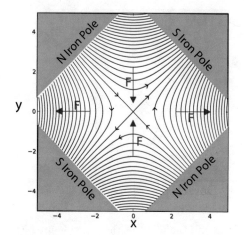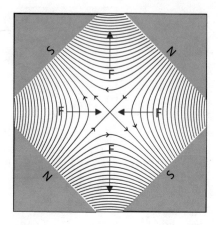

Figure 5.28. The magnetic field lines of focusing quadrupole lenses. The one on the left provides vertical focusing, while the one on the right provides horizontal focusing.

quadrupole magnets, with alternating vertical and horizontal focusing magnets, the beam gets more tightly focused.

1. Create a simulation that plots up the trajectory of an 1 MeV proton passing through a single vertical focusing quadrupole magnet of length 20 cm.

   a. Calculate the initial z component to the proton's velocity based on its energy (it's nonrelativistic).
   b. Assume that the initial beam has a circular cross section of diameter 25 cm, with all of the protons moving in the z direction.
   c. Adjust the value of the field gradient K in (5.84) to obtain 20% focusing in the vertical plane.
   d. Verify with plots that there is vertical focusing of the beam and horizontal defocusing.

2. Extend your program so that it simulates the path of the protons through a back-to-back combination of vertical and horizontal focusing magnets.

   a. Create plots showing the paths through the magnet pair and the final beam cross section.

3. Extend your simulation for an initial beam that has $24°$ angular spread.

4. Extend your simulation to a beam passing through two sets of quadrupoles and estimate the percentage decrease in beam size.

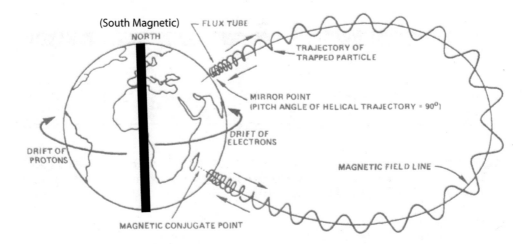

Figure 5.29. A schematic of one line of the earth's magnetic field. The full field resembles that of a bar magnet. Shown is the orbit of one charged particle trapped in a spiral-like orbit between the poles (http://astronauticsnow.com/ENA/ena_rsi_1997.html).

5.7.3 Magnetic Confinement

As illustrated in Figure 5.29, the earth is surrounded by a magnetic field that resembles that of a bar magnet with the magnet's south pole at the earth's north pole. This type of field tends to trap charged particles coming from the solar wind and cosmic rays, and causes them to bounce back and forth between the poles along spiral orbits around the magnetic field lines. (This is called a magnetic mirror effect.) The doughnut-shaped regions in which the charged particles move are known as the Van Allen belts. Because the field gets stronger near the poles, as the particles move closer to the pole they spiral in tighter and tighter orbits, and so their velocity perpendicular to the field rises. However, because a magnetic field cannot change the energy of a particle, the particle's velocity parallel to the field lines decreases, and the particle eventually stops, reverses direction, and spirals out, away from the pole.

In so far as we can model the earth's field as that of a bar magnet, we can use our knowledge of the dipole field given in §5.6.1 to describe the earth's field. Specifically, in terms of the coordinates of Figure 5.26, the radial and azimuthal field components far from the dipole, (5.83), specialized to the earth, are

$$B_r = 2B_0 \left(\frac{R_E}{r}\right)^3 \cos\theta, \qquad B_\theta = B_0 \left(\frac{R_E}{r}\right)^3 \sin\theta, \qquad (5.85)$$

where R_E is the mean radius of the earth ($\simeq 6370$ km) and $B_0 \simeq 3.12 \times 10^{-5}$ T.

1. Develop a simulation that shows the orbits of charged particles around the earth.

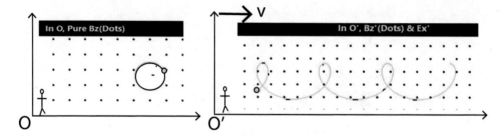

Figure 5.30. Two frames from the animations produced by the program `LorentzFieldVP.py`. *Left:* The B field in frame O causes a charge to move in a circle. *Right:* When viewed in the moving frame O', the motion is no longer circular.

a. Consider the charged particles to be electrons and protons.
b. Consider particles of different speeds and directions projected towards the earth. Determine which ones get captured, deflected, or reach the earth. The solar wind has velocities in the range 300–750 km/s.

5.8 Relativity in E&M

5.8.1 Lorentz Transformation of Fields and Motion

As illustrated on the left of Figure 5.30, an observer in frame O sees a magnetic field in the z direction (the dots coming out of the plane). This observer sets a charge q in motion with a velocity V in the x direction and observes the charge undergoing circular motion in the xy plane. This problem asks you to determine the motion of the charge as viewed in a frame O' moving with velocity v to the right with respect to O (right side of Figure 5.30).

There are at least two ways to solve this problem, and we want you to verify by actual computation that they are equivalent. And, of course, for relativistic effects to matter, the velocities must be large. The first approach is geometric. We have a charge moving in a circle in frame O, which the observer there describes as $[x(t), y(t)]$. The motion in O' is obtained by transforming the circular motion in O into the path $[x'(t'), y'(t')]$. You are given the transformation of positions, times, and velocities:

$$ct' = \gamma(ct - \beta x), \qquad x' = \gamma(x - \beta ct), \tag{5.86}$$

$$y' = y, \quad z' = z, \qquad \gamma = \frac{1}{\sqrt{1 - \beta^2}}, \quad \beta = \frac{v}{c}, \tag{5.87}$$

$$u'_\parallel = \frac{u_\parallel - v}{1 - \frac{\mathbf{v} \cdot \mathbf{u}}{c^2}}, \qquad \mathbf{u}'_\perp = \frac{\mathbf{u}_\perp}{\gamma\left(1 - \frac{\mathbf{v} \cdot \mathbf{u}}{c^2}\right)}. \tag{5.88}$$

Here $u_\parallel$ and $\mathbf{u}_\perp$ denote the components of an observed velocity parallel and perpendicular to velocity of the frames $\mathbf{v}$, and $\mathbf{u}$ is a velocity as observed in the O frame.

The second approach to this problem transforms the electric and magnetic fields in O', and then solves for the motion in O' due to these fields. Physically, the reason for an electric field in O' is clear. Observer O', seeing the charge undergoing translational motion in addition to circular motion, will conclude that there must be an electric field E' causing the translation in addition to the magnetic field B' causing the rotation. You are given that the fields transform as

$$E'_x = E_x, \qquad E'_y = \gamma(E_y - \beta B_z), \tag{5.89}$$
$$E'_z = \gamma(E_z + \beta B_y), \qquad B'_x = B_x \tag{5.90}$$
$$B'_y = \gamma(B_y + \beta E_z), \qquad B'_z = \gamma(B_z - \beta E_y). \tag{5.91}$$

1. Use both ways of solving this problem to determine the motion of the charge as viewed in O'.

   a. To make life simpler, you may want to solve the equations of motion using Euler's integration rule, $\mathbf{y}_{n+1} \simeq \mathbf{y}_n + \frac{d\mathbf{y}}{dt}\big|_n$. Because Euler's rule is of low order in the time step Δt, you must select a smaller step size than you might otherwise with rk4. You can check if you have a small enough time step if the motion in O remains circular, with future orbits falling on top of previous orbits.

   b. In frame O you can use the analytic result for circular motion in a B field, or you can solve the equations of motion numerically.

   c. Solve the equation of motion, $\mathbf{F} = m\mathbf{a}$, in O, and keep repeating the solution until you reach a large value of time t.

   d. Transform the motion in O, $[x(t), y(t)]$ into $[x'(t'), y'(t')]$, as seen in O'.

   e. Is the B in O' larger or smaller than the magnetic field in O? Explain.

   f. Is $\mathbf{E}'$ in O' pointing in the expected direction? Explain.

   g. You can solve the equations of motion in O' as $\mathbf{F}' = m'\mathbf{a}'$, though do remember that m' is no longer the rest mass m.

2. Construct animated plots (movies) of the motion in the O and O' frames. You can do this by adding plotting inside the time loop of Euler's rule.

3. Once your programs are running, investigate a range of values for the velocity v of frame O'. Make sure to include negative as well as positive values, and some very small and some very large ($v \simeq c$) values.

4. Can you find a value of v for which the motion in O' is a closed figure?

5. Repeat some parts of this exercise where relativistic effects seemed largest, only now do it using Galilean transformations.

Our code `LorentzFieldVP.py` that solves this problem by transforming fields is given in Listing 5.18. Some frames from its animations are shown in Figure 5.30.

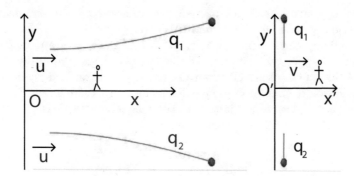

Figure 5.31. Two equal charges start off in frame O (left) with equal velocities **u** and a separation $\mathbf{r}_2 - \mathbf{r}_1$. On the right we see the same two charges as viewed in frame O' moving to the right with velocity **v**.

5.8.2   Two Interacting Charges, the Breit Interaction

A proper relativistic treatment of two interacting particles accounts for the finite amount of time it takes for the field to travel between the particles. This is a hard problem though is necessary to explain the fine structure in hydrogen. When treated to lowest order in v/c, we obtain the Darwin and Breit interactions, which we examine in this section [Jackson(88), Darwin(20), Page & Adams(45), Essen(96)].

As illustrated in Figure 5.31, an observer in frame O sets two identical, spinless charges in motion at the same time, each with a velocity **u** in the x direction. The charges are initially separated by a distance r. The charges repel each other via their **E** fields, yet also attract each other via their **B** fields (use the right hand rule to check that). Since electric forces tend to be stronger than magnetic ones, we expect the observer in O to see the charges move apart in curved trajectories.

As with the previous problem, there are at least two ways to solve this problem, and we want you to verify by actual computation that they are equivalent. And, of course, for relativistic effects to matter, the velocities must be large. The first approach is geometric. We have a charge moving with some trajectory in frame O, which is described as $[x(t), y(t)]$. The solution follows by using (5.86)–(5.88) to Lorentz transform the motion in O into $[x'(t'), y'(t')]$. The second approach uses (5.89)–(5.91) to transform the fields, and then solves the equations of motion in O' using these transformed fields.

In O, the magnetic field produced by charge q_1 with velocity $\mathbf{v_1}$ at the location of charge q_2 a distance **r** away is:

$$\mathbf{B} = \frac{\mu_0}{r^3}\mathbf{v_1} \times \mathbf{r}. \tag{5.92}$$

The force on q_2 due to the magnetic field produced by q_1 is:

$$\mathbf{F_m} = q\mathbf{v_2} \times \mathbf{B} = \frac{\mu_0}{|\mathbf{r_2} - \mathbf{r_1}|^3}\mathbf{v_2} \times [\mathbf{v_1} \times (\mathbf{r_2} - \mathbf{r_1})]. \tag{5.93}$$

The electrostatic force on q_2 due to q_1 is

$$\mathbf{E} = \frac{q_1 q_2(\mathbf{r_2} - \mathbf{r_1})}{4\pi\epsilon_0|(\mathbf{r_2} - \mathbf{r_1})|^3}. \tag{5.94}$$

Because the particles are no longer static, there are retardation corrections to this, which we ignore.

The steps to follow for this problem then are essentially the same as in the previous problem. Figure 5.31 shows some of the results we obtained with the program TwoCharges.py given in Listing 5.16. The program TwoFields.py that transforms fields is given in Listing 5.17.

5.8.3   Field Propagation Effects

Consider again the interaction between the two charged, spinless particles of Figure 5.31. The relativistic two-body problem has its difficulties and is usually treated only approximately. To this end, Jackson derives the Darwin Lagrangian describing the interaction between two charged particles correct to order $1/c^2$ as

$$L_{int} = \frac{q_1 q_2}{r}\left\{-1 + \frac{1}{2c^2}\left[\mathbf{v_1} \cdot \mathbf{v_2} + (\mathbf{v_1} \cdot \hat{\mathbf{r}})(\mathbf{v_2} \cdot \hat{\mathbf{r}})\right]\right\}. \tag{5.95}$$

The full Lagrangian correct to order $1/c^2$ is

$$L_{Darwin} = \sum_{i=1}^{2} \frac{m_i v_i^2}{2}\left(1 + \frac{v_i^2}{4c^2}\right) - \frac{q_1 q_2}{r} + \frac{q_1 q_2}{2c^2 r}\left[\mathbf{v_1} \cdot \mathbf{v_2} + (\mathbf{v_1} \cdot \hat{\mathbf{r}})(\mathbf{v_2} \cdot \hat{\mathbf{r}})\right]. \tag{5.96}$$

The equations of motion are derived from the Lagrangian in the usual way, which we write in several forms as

$$\frac{\partial L(q, \dot{q})}{\partial q_j} - \frac{d}{dt}\frac{\partial L(q, \dot{q})}{\partial \dot{q}} = 0, \tag{5.97}$$

$$\Rightarrow \quad \nabla L(\mathbf{r_1}, \mathbf{v_1}) = \frac{d}{dt}\left[\frac{\partial L(\mathbf{r_1}, \mathbf{v_1})}{\partial \mathbf{v_1}}\right] = \frac{d\mathbf{p_1}}{dt}, \tag{5.98}$$

where $\mathbf{p_1}$ is the generalized momentum associated with particle 1, with a similar equation for particle 2. If we do not include interactions, we obtain the equation of motion for a free particle,

$$\frac{d}{dt}\left[\left(1 + \tfrac{1}{2}\frac{v_1^2}{c^2}\right)m_1\mathbf{v_1}\right] = 0, \qquad \mathbf{p_1} = \left(1 + \tfrac{1}{2}\frac{v_1^2}{c^2}\right)m_1\mathbf{v_1}. \tag{5.99}$$

As advertised, the Lagrangian includes relativistic effects up to order v^2/c^2. Finally, we obtain the equation of motion and the generalized momentum, which includes a contribution from the electromagnetic field:

$$\frac{d\mathbf{p}_1}{dt} = \frac{q_1 q_2}{r^2} \left\{ \hat{\mathbf{r}} + \frac{\mathbf{v}_1(\hat{\mathbf{r}} \cdot \mathbf{v}_2) + \mathbf{v}_2(\hat{\mathbf{r}} \cdot \mathbf{v}_1) - \hat{\mathbf{r}}[\mathbf{v}_1 \cdot \mathbf{v}_2 + 3(\mathbf{v}_1 \cdot \hat{\mathbf{r}})(\mathbf{v}_2 \cdot \hat{\mathbf{r}})]}{2c^2} \right\}, \quad (5.100)$$

$$\mathbf{p}_1 = m_1 \mathbf{v}_1 + \frac{v_1^2}{2c^2} m_1 \mathbf{v}_1 + \frac{q_1}{c} \mathbf{A}, \quad (5.101)$$

$$\mathbf{A} = \frac{q_2}{2cr}[\mathbf{v}_2 + \hat{\mathbf{r}}(\hat{\mathbf{r}} \cdot \mathbf{v}_2)], \qquad \mathbf{r} = \mathbf{r}_1 - \mathbf{r}_2. \quad (5.102)$$

Note: While we have given several references that study the effects of the Darwin interaction, none of them appear to compute numerical values that indicate the size of the effects. So consider the problems here exploratory in nature and possibly worthy of serious study.

1. Consider again Figure 5.31 showing two charges initially separated by a distance r. Since we start off with a symmetrical configuration for both charges, and since the equations of motion are symmetric, assume that the paths of the charges remain symmetric.

   a. Program up the equation of motion (5.100) so that you can solve it with a standard ODE solver such as rk4.
   b. The point of this problem is to explore the size of the relativistic effects in the interaction term for two particles. So you need to have v^2/c^2 large enough to be significant, but not so large that the ignored v^4/c^4 terms are significant. In particular, see if you can find any differences from the treatment in §5.8.2.

2. A physical system to which the Darwin interaction has been applied is positronium, a bound state of an electron and a positron [Page & Adams(45)]. Consider a situation in which the charges are moving in a plane, in a circular orbit, and with a common velocity.

   a. Explore the solution to the equation of motion and identify effects due to field propagation.

5.9   Code Listings

```
# LaplaceTri.py: Matplotlib, Laplace Eq field triangular capacitor

import matplotlib.pylab as p
from mpl_toolkits.mplot3d import Axes3D;   from numpy import *

#i = 0;       x = 0.0;      y = 0.3;    a1 = -0.5;   b1 = -0.433
#a2 = 0.5;    b2 = -0.433;  a3 = 0.0;   b3 = 0.433   vertices
```

```
Nmax = 100;   Niter = 600
V = zeros ((Nmax, Nmax), float);   grid = ones ((Nmax,Nmax))
sq3 = sqrt (3.)
y0 = -sq3/4.

def contour ():                                      # Potential on one side
    for j in range(0,Nmax):  V[0,j] = 1000                      # Columns
for i in range(0,Nmax):                       # Set interior grid points to 0
    y = y0 +i*0.01
    x0 = -0.5
    for j in range(0,Nmax):
        x = x0+j*0.01
        if (y <= sq3*(x+0.25) and x<0.) or (y<-sq3*(x-0.25) and x >= ↩
            0):
            grid[i,j] =  0
        else:   if(y <= sq3/4. + 0.01): V[i,j] = 0.0 # Triangle tip
for iter in range(1,Niter):                               # Iterate
    if(iter%50==0): print('On iteration', iter,'out of',Niter)
    contour () # keep one side at 1000V
    for i in range(1, Nmax-2):
        for j in range(1,Nmax-2):
            if grid[i,j]==0.: # interior points
                V[i,j] = 0.25*(V[i+1,j]+V[i-1,j]+V[i,j+1]+V[i,j-1])
x = range(0, Nmax-1, 2)
y = range(0, Nmax, 2)                              # Plot every other point
X, Y = p.meshgrid (x,y)

def functz(V):                                  # Function returns V(x, y)
    z = V[X,Y]
    return z

Z = functz(V)
fig = p.figure()                                        # Create figure
ax = Axes3D(fig)                                          # Plot axes
ax.plot_wireframe(X, Y, Z, color = 'r')               # Red wireframe
ax.set_title('Potential within Triangle (Rotatable)')
ax.set_xlabel('X')                                     # Label axes
ax.set_ylabel('Y')
ax.set_zlabel('Potential')
p.show()                                               # Display fig
```

Listing 5.1. LaplaceTri.py solves Laplace's equation within a triangular conductor. Various parameters should be adjusted for an accurate solution.

```
# LaplaceLine.py: Matplotlib, Solve Laplace's eqtn in square

import matplotlib.pylab as p, numpy
from mpl_toolkits.mplot3d import Axes3D;   from numpy import *;

Nmax = 100; Niter = 50
V = zeros ((Nmax, Nmax), float)
print ("Working hard, wait for the figure while I count to 60")

for k in range(0, Nmax-1):  V[0,k] = 100.0        # Line at 100V
for iter in range(Niter):
    if iter%10 == 0: print(iter)
    for i in range(1, Nmax-2):
        for j in range(1,Nmax-2):
```

```
            V[i,j] = 0.25*(V[i+1,j]+V[i-1,j]+V[i,j+1]+V[i,j-1])
        print ("iter, V[Nmax/5,Nmax/5]", iter, V[Nmax/5,Nmax/5])
x = range(0, 50, 2);   y = range(0, 50, 2)
X, Y = p.meshgrid(x,y)

def functz(V):                                              # V(x, y)
    z = V[X,Y]
    return z

Z = functz(V)
fig = p.figure()                                       # Create figure
ax = Axes3D(fig)                                         # Plot axes
ax.plot_wireframe(X, Y, Z, color = 'r')         # Red wireframe
ax.set_xlabel('X');   ax.set_ylabel('Y');   ax.set_zlabel('V(x,y)')
ax.set_title('Potential within Square V(x=0)=100V (Rotatable)')
p.show()                                                # Show fig
```

Listing 5.2. **LaplaceLine.py** solves Laplace's equation via relaxation. Various parameters should be adjusted for an accurate solution.

```
# LaplaceDisk.py:  Laplace's eqtn in cyclinder, Matplotlib 3D plot

from scipy import special; from mpmath import *       # for hypergeo
from matplotlib.pyplot import figure, show, rc; import numpy as np

V = pi;   R = 1;   r = 1.1;   const = 2*V*R/(pi*r);   st = [1,1]
incr = 2*pi/100.;   thView = [pi/2, 3*pi/2];   Mmax = 10
x = [0]*(100);   xx = [0]*(100);   xxx = [0]*(100)  # Declare arrays

def legendre(n,x):                              # Legendre polymonial
    if(n==0):     p = 1
    elif(n==1):   p = x
    else:
        p0 = 1;   p1 = x
        for m in range(1,n):
            p2 = (((2*m+1)*x*p1-m*p0)/(m+1))           # Recurrence
            p0 = p1;   p1 = p2;   p = p2
    return p

def U(r,theta):
    summ = 0
    for m in range(0,Mmax):
        twom = 2*m;   w = cos(theta)
        leg = legendre(twom,w)
        term = ((-1)**m) *(R/r)**twom*leg/(twom+1)
        summ = summ + term
    pot = summ*const
    return pot

i = 0
for theta in np.arange(0,2*pi,incr):
    x[i] = U(r,theta);   xx[i] = U(3,theta);   xxx[i] = U(1.3,theta)
    i += 1
fig = figure()
ax = fig.add_subplot(111, projection = 'polar')
theta =  np.arange(0,2*pi,incr)
ax.plot(theta,x,label='r = 1.1',linewidth=3)
ax.plot(theta,xx,color='r',label='r = 3',linewidth=3)
```

```
ax.plot(theta,xxx,linewidth=3,label='r = 1.3',color=(1,0.8,0))
ax.plot(thView,st,linewidth=6,color=(0.4,0.4,0.4),label="Edge")
ax.legend(loc='best')
show()
```

Listing 5.3. **LaplaceDisk.py** creates a polar plot in θ of the potential $U(r,\theta,\phi)$ due to a charged disc. The curves are for three fixed values of r, with there being no ϕ dependence, and uses Matplotlib for plotting.

```
# LaplaceCyl.py:   Laplace's eqtn in cyclinder, Matplotlib 3D plot

from scipy import special; from mpl_toolkits.mplot3d import Axes3D
import numpy as np, matplotlib.pyplot as plt
from matplotlib import cm

a = 20.;   L = 20;   z = 18               # Cylinder size, U at z
rhop = 12.0; phip = 7*np.pi/4.;   zp = 15      # Charge location
Nzeros = 80; Nzeros2 = int(Nzeros/2)

def potential(rho,phi):
    suma = 0
    for m in range (-Nzeros2, Nzeros2+1):
        xmn = special.jn_zeros(m,Nzeros)                # Jm zeros
        xmnr = xmn*rho/a;   xmnp = xmn*rhop/a
        jm1 = special.jn(m,xmnr);   jm2 = special.jn(m,xmnp)
        sh = np.sinh(xmn*L/a);  sh2 = np.sinh(xmn*zp/a);
        sh3 = np.sinh(xmn*(L-z)/a)
        ex = np.cos(m*(phi-phip))             # Re exp[im(f-f')]
        jmp = special.jn(m+1,xmn)                       # J_m+1
        for n in range(0,Nzeros):            # Sums over zeros
            num = ex*jm1[n]*jm2[n]*sh2[n]*sh3[n]
            den = xmn[n]*sh[n]*jmp[n]**2
            pot = num/den
        poten = pot                      # Potential from one m
        suma = suma + poten              # Sum all m
    return suma

fig = plt.figure(figsize=(8,8))
ax = fig.add_subplot(111, projection='3d')
rho = np.linspace(0,a,Nzeros)
phi = np.linspace(0,2*np.pi,Nzeros)
R, P = np.meshgrid(rho, phi)             # Polar coords
X, Y = R*np.cos(P), R*np.sin(P)          # Cartesian coord
Z = potential(R,P)                       # U(z,r,phi)
ax.plot_surface(X, Y, Z, rstride=1, cstride=1, cmap=cm.coolwarm)
ax.set_xlabel('X'); ax.set_ylabel('Y'); ax.set_zlabel('U(x,y)')
plt.show()
```

Listing 5.4. **LaplaceCyl.py** evaluates the series expansion for the potential due to a charge within a grounded cylinder.

```
# ImagePlaneMat.py: E for charge left of plane plus image

import numpy as np,   matplotlib.pyplot as plt
from matplotlib.patches import Circle

Nx = 50;                         Ny = 50                        # x,y 50 grid
x = np.linspace(-5,5,Nx);        y = np.linspace(-5,5,Ny)
X,Y = np.meshgrid(x,y)                             # Transform coordinates
Ex = np.zeros((Nx,Ny));          Ey = np.zeros((Nx,Ny))    # Ex,Ey(x,y)

def E(xx,x,y):                                # E due to charge q at xx
    r = np.sqrt(x**2+y**2)                              # Distance
    dm = (x-xx)                               # Position q to xx
    d1 = np.sqrt((dm**2+y**2))                # Position q to x
    dp = (x+xx)                               # x component q
    d2 = np.sqrt((dp**2+y**2))                # Distance -q to (x,y)
    Ex = dm/d1**3-dp/d2**3
    Ey = y/d1**3 -y/d2**3
    return Ex,Ey

Ex,Ey = E(2.5,X,Y)
fig = plt.figure()
ax = fig.add_subplot(111)
circle1 = plt.Circle((2.5, 0),0.2, color='r')
circle2 = plt.Circle((-2.5, 0),0.2, color='b')
ax.add_artist(circle1)
ax.add_artist(circle2)
ax.streamplot(x,y,Ex,Ey)
ax.set_aspect('equal')
ax.set_title('E Field Due to Charge Left of Plane (Red Image)')
ax.set_xlabel('x')
ax.set_ylabel('y')
l = plt.axvline(x=0, linewidth=2, color='g')
plt.show()
```

Listing 5.5. **ImagePlaneMat.py** uses Matplotlib to plot in 3-D the electric field due to a charge to the left of a conducting plane determined via the image method.

```
# ImagePlaneVP.py: E field lines, charge plus image wi VPython

from vpython import *
from numpy import *

scene = canvas(width=500, height=500,range=100,
   title="E of Charge Left of Plane (Red Image)")
plane = box(pos=vector(0,0,0),length=2,height=130,width=130,
   color=vector(0.9,0.9,0.9),opacity=0.5)
gridpts = points(radius=4, color=color.cyan)
PlusCharge = sphere(radius=5,color=color.red, pos=vector(40,0,0))
NegCharge = sphere(radius=5,color=color.green,pos=vector(-40,0,0))

def grid3d():
    for z in range(-60,80,20):
        for y in range(-60,80,20):
            for x in range(-60,80,20): gridpts.append(pos=vector(x,y,z))

def electricF():
    for y in range(-60,80,20):
```

```
        for z in range(-60,80,20):
          for x in range(-60,80,20):
            r  = vector(x,y,z)                      # E vector here
            xx = vector(40.,0,0)                        # q location
            d  = vector(r-xx)                      # Vector q to r
            dm = vector(r+xx)                     # Vector q' to r
            dd  = mag(d)                            # Mag vector d
            ddp = mag(dm)                          # Mag vector dm
            if(x==40 and y==0 and z==0):     continue
            if ddp !=0:
              E1 =  d/dd**3                          # E due to q
              E2 = -dm/ddp**3                       # E due to -q
              E =  E1 + E2                            # Total E
              elecF = arrow(pos=r,color=color.orange)
              elecF.axis = 10*E/mag(E)          # 10 x unit vector
  grid3d()
  electricF()
```

Listing 5.6. ImagePlaneVP.py uses VPython to plot in 3-D the electric field due to a charge to the left of a conducting plane determined via the image method.

```
# ImageSphereVP.py: E field lines for charge + image wi VPython

from vpython import *
scene= canvas(width=500, height=500,range=100,
  title="E of Charge in Sphere (Red Image)")
gridpts = points(radius=4, color=color.cyan)

def grid3d():
  for z in range(-60,80,20):
    for y in range (-60,80,20):
      for x in range(-60,80,20): gridpts.append(pos = (x,y,z))

grid3d()
xp = 60;  yp = 40;  zp = 0;  a = 30;  q = 1
xx = vector(xp, yp, zp)                        # Charge location
xxp = xx*a**2/(mag(xx))**2                       # Image location
qp = -q*a/mag(xx)                                # Image charge
ball = sphere(pos=vector(0,0,0),radius=a, opacity=0.5)
poscharge = sphere(radius=5,color=color.red, pos=vector(xp,yp,zp))
negcharge = sphere(radius=5,color=color.blue, pos=xxp)

def electricF():
  for y in range(-60,80,20):
    for z in range(-60,80,20):
      for x in range(-60,80,20):
        r = vector(x,y,z)                           # E here
        d = vector(r-xx)                       # Vector q to r
        dm = vector(r-xxp)
        dd = mag(d)                            # Magnitude d
        ddp = mag(dm)                          # Magnitude dm
        if dd !=0:
          E1 =  d/dd**3                          # E due to q
          E2 = -dm/ddp**3                       # E due to -q
          E =  E1 + E2
          elecF = arrow(pos=r,color=color.orange) # E
          elecF.axis = 10*E/mag(E)          # 10 x unit vector
electricF()
```

Listing 5.7. ImageSphereVP.py uses VPython to plot in 3-D the electric field due to a charge within a conducting sphere, determined via the image method.

```
# ImageSphereMat.py: E field lines for charge plus image

import numpy as np,    matplotlib.pyplot as plt
from matplotlib.patches import Circle

Nx = 50;    Ny = 50; q = 1              # x, y grids, charge strength
x = np.linspace(-5,5,Nx);      y = np.linspace(-5,5,Ny)
X,Y = np.meshgrid(x,y)                  # Transform coordinates
Ex = np.zeros((Nx,Ny));      Ey = np.zeros((Nx,Ny))   # Arrays
xx = 3.0;    yy = 3.0                    # Charge coordinates
dq = np.sqrt(xx**2+yy*2);    a = 1.   # Origin, sphere radius
qp = -a*q/dq                           # Magnitude image charge
xp =  a**2*xx/dq**2;          yp =  a**2*yy/dq**2

def E(xx,yy,x,y):                       # xx,yy coord image q
    r = np.sqrt(x**2+y**2)
    dx = (x-xx);            dy = (y-yy)
    d1 = np.sqrt((dx**2 + dy**2))        # Distance q to (x,y)
    dpx = x-xp;            dpy = y-yp
    d2 = np.sqrt((dpx**2 + dpy**2))      # Distance -q to (x,y)
    Ex =  dx/d1**3-dpx/d2**3;       Ey = dy/d1**3 -dpy/d2**3
    return Ex,Ey

Ex,Ey = E(xx,yy,X,Y)
fig = plt.figure()
ax = fig.add_subplot(111)
circle1 = plt.Circle((xx, yy),0.2, color='r')
circle2 = plt.Circle((xp,yp),0.2, color='b')
sphere = plt.Circle((0,0),a, color='y',alpha=0.5)
ax.add_artist(circle1);        ax.add_artist(circle2)
ax.add_artist(sphere);        ax.streamplot(x,y,Ex,Ey)
ax.set_aspect('equal')
ax.set_title('E Field for a Charge (in Sphere) & Image')
ax.set_xlabel('x');        ax.set_ylabel('y')
plt.show()
```

Listing 5.8. ImageSphereMat.py visualizes the *E* field lines for a charge plus its image using Matplotlib.

```
# FDTD.py  FDTD Maxwell's equations in 1-D wi Visual

from visual import *

Xm = 201;  Ym = 100; Zm = 100; ts = 2;   beta = 0.01
Ex = zeros((Xm,ts),float);    Hy = zeros((Xm,ts),float)  # Arrays
scene = display(x=0,y=0,width= 800, height= 500,
        title= 'E: cyan, H: red. Periodic ←
```

```
              BC', forward=(-0.6,-0.5,-1))
Eplot = curve(x=list(range(0,Xm)),color=color.cyan,
   radius=1.5,display=scene)
Hplot = curve(x=list(range(0,Xm)),color=color.red,
   radius=1.5,display=scene)
vplane = curve(pos=[(-Xm,Ym),(Xm,Ym),(Xm,-Ym),(-Xm,-Ym),
                    (-Xm,Ym)],color=color.cyan)
zaxis = curve(pos=[(-Xm,0),(Xm,0)],color=color.magenta)
hplane = curve(pos=[(-Xm,0,Zm),(Xm,0,Zm),(Xm,0,-Zm),(-Xm,0,-Zm),
                    (-Xm,0,Zm)],color=color.magenta)
ball1 = sphere(pos = (Xm+30, 0,0), color = color.black, radius = 2)
ExLabel1 = label( text = 'Ex', pos = (-Xm-10, 50), box=0)
ExLabel2 = label( text = 'Ex', pos = (Xm+10, 50), box=0)
HyLabel  = label( text = 'Hy', pos = (-Xm-10, 0,50), box=0)
zLabel   = label( text = 'Z',  pos = (Xm+10, 0), box=0)

def PlotFields():
    z = arange(Xm)
    Eplot.x = 2*z-Xm                        # World to screen coords
    Eplot.y = 800*Ex[z,0]
    Hplot.x = 2*z-Xm
    Hplot.z = 800*Hy[z,0]

z = arange(Xm)
Ex[:Xm,0]  = 0.1*sin(2*pi*z/100.0)          # Initial field
Hy[:Xm,0]  = 0.1*sin(2*pi*z/100.0)
PlotFields()

while True:
    rate(600)
    Ex[1:Xm-1,1] = Ex[1:Xm-1,0] + beta*(Hy[0:Xm-2,0]-Hy[2:Xm,0])
    Hy[1:Xm-1,1] = Hy[1:Xm-1,0] + beta*(Ex[0:Xm-2,0]-Ex[2:Xm,0])
    Ex[0,1]      = Ex[0,0] + beta*(Hy[Xm-2,0]  -Hy[1,0])    # BC
    Ex[Xm-1,1]   = Ex[Xm-1,0] + beta*(Hy[Xm-2,0]  -Hy[1,0])
    Hy[0,1]      = Hy[0,0] + beta*(Ex[Xm-2,0]  -Ex[1,0])    # BC
    Hy[Xm-1,1]   = Hy[Xm-1,0] + beta*(Ex[Xm-2,0] - Ex[1,0])
    PlotFields()
    Ex[:Xm,0]  = Ex[:Xm,1]                   # New -> old
    Hy[:Xm,0]  = Hy[:Xm,1]
```

Listing 5.9. FDTD.py solves Maxwell's equations via FDTD time stepping for linearly polarized wave propagating in the z direction.

```
# Dielect.py;  Matplotlib Animated E & B space to dielectric

from numpy import *
import numpy as np;  import matplotlib.pyplot as plt
import matplotlib.animation as animation

Xmax = 401;  Ymax = 100;  Zmax = 100
eps = 4; dd = 0.5; Xmax = 401        # Dielectric, stability param
Ex = zeros((Xmax),float);  Hy = zeros((Xmax),float) # Declare E,H
beta = zeros((Xmax),float)

for i in range(0,401):
    if i<201:  beta[i] = dd          # Free space stability cond
    else:      beta[i] = dd/eps                  # In dielectric
z = arange(201)                      # Initial outside dielectric
```

```python
xs = np.arange(1,Xmax-1)
Ex[:201] = 0.5*sin(2*pi*z/100.)                    # Slice entire range
Hy[:201] =  0.5*sin(2*pi*z/100.)
fig = plt.figure()
ax = fig.add_subplot(111,autoscale_on=False,
                 xlim=(1,Xmax-1),ylim=(-1.5,1.5))
ax.grid()
line, = ax.plot(xs,Ex[1:Xmax-1], lw=2)

def animate(dum):
    for x in range (1,Xmax-1):
        Ex[x] = Ex[x] + beta[x]*(Hy[x-1]-Hy[x])
        Hy[x] = Hy[x]+ dd*(Ex[x]-Ex[x+1])
    line.set_data(xs,Ex[1:Xmax-1])
    return line,

plt.title('Refraction & Reflection at Dielectric (right)')
plt.xlabel('z')
plt.ylabel('Ex')
p = plt.axvline(x=200,color='r')    # Vert line separator
ani = animation.FuncAnimation(fig, animate,1,blit=True)
plt.show()
```

Listing 5.10. Dielect.py solves Maxwell's equations via FDTD and time stepping for linearly polarized wave incident from the left and produces a 2-D animation using Matplotlib.

```python
# CircPolarztn.py: Maxwell eqs. circular polariz via FDTD

from visual import *

scene = display(x=0,y=0,width=600,height=400,range=200,
   title='Circular Polarized E(white)& H(yellow)Fields')
global phy, pyx
max = 201; c = 0.01                      # Courant stable if c < 0.1
Ex = zeros((max+2,2),float);    Ey = zeros((max+2,2),float)
Hy = zeros((max+2,2),float);    Hx = zeros((max+2,2),float)
arrowcol= color.white
Earrows = [];                      Harrows  = []
for i in range(0,max,10):
    Earrows.append(arrow(pos=(0,i-100,0), axis=(0,0,0),
           color=arrowcol))
    Harrows.append(arrow(pos=(0,i-100,0), axis=(0,0,0),
           color=color.yellow))

def plotfields(Ex,Ey,Hx,Hy):
    for n, arr in enumerate(Earrows):
        arr.axis = (35*Ey[10*n,1],0,35*Ex[10*n,1])
    for n, arr in enumerate(Harrows):
        arr.axis = (35*Hy[10*n,1],0,35*Hx[10*n,1])

def inifields():                          # Initial E & H
    phx = 0.5*pi;   phy = 0.0
    z = arange(0,max)
    Ex[:-2,0] = cos(-2*pi*z/200+phx)
    Ey[:-2,0] = cos(-2*pi*z/200+phy)
    Hx[:-2,0] = cos(-2*pi*z/200+phy+pi)
    Hy[:-2,0] = cos(-2*pi*z/200+phx)
```

```
def newfields():
    while True:                                    # Time stepping
        rate(1000)
        Ex[1:max-1,1] =Ex[1:max-1,0] + c*(Hy[:max-2,0]-Hy[2:max,0])
        Ey[1:max-1,1] =Ey[1:max-1,0] + c*(Hx[2:max,0]-Hx[:max-2,0])
        Hx[1:max-1,1] =Hx[1:max-1,0] + c*(Ey[2:max,0]-Ey[:max-2,0])
        Hy[1:max-1,1] =Hy[1:max-1,0] + c*(Ex[:max-2,0]-Ex[2:max,0])
        Ex[0  ,1] = Ex[0  ,0]   + c*(Hy[200-1,0]-Hy[1,0]) # Periodic
        Ex[200,1] = Ex[200,0]   + c*(Hy[200-1,0]-Hy[1,0])
        Ey[0  ,1] = Ey[0  ,0]   + c*(Hx[1,0] - Hx[200-1,0])
        Ey[200,1] = Ey[200,0]   + c*(Hx[1,0] - Hx[200-1,0])
        Hx[0  ,1] = Hx[0  ,0]   + c*(Ey[1,0] - Ey[200-1,0])
        Hx[200,1] = Hx[200,0]   + c*(Ey[1,0] - Ey[200-1,0])
        Hy[0  ,1] = Hy[0  ,0]   + c*(Ex[200-1,0]-Ex[1,0])
        Hy[200,1] = Hy[200,0]   + c*(Ex[200-1,0]-Ex[1,0])
        plotfields(Ex,Ey,Hx,Hy)
        Ex[:max,0] = Ex[:max,1];   Ey[:max,0] = Ey[:max,1] # Update
        Hx[:max,0] = Hx[:max,1];   Hy[:max,0] = Hy[:max,1]

inifields()                                        # Initial
newfields()                                        # New
```

Listing 5.11. **CircPolartzn.py** solves Maxwell's equations via FDTD and time-stepping for circularly polarized wave propagation in the z direction.

```
# QuarterPlate.py  FDTD solution of Maxwell's equations in 1-D

from visual import *

xmax = 401; ymax = 100;  zmax = 100; ts = 2; beta = 0.01

Ex = zeros((xmax,ts),float); Ey = zeros((xmax,ts),float)
Hx = zeros((xmax,ts),float); Hxx = zeros((xmax,ts),float)
Hy = zeros((xmax,ts),float); Hyy = zeros((xmax,ts),float)
Exx = zeros((xmax,ts),float);  Eyy = zeros((xmax,ts),float)

scene = display(x=0,y=0,width= 800, height= 500, title= 'Ey:
  cyan, Ex:yellow. periodic BC', forward=(-0.8,-0.3,-0.7))
Exfield = curve(x=list(range(0,xmax)),color= color.yellow,
  radius=1.5,display=scene)
Eyfield = curve(x=list(range(0,xmax)),color=color.cyan,
  radius=1.5,display=scene)
vplane= curve(pos=[(-xmax,ymax),(xmax,ymax),(xmax,-ymax),
  (-xmax,-ymax), (-xmax,ymax)],color=color.cyan)
zaxis = curve(pos=[(-xmax,0),(xmax,0)],color=color.magenta)
hplane = curve(pos=[(-xmax,0,zmax),(xmax,0,zmax),(xmax,0,-zmax),
  (-xmax,0,-zmax),(-xmax,0,zmax)],color=color.magenta)
ball1 = sphere(pos=(xmax+30,0,0), color=color.black, radius = 2)
ba2 = sphere(pos=(xmax-200,0),color=color.cyan,radius=3)
plate = box(pos=(-100,0,0),height=2*zmax,width=2*ymax,
  length=0.5*xmax, color=(1.0,0.6,0.0),opacity=0.4)
Exlabel1 = label( text = 'Ey', pos = (-xmax-10, 50), box = 0 )
Exlabel2 = label( text = 'Ey', pos = (xmax+10, 50), box = 0 )
Eylabel = label( text = 'Ex', pos = (-xmax-10, 0,50), box = 0 )
zlabel  = label( text =  'Z', pos =  (xmax+10, 0), box = 0 )
polfield = arrow(display = scene)
polfield2 = arrow(display = scene)
ti = 0
```

```
def  InitField ():
     kar = arange(xmax)
     phx = 0.5*pi
     Hyy[:xmax,0] = 0.1*cos(-2*pi*kar/100)
     Exx[:xmax,0] = 0.1*cos(-2*pi*kar/100)
     Eyy[:xmax,0] = 0.1*cos(-2*pi*kar/100)
     Hxx[:xmax,0] = 0.1*cos(-2*pi*kar/100)
     Ey[:xmax,0]  = 0.1*cos(-2*pi*kar/100)
     Hx[:xmax,0]  = 0.1*cos(-2*pi*kar/100)

def  InitExHy():
     k  =  arange(101)
     Ex[:101,0] = 0.1*cos(-2*pi*k/100)
     Hy[:101,0] = 0.1*cos(-2*pi*k/100)
     kk = arange(101,202)          # Inside plate, delay lambda/4
     Ex[101:202,0]  =  0.1*cos(-2*pi*kk/100.0-0.005*pi*(kk-101))
     Hy[101:202,0]  =  0.1*cos(-2*pi*kk/100.0-0.005*pi*(kk-101))
     kkk = arange(202,xmax)        # After plate, phase diff pi/2
     Ex[202:xmax,0]  =  0.1*cos(-2*pi*kkk/100-0.5*pi)
     Hy[202:xmax,0]  =  0.1*cos(-2*pi*kkk/100-0.5*pi)

def  PlotFields(ti):                      # screen coordinates
     k  =  arange(xmax)
     Exfield.x  =  2*k-xmax                # world to screen coords.
     Exfield.y  =  800*Ey[k,ti]
     Eyfield.x  =  2*k-xmax                # world to screen coords.
     Eyfield.z  =  800*Ex[k,ti]

InitField()
InitExHy()
PlotFields(ti)
j = 0
end = 0
while end < 5:
     rate(150)
     Exx[1:xmax-1,1] = Exx[1:xmax-1,0] + beta*(Hyy[0:xmax-2,0]
          - Hyy[2:xmax,0])
     Eyy[1:xmax-1,1] = Eyy[1:xmax-1,0] + beta*(Hxx[0:xmax-2,0]
          - Hxx[2:xmax,0])
     Hyy[1:xmax-1,1] = Hyy[1:xmax-1,0] + beta*(Exx[0:xmax-2,0]
          - Exx[2:xmax,0])
     Hxx[1:xmax-1,1] = Hxx[1:xmax-1,0] + beta*(Eyy[0:xmax-2,0]
          - Eyy[2:xmax,0])
     Ex[1:xmax-1,1] =  Ex[1:xmax-1,0]  + beta*(Hy[0:xmax-2,0]
          - Hy[2:xmax,0])
     Ey[1:xmax-1,1] =  Ey[1:xmax-1,0]  + beta*(Hxx[0:xmax-2,0]
          - Hxx[2:xmax,0])
     Hy[1:xmax-1,1] =  Hy[1:xmax-1,0]  + beta*(Ex[0:xmax-2,0]
          - Ex[2:xmax,0])
     Hx[1:xmax-1,1] =  Hx[1:xmax-1,0]  + beta*(Eyy[0:xmax-2,0]
          - Eyy[2:xmax,0])
     polfield.pos = (-280,0,0)
     polfield.axis = (0,700*Exx[60,1],700*Eyy[60,1])
     polfield2.pos = (380,0,0)
     polfield2.axis = (0,700*Ex[360,1],-700*Ey[360,1])
  # Periodic BC
     Exx[0,1] = Exx[0,0] + beta*(Hyy[xmax-2,0] - Hyy[1,0])
     Eyy[0,1]=  Eyy[0,0] + beta*(Hxx[xmax-2,0] - Hxx[1,0])
     Hyy[0,1] = Hyy[0,0] + beta*(Exx[xmax-2,0] - Exx[1,0])
     Hxx[0,1] = Hxx[0,0] + beta*(Eyy[xmax-2,0] - Eyy[1,0])
     Hyy[xmax-1,1] = Hyy[xmax-1,0] + beta*(Exx[xmax-2,0]
          - Exx[1,0])
     Hxx[xmax-1,1] = Hxx[xmax-1,0] + beta*(Eyy[xmax-2,0]
```

```
                        − Eyy[1,0])
        Exx[xmax−1,1] = Exx[xmax−1,0] + beta*(Hyy[xmax−2,0]
                        − Hyy[1,0])
        Eyy[xmax−1,1] = Eyy[xmax−1,0] + beta*(Hxx[xmax−2,0]
                        − Hxx[1,0])
        Ex[0,1] = Exx[0,0] + beta*(Hyy[xmax−2,0] − Hyy[1,0])
        Ey[0,1] = Eyy[0,0] + beta*(Hxx[xmax−2,0] − Hxx[1,0])
        Hy[0,1] = Hyy[0,0] + beta*(Exx[xmax−2,0] − Exx[1,0])
        Hx[0,1] = Hxx[0,0] + beta*(Eyy[xmax−2,0] − Eyy[1,0])
        Hy[xmax−1,1] = Hy[xmax−1,0] + beta*(Ex[xmax−2,0]
                        − Ex[xmax−100,0])
        Hx[xmax−1,1] = Hx[xmax−1,0] + beta*(Ey[xmax−2,0]
                        − Ey[1,0])
        Ex[xmax−1,1] = Ex[xmax−1,0] + beta*(Hy[xmax−2,0]
                        − Hy[xmax−100,0])
        Ey[xmax−1,1] = Ey[xmax−1,0] + beta*(Hxx[xmax−2,0]
                        − Hxx[1,0])
        PlotFields(ti)
        k = arange(101,202)
        Ex[101:202,1] = 0.1*cos(−2*pi*k/100−0.005*pi*(k−101)
                        + 2*pi*j/4996.004)
        Hy[101:202,1] = 0.1*cos(−2*pi*k/100−0.005*pi*(k−101)
                        + 2*pi*j/4996.004)
        Exx[:xmax,0]  = Exx[:xmax,1]
        Eyy[:xmax,0]  = Eyy[:xmax,1]
        Hyy[:xmax,0]  = Hyy[:xmax,1]
        Hxx[:xmax,0]  = Hxx[:xmax,1]
        Ex[:xmax,0]  =   Ex[:xmax,1]
        Ey[:xmax,0]  =   Ey[:xmax,1]
        Hx[:xmax,0]  =   Hx[:xmax,1]
        Hy[:xmax,0]=     Hy[:xmax,1]
        if j%4996 == 0:
            j = 0
            end += 1
        j = j+1
```

Listing 5.12. **QuarterPlate.py** solves Maxwell's equations via FDTD and time stepping for transmission through a quarter-wave plate.

```
# TeleMat.py:    Lossless transmission line animation, Matplotlib

from numpy import *; from matplotlib import animation
import numpy as np, matplotlib.pyplot as plt

L = 0.1; C = 2.5; c = 1/sqrt(L*C); dt = 0.025;dx = 0.05
R = (c*dt/dx)**2
V = np.zeros( (101, 3), float)                    # (Nx, Nt)
xx = 0
fig = plt.figure()                                # Figure to plot
ax = fig.add_subplot(111,autoscale_on=False, xlim=(0,100),
    ylim=(−40,40))
ax.grid()
plt.title("Wave Transmission Via Telegrapher's Equations")
plt.xlabel("x");   plt.ylabel ("V(x,t)")
line , = ax.plot([],[], lw=2)     # x axis, y values, linewidth=2

def init():
    line.set_data([],[])
```

```
        return line ,

for i in arange (0,100):
    V[i,0] =  10*exp(-(xx**2)/0.1)
    xx = xx + dx

for i in range(1, 100):
   V[i,2] = V[i,0] + R*(V[i+1,1] + V[i-1,1] - 2*V[i,1])
V[:,0] = V[:,1];  V[:,1] = V[:,2]                 # Recycle array

def animate(dum):
        i =    arange(1, 100)
        V[i, 2] = 2.*V[i,1]-V[i,0]+R*(V[i+1,1]+V[i-1,1]-2*V[i,1])
        line.set_data(i,V[i,2])
        V[:,0] = V[:,1]; V[:,1] = V[:,2]          # Recycle array
        return line ,

# Plot 10000 frames , delay 10ms, blit=True redraw only changes
ani = animation.FuncAnimation(fig , ↩
    animate,init_func=init ,frames=10000,
  interval=20,blit=True)
plt.show()
```

Listing 5.13. **TeleMat.py** solves the telegraph equations for wave transmission on a telegraph line and produces an animation with Matplotlib.

```
# TelegraphVis.py:   Lossless  transmission  line  animation ,  Visual

from visual import *

g = display(width = 600, height = 300, title='Telegrapher's Eqnt')
vibst = curve(x=list(range(0,101)), color=color.yellow,radius=0.5)
L = 0.1; C = 2.5; c = 1/sqrt(L*C); dt = 0.025; dx = 0.05
R = (c*dt/dx)**2                        # R = 1 for stabiity
V = zeros( (101, 3), float)             # Declare array
xx = 0

for i in arange (0,100):
    V[i,0] =  10*exp(-(xx**2)/0.1)
    xx = xx+dx
    vibst.x[i] = 2.0*i - 100.0    # i=0-> x=-100;  i =100, x=100
    vibst.y[i] = 0.0                      # Eliminate a curve

for i in range(1, 100):
   V[i, 2] = V[i,0] + R*(V[i+1,1] + V[i-1,1] - 2*V[i,1])
   vibst.x[i] = 2.0*i - 100.0                   # x scale again
   vibst.y[i] = V[i, 2]
V[:,0] = V[:,1];  V[:,1] = V[:,2]              # Recycle array

while 1:
    rate(20)                          # Delay plot, large = slow
    for i in range(1, 100):
        V[i, 2] = 2.*V[i,1]-V[i,0]+R*(V[i+1,1]+V[i-1,1]-2*V[i,1])
        vibst.x[i] = 2.*i - 100.0                  # Scale x
        vibst.y[i] = V[i, 2]
    V[:,0] = V[:,1];  V[:,1] = V[:,2]             # Recycle array
```

Listing 5.14. TeleVis.py solves the telegraph equations for wave transmission on a telegraph line and produces an animation with Visual.

```python
# ThinFilm.py: Thin film interference by reflection (AJP ↩
    72,1248-1253)

from vpython import *
from numpy import *

escene = ↩
    canvas(width=500,height=500,range=400,background=color.white,
    foreground=color.black,title='Thin Film Interference')
Rcurve = curve(color=color.red,radius=2)
Gcurve = curve(color=color.green,radius=2)
Bcurve = curve(color=color.blue,radius=2)
title  = label(pos=vector(-20,350,0),
    text='Intensity vs Thickness (nA)',box=0)
waves  = label(pos=vector(-30,320,0),
    text='Red, Green, and Blue Intensities',box=0)
trans  = label(pos=vector(-280,300,0),text='Transmission',box=0)
refl   = label(pos=vector(210,300,0),text='Reflection',box=0)
lamR   = 572;    lamB = 430; lamG = 540; i = 0      # R,B, G
film = curve(pos=[(-150,-250,0),(150,-250,0),(150,250,0),
        (-150,250,0),(-150,-250,0)])
Rc = [];   Gc = []; Bc = []                        # R,G,B intensity

nA = arange(0,1250,10)
delR = 2*pi*nA/lamR+pi; delG = 2*pi*nA/lamG+ pi
delB = 2*pi*nA/lamB+pi
intR = cos(delR/2)**2; intG = cos(delG/2)**2; intB = cos(delB/2)**2
xrp = 300*intR-150; xbp = 300*intB-150;  xgp = 300*intG-150
ap = -500*nA/1240 +250
Rc = Rc+[intR];   Gc = Gc+[intG];   Bc = Bc+[intB]
Rt = [];   Gt = [];      Bt = []
DelRt = 4*pi*nA/lamR;   DelGt = 4*pi*nA/lamG; DelBt = 4*pi*nA/lamB
IntRt = cos(DelRt/2)**2; IntGt = cos(DelGt/2)**2
IntBt = cos(DelBt/2)**2
xRpt = 300*intR-150;   xBpt = 300*intB-150;  xGpt = 300*intG-150
Rt = Rt + [intR];      Gt = Gt + [intG];       Bt = Bt + [intB]
ap =  -500*nA/1240 +250                         # Film height

for nA in range (0,125):
    col = vector(intR[nA],intG[nA],intB[nA])           # RGB reflection
    reflesc = -500*nA/125+250
    box(pos=vector(205,reflesc,0),width=0.1,height=10,
        length=50,color=col)
    colt = vector(IntRt[nA],IntGt[nA],IntBt[nA])
    box(pos=vector(-270,reflesc,0),width=0.1,height=10,
        length=50,color=colt)
    if (nA%20==0):
        prof = nA*10
        escal = -500*nA/125+250
        depth = label(pos=vector(-200,escal,0),text='%4d'%prof,box=0)
for i in range(0,1250,10):
    dlR =  2*pi*i/lamR+pi
    dlG =  2*pi*i/lamG+ pi
```

```
dlB  =    2*pi*i/lamB+pi
inR  =  ( cos ( dlR/2 ) )**2
inG  =  ( cos ( dlG/2 ) )**2
inB  =  ( cos ( dlB/2 ) )**2
xr   =    300*inR−150
xb   =    300*inB−150
xg   =    300*inG−150
ar   =   −500*i/1240  +250
#print ( 'dlR ', inR , xr , ar )
Rcurve . append ( pos=vector (  xr , ar ,0 ) )
Bcurve . append ( pos=vector (  xb , ar ,0 ) )
Gcurve . append ( pos=vector (  xg , ar ,0 ) )
```

Listing 5.15. ThinFilm.py computes the intensity distributions and the spectra shown in Figure 5.19 right of light reflected from and transmitted through a thin film. The colors arise from the interference of reflected and transmitted waves.

```
# TwoCharges.py:   Motion of 2 charges in 2 frames wi Visual

from vpython import *

scene = canvas ( width=700 , height=300 , range=1 , background
   =vector ( 1 , 1 , 1 ) , title= "In O: Initial Parallel Velocities" )
graf = curve ( color=color . red )
r1 = vector ( −.9 , .2 , 0 )
charge = sphere ( pos=r1 , color=color . red ,  radius =.02 , make_trail=True )
r2 = vector ( −.9 , −.2 , 0 )
charge2 = sphere ( pos=r2 , color=color . red , radius =.02 , make_trail=True )
scene2 = canvas ( y=700 , width=700 , height=300 ,  range=1 ,
   background=vector ( 1 , 1 , 1 ) , title= "In O' (Moving Frame)" )
r1p = r1 ;      r2p = r2
charge3 = sphere ( color=color . red , radius =.02 , make_trail=True ,
   display=scene2 )
charge4 = sphere ( color=color . red , radius =.02 , make_trail=True ,
   display=scene2 )
mu0 = 1 ;        e0 = 1 ;           q1 = q2 = 2 ;      m0 = 50 .
beta = 0.3 ;     gamma = 1/sqrt ( 1−beta**2 )
m = m0*gamma ;   u = vector ( 0.3 , 0 , 0 )                   # Vo in O of q1
dt = 0.01 ;       dtp = dt*gamma                     # Time step in O, O'
r1 = vector ( −.9 , .2 , 0 ) ;    r2 = vector ( −.9 , −.2 , 0 )   # Initial 1, 2

def EulerPlusTF ( u , r1 , r2 , beta ) :   # Euler ODE solve + Lorentz TF
   v2 = u ;      v1 = u
   for i in range ( 0 , 300 ) :                      # Motion loop
      rate ( 100 )
      rr1 = r2−r1                                    # q2 wrt r1
      rr2 = −rr1                                     # q1 wrt r2
      rr = mag ( rr1 )
      B1 =    q1*cross ( v1 , rr1 )/(4*pi *rr**3 )          # B at q2
      B2 =    q2*cross ( v2 , rr2 )/(4*pi *rr**3 )          # B at q1
      E1 = q1*rr1/(4*pi*rr**3 )                         # E at q2
      E2 = q2*rr2/(4*pi*rr**3 )                         # E at q1
      F1 = q2*(E1 +  cross ( v2 , B1 ) )                # Force on q2
      F2 = q1*(E2 +  cross ( v1 , B2 ) )                # Force on q1
      a2 = F1/m0 ;        a1 = F2/m0
      v1 = v1 + a2*dt ;    v2 = v2 + a1*dt
      x1 = r1.x ;
```

```
                  r1 = r1 + v2*dt;        r2 = r2 + v1*dt               # Update
                  x1 =r1.x;               x2 = r2.x
                  y1 = r1.y;              y2 = r2.y
                  t = i*dt                                              # Time in O
        #         Now transform to O'
                  x1p = (x1-beta*t)/(sqrt(1-beta*beta))        # TF x to x'
                  y1p = y1                                      # TF y to y'
                  charge3.pos = vector(x1p,y1p,0)
                  x2p = (x2-beta*t)/(sqrt(1-beta*beta))
                  y2p = y2
                  charge3.pos = vector(x1p,y1p,0)
                  charge4.pos = vector(x2p,y2p,0)
                  charge2.pos = r2
                  charge.pos = r1

        EulerPlusTF(u, r1, r2, beta)                         # Call animation
```

Listing 5.16. TwoCharges.py The motion of two charges as viewed in different reference frames.

```
# TwoFields.py:   Motion of charges in 2 frames wi Field TF

from vpython import *

scene = canvas(width=700, height=300, range=1,
               title="Frame S with moving charges")
graf = curve(color=color.red)
r1 = vector(-.9,.2,0)  # charge up
charge = sphere(pos=r1,color=color.red, radius=.02,make_trail=True)
r2 = vector(-.9,-.2,0)  # charge down
charge2 = sphere(pos=r2,color=color.red,radius=.02,make_trail=True)
scene2 = canvas(y=300,width=700,height=300, range=1,
                title= "Frame S' with velocity u to the right")
r1p = r1
r2p = r2
charge3 = sphere(color=color.red, radius=.02,
    make_trail=True,display=scene2)
charge4 = sphere(color=color.red, radius=.02
    ,make_trail=True,display=scene2)
mu0 = 1                          # permeability vacuum
e0 = 1                           # in vacuum
q1 = 2                           # charge for   stronger interactions
q2 = 2
m0 = 50.                         #   convenient for the units used
dt = 0.01                        # time step in S
beta = 0.4                       # beta = v/c, v of S' wrt S, c=1
gamma = 1/sqrt(1-beta**2)        # gamma from special relativity
m = m0*gamma
dtp = dt*gamma                   # time increment   in S' approx.1st time
facv = 1./(1.-beta**2)           # factor needed
ux = beta                        # velocity of S' wrt to S

def EBtransform(E,B):            # parameters: E, B in S
    Exp = E.x                    # transformation of E and B to S'
    Eyp = gamma*(E.y-ux*B.z)
    Ezp = gamma*(E.z+ux*B.y)
    Bxp = B.x
```

```
        Byp = gamma*(B.y+ux*E.x)
        Bzp = gamma*(B.z−ux*E.y)
        Ep = vector(Exp,Eyp,Ezp)
        Bp = vector(Bxp,Byp,Bzp)
        return Ep,Bp                    # returns E, B fields in S'


    def EulerPlusTF():                  # Euler ODE solve + Lorentz TF

        u = vector(0.4,0,0)        # initial velocity of q1 in S
        r1 = vector(−.9,.2,0)      # initial position q1 in S  wrt S
        r2 = vector(−.9,−.2,0)     # initial position q2 in S wrt S
        v2 = u                     # each charge has the same velocity
        v1 = u                     # velocities of charges in S
        fcv1 = 1./(1−u.x*mag(v1))  # factor
        v1xp = (v1.x−u.x)*fcv1
        v1yp = v1.y*fcv1/gamma
        v1zp = v1.z*fcv1/gamma
        v1p = vector(v1xp,v1yp,v1zp)
        v2p = v1p
        r1p = r1                   # initial positions coincide at t=0 in
        r2p = r2                   # both reference systems S and S'

        for i in range (0,300):    # loop for  motion
            rate(100)              # slow  the  motion
            rr1 = r2−r1            # relative position q2 wrt r1
            rr2 = −rr1             # relative position q1 wrt r2
            rr = mag(rr1)          # magnitud of vector rr1
            B1 =  q1*cross(v1,rr1)/(4*pi*rr**3) # B field at q2
            B2 =  q2*cross(v2,rr2)/(4*pi*rr**3) # B field at q1
            E1 = q1*rr1/(4*pi*rr**3) # E field of q1 at  q2
            E2 = q2*rr2/(4*pi*rr**3) # E field of q2 at q1
            E1p,B1p = EBtransform(E1,B1)
            E2p,B2p = EBtransform(E2,B2)
            ux = u.x                   # x component of velocity u
            FB1 = cross(v2,B1)         # magnetic force on q2
            FB2 = cross(v1,B2)         # magnetic force on q1
            F2 = q1*(E2+FB2)           # Lorentz force on q1
            F1 = q2*(E1+FB1)           # Lorentz force on q2
            FB1p = cross(v2p,B1p)      # magnetic force on q2
            FB2p = cross(v1p,B2p)      # magnetic force on q2
            F1p = q2*(E1p+FB1p)        # Lorentz force on q2 in S'
            F2p = q1*(E2p+FB2p)        # Lorentz force on q1  in S'
            a2 = F1/m0                 # acceleration q2 in S
            a1 = F2/m0                 # acceleration q1 in S
            a2p = F1p/m                # acceleration q2 in S'
            a1p = F2p/m                # acceleration q1 in S'
            v2 = v2+a1*dt              # velocity q1 in S
            v1 = v1+a2*dt              # velocity q2 in S
            x1 = r1.x                  # x component of r (x1)
            r1 = r1+v2*dt
            x2 = r1.x                  # x component of r (x2) after dt
            dx = x2−x1
            dtp = (dt−dx*u.x)*gamma    # time increment in S'
            r2 = r2+v1*dt
            v1p = v1p+a2p*dtp          # velocity in S'
            v2p = v2p+a1p*dtp          # velocity in S'
            r1p = r1p+v2p*dtp
            r2p = r2p+v1p*dtp
            charge3.pos = r1p
            charge4.pos = r2p
            charge2.pos = r2
            charge.pos = r1
```

```
EulerPlusTF ()                          # call to begin animation
```

Listing 5.17. **TwoFields.py** The fields of two charges as viewed in different reference frames.

```
# LorentzFieldVP.py:   Lorentz TF of E, B, & V with Visual

from vpython import *

scene = canvas(width=700,height=400,range=100,title="Dots: Bz in ←
    O")
graf = curve(color=color.red)
rr = vector(40,35,0)
charge = sphere(pos=rr,color=color.red, radius=2,make_trail=True)
B = vector(0,0,.1);   Bz = B.z                # 3-D B in O, Bz
m0 = 1; q = 1                                 # Mass, charge
beta =  0.9;   dt = 0.001                      # v/c, Time step in O
gamma = 1/sqrt(1.-beta**2)

def plotB():                                  # Plot B as dots
    for i in range(-100,110,10):
        for j in range (-50,60,10):
            points(pos =vector(i,j,0),radius=4,display=scene)

def plotBp():                                 # Plot B' as dots
    for i in range(-500,501,8):
        for j in range (-50,60,8):
            points(pos=vector(i,j,0),radius=1,display=scene2)

def Euler():                    # Euler method, solve Eq Mtn in O'
    V = vector(0.9,0,0);   Vx = V.x ;  Vy = V.y    # Vo in O
    v = vector(beta,0,0)                        # Relative v
    r = vector(40,35,0)                            # R(t=0)
    den = (1-Vx*beta)
    Vxp = (V.x-beta)/den; Vyp = Vy/den/gamma
    Vp = vector(Vxp,Vyp,0)
    rp = r                                      # Initially aligned
    for i in range (0,200000):                  # Motion loop
        Bzp = gamma*Bz;  Bp = vector(0,0,Bzp)          # B'
        Eyp = -gamma*beta*Bz;   Ep = vector(0,Eyp,0)   # E'
        F = V.cross(B)                             # Force
        a = F/m0                                 . # Acceleration
        V = V + a*dt
        r = r + V*dt
        Fp = Vp.cross(Bp)+q*Ep                    # Force in O'
        m = m0*gamma                              # Mass in S'
        ap = Fp/m                            # Acceleration in O'
        dtp = dt*gamma
        Vp = Vp + ap*dtp                                # V'
        rp = rp + Vp*dtp - v*dtp                        # R'
        charge2.display = scene2
        charge2.pos = vector(rp)                   # O' plot
        charge.display = scene                     # O plot
        charge.pos = vector(r)
        rate(5000)
plotB();
scene2 = canvas(y=400,width=700,height=400,range=500,
    title="In O', Bz'(Dots) & Ex'")
```

```
charge2 = sphere(pos=rr, color=color.red, radius=2,make_trail=True)
plotBp()                                    # Call plots
Euler()                                     # Begin animation
```

Listing 5.18. **LorentzFieldVP.py** A Vpython visualization of the effect of a Lorentz transformation on electric fields.

6

Quantum Mechanics

6.1 Chapter Overview

Quantum mechanics lends itself to a variety of problems, and so this too is a long chapter. Section 6.2 contains the traditional, semi-analytic bound state in a 1-D box problem that requires a trial-and-error search algorithm from §2.2. Then we examine two techniques for finding the bound states for an arbitrary potential as a numerical eigenvalue problem. These solve an ODE using the rk4 algorithm of §1.7.1 along with a trial-and-error search. In §6.2.4 we use similar techniques for the relativistic hydrogen states of the Klein-Gordon equation.

Section 6.3 employs a Monte Carlo algorithm to simulate the spontaneous decay of atoms or nuclei. Next, §6.3.1 covers a simple fitting of the black body spectrum, in this case to cosmic microwave background. Section 6.4 starts with the numerical solution of the time-independent Schrödinger equation. Then partial wave expansions are examined, which leads to methods of computing the spherical harmonics and the associated Legendre functions, and ultimately the 3-D hydrogenic wave functions. Section 6.7 contains several problems dealing with wave packets, some solving the time-dependent Schrödinger PDE directly, and others studying packet properties. Then §6.8 presents scattering problems: a 3-D square well potential, a Coulomb potential, and three stationary disks on a billiard table. Billiard scattering may be chaotic, and is best understood after studying classical billiards in §3.10.

Section 6.9 deals with the numerical solution of the integral-equation form of the Schrödinger equation expressed in terms of matrix equations. The solution of integral equations may be challenging to some students who have never seen them before. Then on to the hyperfine structure of hydrogen, which requires a symbolic manipulation package, and the SU(3) symmetry of quarks is probed via a linear algebra subroutine library. Section 6.10 deals with coherent (Glauber) and entangled states. Then there are problems dealing with the changing superposition of kaon states and their relation to oscillations between double wells. The last section in the chapter deal with the Feynman path integral formulation of quantum mechanics, and its solution via the Metropolis algorithm.

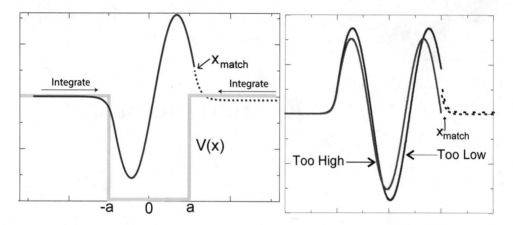

Figure 6.1. *Left:* The solution of the Schroödinger found by integrating out from the left is matched at x_{match} to the solution found by integrating in from the right. *Right:* The wave functions for energies that are either too low or too high to be an eigenvalue.

6.2   Bound States

6.2.1   Bound States in 1-D Box (Semianalytic)

Conduct a trial-and-error search technique to solve a transcendental equation to determine the energies of a particle bound within a 1-D square well of radius R:[1]

$$V(x) = \begin{cases} -V_0, & \text{for } |x| \leq R, \\ 0, & \text{for } |x| \geq R. \end{cases} \tag{6.1}$$

The energies of the bound states $E = -E_B < 0$ within this well are solutions of the transcendental equations [Gottfried(66)]

$$\sqrt{V_0 - E_B} \, \tan\left(\sqrt{V_0 - E_B}\right) = \sqrt{E_B} \quad \text{(even)}, \tag{6.2}$$

$$\sqrt{V_0 - E_B} \, \cotan\left(\sqrt{V_0 - E_B}\right) = \sqrt{E_B} \quad \text{(odd)}, \tag{6.3}$$

where even and odd refer to the symmetry of the wave function. Here we have chosen units such that $\hbar = 1$, $2m = 1$, $a = 1$. You can solve this problem by stepping through values for E_B and observing when the LHS and the RHS are equal. That is essentially what we automate using the bisection algorithm, as illustrated in Figure 2.1 left, and programmed in `Bisection.py` in Listing 2.6.

1. Rewrite (6.2)–(6.3) in the standard form for a search: $f(x) = 0$.

---

[1]We solve this same problem in §6.2.2 using a completely numerical approach that is applicable to almost any potential and which also provides the wave functions.

2. Choose a value for V_0 and a guess for the energy $|E_0| < |V|$.

3. Adjust x_{min} and x_{max} for your search until you can find several bound-state energies E_B for even wave functions. Ensure that your answer is accurate to six decimal places.

4. Vary x_{min}, x_{max}, and E_0 as a check that you have not missed any states.

5. For the same value for V_0, find several bound-state energies E_B for odd wave functions.

6. Increase the depth of the well until you find additional bound states.

7. What is the relation in your results between the energy of a particular bound state and the well depth?

6.2.2 Bound States in Arbitrary Potential (ODE Solver + Search)

A quantum particle in stationary state of definite energy E is bound by a 1-D potential. Its wave function is determined by an ordinary differential equation (ODE), the time-independent Schrödinger equation:[2]

$$\frac{d^2\psi(x)}{dx^2} - \frac{2m}{\hbar^2}V(x)\psi(x) = \kappa^2\psi(x), \qquad \kappa^2 = -\frac{2m}{\hbar^2}E = \frac{2m}{\hbar^2}|E|. \qquad (6.4)$$

When a particle is bound it is confined to some finite region of space, and this implies that $\psi(x)$ is normalizable. This, in turn, tells us that the energy must be negative in order to have $\psi(x)$ decay exponentially as $x \to \pm\infty$:

$$\psi(x) \to \begin{cases} e^{-\kappa x}, & \text{for } x \to +\infty, \\ e^{+\kappa x}, & \text{for } x \to -\infty. \end{cases} \qquad (6.5)$$

Although it is straightforward to solve the ODE (6.4) with an rk4 algorithm, the catch here is that we must *also* require that the solution $\psi(x)$ satisfies the boundary conditions (6.5). This extra condition turns the ODE problem into an *eigenvalue problem* for which solutions exist only for certain values of the energy E, the *eigenvalues*. The solution, if it exists, follows by guessing an energy, solving the ODE, and then varying the energy as part of a trial-and-error search for a wave function that satisfies the boundary conditions (6.5).

1. Start your program on the far *left* at $x = -X$, where $X \gg a$. Here X is your numerical approximation to infinity at which $V(-X) \simeq 0$ and $\psi \sim e^{\pm\kappa x}$.

$$\psi_L(x = -X) = e^{+\kappa(x=-X)} = e^{-\kappa X_{max}}. \qquad (6.6)$$

---

[2]For a time-dependent problem, even in 1-D, we would need to solve a PDE, as done in §6.7.3.

2. Use an ODE solver to step $\psi_L(x)$ in toward the origin (to the right) from $x = -X$ until you reach the *matching radius* x_m near the potential region. The exact value of this matching radius is not important, and your final solution should be independent of it. In Figure 6.1 we show a sample solution with $x_m = a$; that is, we match at the right edge of the potential well (since the potential vanishes for $x < a$, the wave function has its asymptotic form).

3. Start at the extreme *right*, that is, at $x = X \simeq +\infty$, with a wave function that satisfies the right-hand boundary condition:

$$\psi_R(x = \kappa X) = e^{-\kappa(x=X)} = e^{-\kappa X}. \tag{6.7}$$

4. Use your ODE solver to step ψ_R in toward the origin (to the left) from $x = X$ until you reach the *matching radius* x_m (Figure 6.1).

5. In order for probability and current to be continuous at $x = x_m$, $\psi(x)$ and $\psi'(x)$ must be continuous there. Requiring continuity of the *logarithmic derivative* $\psi'(x)/\psi(x)$ combines both requirements into a single condition that is independent of ψ's normalization. Incorporate expressions in your program for the logarithmic derivative of both the left and right wave functions.

6. A good guess for ground-state energy would be a value somewhat up from the bottom of the well, $|E| < V_0$. It is unlikely the left and right wave functions will match for an arbitrary guess, and so we measure the mismatch in logarithmic derivatives as:

$$\Delta(E, x) = \frac{\psi'_L(x)/\psi_L(x) - \psi'_R e\psi(x)/\psi_R e\psi(x)}{\psi'_L(x)/\psi_L(x) + \psi'_R e\psi(x)/\psi_R e\psi(x)}\Bigg|_{x=x_m}, \tag{6.8}$$

where the denominator is included in order to limit the magnitude of Δ. Use this measure to search for an energy at which $\Delta = 0$ within some set tolerance. In Figure 6.1 we show some guesses that do not match.

Our solution for a square well, `QuantumEigenCall.py`, which can be generalized for an arbitrary potential, uses rk4 and the bisection algorithm, and is given in Listing 6.3. Here is its pseudocode:

```
# Pseudocode for QuantumEigenCall.py: Finds E & psi via rk4 + bisection

Import library functions
Define Constants
Define the RHS function f(x,y) for ODE
Define potential function V(x)
Define ODE solution  function diff(h)
Define plotting function plot(h)
     Integrate left Psi   -> from -Inf to Rmatch
     Integrate right Psi  <- from Inf to Rmatch
Normalize left Psi to match right Psi
Use bisection algorithm search on E
     Quit when L & R  log(deriv)  <  eps or after Nmax iterations
```

For ODEs not containing any first-order derivatives, a more efficient algorithm than rk4 is the Numerov algorithm,

$$\psi(x+h) \simeq \frac{2[1 - \frac{5}{12}h^2 k^2(x)]\psi(x) - [1 + \frac{h^2}{12}k^2(x-h)]\psi(x-h)}{1 + h^2 k^2(x+h)/12}. \tag{6.9}$$

We see that the Numerov algorithm uses the values of ψ at the two previous steps x and $x - h$ to move ψ forward to $x + h$. To step backward in x, we need only to reverse the sign of h. Our implementation of this algorithm for bound-state solutions of the Schrödinger equation, `QuantumNumerov.py`, is given in Listing 6.4.

6.2.3   Bound States in Arbitrary Potential (Sloppy Shortcut)

A bound state by definition is confined to a region of space. In most cases this means that the wave function far from the influence of the potential falls off exponentially. Accordingly, a quick, though sloppy, way to find a bound state is to integrate out from the origin and check if the external wave function is a decaying exponential. The problem with approach, and why we call it "sloppy", is that it can lead to very large error accumulation and for this reason unreliable results. For large x both $\exp(-\kappa x)$ and $\exp(+\kappa x)$ are valid solutions of the ODE, though not of the eigenvalue problem. So if you start with a numerical solution that behaves like $\exp(-\kappa x)$ and integrate it out to larger x, the small amount of $\exp(+\kappa x)$ present due to numerical error will grow as x increases, while the desired $\exp(-\kappa x)$ solution will become relatively smaller. Ultimately, the error may dominate, which leaves you with garbage.

1. Choose a potential whose bound states you know analytically or numerically.

2. Apply your ODE solver so that it integrates out from the origin to region where the potential vanishes or is vanishingly small.

3. Verify that the wave function is a decaying exponential, and , accordingly, a bound state.

4. Vary the energy in small, continuous, negative steps and note at which energies the external wave function is a decaying exponential, and , accordingly, a bound state.

5. Compare those energies to previously found using matching and searching.

6. To see if there is a serious weakness in this method arising from integrating outward on a decaying exponential, increase the x value at which you look for the decaying exponential, and note if you can still find a decaying exponential and if the deduced eigenenergies change.

6.2.4   Relativistic Bound States of Klein-Gordon Equation

Incorporating relativity into quantum mechanics presents challenges, one of which is the inconsistent possibility of multiple particles within a single-particle theory. Inconsistencies aside, relativity is a more accurate description of nature than is Newtonian dynamics, and so should be incorporated into a proper quantum mechanics. Here we look at solutions of the Klein-Gordon equation (KGE) that intrinsically include relativity. Although spin effects arise naturally in relativistic quantum mechanics, that requires the more complex Dirac equation, and so the problem here is more appropriate as description of spinless particles, like pions, than of electrons [Landau(96)]. Furthermore, since a π^- forms an exotic atom with a radius some 280 times smaller than that of an electron, relativistic effects are more important for pionic atoms than for electrons.

The KGE starts with the relation

$$E^2 = p^2 + m^2, \tag{6.10}$$

where we use units in which $\hbar = c = 1$. As did Schrödinger with $E = KE + PE$, we makes the canonical associations

$$E \;\rightarrow\; \mathbf{H}_0 \;\rightarrow\; i\partial/\partial t, \qquad \mathbf{p} \;\rightarrow\; \boldsymbol{\nabla}/i, \tag{6.11}$$

$$\Rightarrow \quad -\frac{\partial^2 \psi(\mathbf{x},t)}{\partial t^2} = -\nabla^2 \psi(\mathbf{x},t) + m^2 \psi(\mathbf{x},t). \tag{6.12}$$

The interaction with a static electric field ϕ follows by the minimal coupling postulate:

$$(i\frac{\partial}{\partial t} - q\Phi)^2 \psi(\mathbf{x},t) = -\nabla^2 \psi(\mathbf{x},t) + m^2 \psi(\mathbf{x},t). \tag{6.13}$$

We solve (6.13) in the partial-wave basis,[3]

$$\psi(\mathbf{x},t) = e^{-iEt} \sum_{\ell=0}^{\infty} \sum_{m=-\ell}^{\ell} \frac{u_l(kr)}{r} Y_{lm}(\theta,\phi), \tag{6.14}$$

with the external electric field generated by a nuclear charge Ze, $q\Phi = -Ze^2/r$.

1. Show that the resulting radial KGE is

$$\frac{d^2 u_l(kr)}{dr^2} + \left[\frac{2EZ\alpha}{r} - (m^2 - E^2) - \frac{l(l+1) - (Z\alpha)^2}{r^2} \right] u_l(kr) = 0, \tag{6.15}$$

$$\alpha = e^2 \equiv \frac{e^2}{\hbar c} \simeq \frac{1}{137} \quad \text{(fine-structure constant).} \tag{6.16}$$

[3]Note that the partial wave basis used previously for the hydrogen bound state did not have the $1/r$ division.

2. Solve (6.15) numerically for the lowest two bound states of pionic barium, $Z = 56$, $m = 139 MeV/c^2$ with $\ell = 0$ and $\ell = 1$. *Beware:* the $l = 0$ wave function (though not probability) is singular at the origin, and so may require some care.

3. Already evident in (6.15) is a relativistic correction to the angular momentum barrier, $l(l+1) \to l(l+1) - (Z\alpha)^2$. This makes the orbits precess and removes the l degeneracy in the energy. Accordingly, be careful to evaluate the amount by which the $2S$ and $2P$ states are no longer degenerate.

4. Compare your results to the nonrelativistic answers. Since

$$\sqrt{p^2 + m^2} - m \simeq \frac{p^2}{2m} - \frac{p^4}{8m} + \cdots ,\tag{6.17}$$

we would expect slightly lower levels as compared to the nonrelativistic results.

5. There is actually an analytic solution to (6.15) obtained by following the same path as used for the Schrödinger hydrogen atom. It yields:

$$E_{n,l} = \frac{m}{\left(1 + (Z\alpha)^2/\left[n - l - \frac{1}{2} + \sqrt{(l+\frac{1}{2})^2 - (Z\alpha)^2}\right]^2\right)^{1/2}},\tag{6.18}$$

where n is the principal quantum number. Compare the answers you obtained by solving the KGE as an ODE to (6.18). (See too next question.)

6. *Warning:* The radical in (6.18) contains the addition of some very small numbers to some very large numbers, a situation fraught with numerical dangers. One way to moderate the dangers is to expand E in powers of $Z\alpha$ and evaluate the series term by term:

$$E_{n,l} \simeq m - \frac{m(Z\alpha)^2}{2n^2} - \frac{m(Z\alpha)^4}{2n^4}\left(\frac{n}{l+\frac{1}{2}} - \frac{3}{4}\right) + O(\alpha^6).\tag{6.19}$$

We see that the total energy consists of the electron's rest energy plus Bohr's result for the binding energy with a relativistic correction that produces a fine structure and removes the l degeneracy. Compare your evaluation of (6.18) to the expansion (6.19) and to your result from the solution of the ODE.

7. Although the Klein-Gordon equation does not contain spin, it is still a good guide to the size expected for relativistic kinematics corrections in hydrogen. Calculate the relativistic corrections to the hydrogen bound state energies predicted by the Klein-Gordon equation.

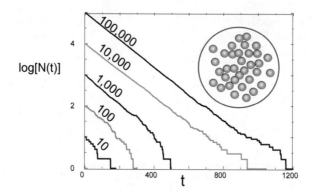

Figure 6.2. The circle shows a sample of N atoms spontaneously decaying. The semilog plots show results of five simulations, all with the same decay constant, but differing initial numbers of atoms.

6.3   Spontaneous Decay Simulation

Quantum textbooks often start off with the history of how quantum levels within atoms were implied by the discrete frequencies of light emitted in the decay of excited states of atoms. However, those books seldom examine the physics of spontaneous decay in which an excited particle decays into other particles without any external stimulation. Furthermore, the probability $\mathcal{P}$ of decay of an unstable state within a given time interval is constant, yet just when, or if it decays in that time interval, is a random event.[4] Accordingly, if you start with a collection of N particles, that number will decrease in time, though the probability of any one particle decaying in some time interval remains the same for as long as that particle exists.

We express these ideas in equation form as

$$\mathcal{P} = \frac{\Delta N(t)/N(t)}{\Delta t} = -\lambda, \quad \Rightarrow \quad \frac{\Delta N(t)}{\Delta t} = -\lambda N(t), \tag{6.20}$$

where λ is the *decay rate constant*. Because $N(t)$ decreases in time, the *activity* $\Delta N(t)/\Delta t$ (sometimes called decay rate) also decreases with time. Equation (6.20) is a *finite-difference equation* relating the quantities $N(t)$, $\Delta N(t)$, and Δt, and is the basic, experimental law describing spontaneous decay. In the limit of infinitesimal Δt and ΔN, (6.20) can be approximated as an ODE with an exponential solution:

$$\frac{dN(t)}{dt} \simeq -\lambda N(t) \quad \Rightarrow \quad N(t) \simeq N(0)e^{-\lambda t}. \tag{6.21}$$

However, since in nature Δt and ΔN are finite, (6.20) is actually more exact, and is precisely what you are asked to simulate.

---

[4]Stimulated emission within a reflecting cavity is a different story.

The algorithm for simulating radioactive decay is simple, with time advancing in steps of Δt. After each time step we decide whether each individual atom decays based on a comparison of λ to a random number, and keep a running count of the number of decays during that Δt. The process continues until all atoms are gone. DecaySound.py is our Python simulation code which also creates a sound each time a particle decays and plots the results (Figure 6.2). Its output sounds very much like a Geiger counter, which lends some credence to the model and to the simulation. Here's its pseudocode:

```
input N, lambda
t = 0
while N > 0                          # Time Loop
        DeltaN = 0
             for i = 1..N            # Number Loop
                   if  r_i < lambda  then  DeltaN = DeltaN + 1
        t = t +1
        N = N - DeltaN
        Output t, DeltaN, N
```

1. Using our pseudocode as a guide, write your own spontaneous decay code. You should obtain results like those in Figure 6.2.

   a. Note, your pick for the decay constant $\lambda = 1/\tau$ sets the time scale for the simulation.
   b. Plot the logarithm of the number left $\ln N(t)$ and the logarithm of the decay rate $\ln \Delta N(t)/\Delta t$ versus time.
   c. Why does there appear to be larger statistical fluctuations in $\Delta N(t)$ than in $N(t)$?
   d. Confirm that you obtain what looks like exponential decay when you start with large values for $N(0)$, but that inevitably the decay displays its stochastic nature as $N(t)$ becomes small.
   e. Show that you obtain the same slopes for $N(t)$ versus t for different values of $N(0)$, and that the slopes are proportional to the value chosen for λ.
   f. Explain in your own words how a process that is spontaneous and random at its very heart can be modeled by an exponential function.

2. Devise and test a model for spontaneous decay including the effect of stimulated emission. This might describe decaying particles trapped within a cavity. One approach is to have the decay rate parameter increase as the number of atoms increases, say as a modification of (6.20):

$$\mathcal{P} = \frac{\Delta N(t)/N(t)}{\Delta t} = -\lambda \left(1 + \alpha \frac{\Delta N(t)}{\Delta N(0)} \right), \tag{6.22}$$

where α is a parameter you may adjust as desired. Note, this addition makes the fundamental law nonlinear, and so may lead to some interesting effects.

3. Investigate some simulations in which there is growth instead of decay within each time interval, that is, for which the fundamental law is

$$\mathcal{P} = \frac{\Delta N(t)/N(t)}{\Delta t} = +\lambda. \tag{6.23}$$

  a. Extend this model to include a limit on the growth rate. One way of doing that is to make the growth rate parameter decrease as the number of particles increases, such as

$$\mathcal{P} = \frac{\Delta N(t)/N(t)}{\Delta t} = +\lambda \left(1 - \alpha \frac{N(t)}{N(0)} \right), \tag{6.24}$$

  where α is a parameter you may adjust as desired. Note, this addition makes the fundamental law nonlinear, and so may lead to some interesting effects.

6.3.1   Fitting a Black Body Spectrum

Quantum mechanics began with Planck's fit to the spectrum of black body radiation:

$$\mathrm{I}(\nu, T) = \frac{2h\nu^3}{c^2} \frac{1}{e^{\frac{h\nu}{kT}} - 1}. \tag{6.25}$$

Here $\mathrm{I}(\nu, T)$ is the energy per unit time of radiation with frequency ν emitted per unit area of emitting surface, per unit solid angle, and per unit frequency by a black body at temperature T. The parameter h is Planck's constant, c is the speed of light in vacuum, and k is Boltzmann constant. The [COBE(16)] project measured the cosmic background radiation and obtained the results given in Table 6.1.

1. Plot the COBE data and see if it has a shape similar to the black body spectrum first explained by Planck.

2. Use these data to deduce the temperature T of the cosmic microwave background radiation.

3. *Hint:* Applying a logarithm to (6.25) will permit a linear, least-squares fit.

6.4   Wave Functions

6.4.1   Harmonic Oscillator Wave Functions

In dimensionless form, the time-independent Schrödinger equation for stationary states of the harmonic oscillator is:

$$\frac{d^2u}{dx^2} + (2n + 1 - x^2)u = 0, \quad n = 0, 1, 2, \ldots, \tag{6.26}$$

Table 6.1. The energy spectrum of microwave radiation measured by COBE.

ν 1/cm	$I(\nu,T)$ MJy/sr	Error kJy/sr	ν 1/cm	$I(\nu,T)$ MJy/sr	Error kJy/s	ν 1/cm	$I(\nu,T)$ MJy/sr	Error kJy/sr
2.27	200.723	14	2.72	249.508	19	3.18	293.024	25
3.63	327.770	23	4.08	354.081	22	4.54	372.079	21
4.99	381.493	18	5.45	383.478	18	5.90	378.901	16
6.35	368.833	14	6.81	354.063	13	7.26	336.278	12
7.71	316.076	11	8.17	293.924	10	8.62	271.432	11
9.08	248.239	12	9.53	225.940	14	9.98	204.327	16
10.44	183.262	18	10.89	163.830	22	11.34	145.750	22
11.80	128.835	23	12.25	113.568	23	12.71	99.451	23
13.16	87.036	22	13.61	75.876	21	14.07	65.766	20
14.52	57.008	19	14.97	49.223	19	15.43	42.267	19
15.88	36.352	21	16.34	31.062	23	16.79	26.580	26
17.24	22.644	28	17.70	19.255	30	18.15	16.391	32
18.61	13.811	33	19.06	11.716	35	19.51	9.921	41
19.97	8.364	55	20.42	7.087	88	20.87	5.801	155
21.33	4.523	282						

where n labels the bound state energies. Even though finding bound states is an eigenvalue problem, we do not have to search because we have already built in the energy eigenvalue into (6.26). Accordingly, rather matching logarithmic derivatives of the right and left wave function, as we did in §6.2.2, we can just integrate the ODE directly out from the origin. Our sample code is given in Listing 6.1, with typical output given in Figure 6.4. In Listing 6.2 we give the code HOanal.py that solves the ODE using the Sympy symbolic manipulation package.

1. Use a Runge-Kutta 4 algorithm to solve (6.26) by integrating from $x = 0$ to a large positive x:

$$0 \leq x \leq 5, \quad n = 0, 1, 2, \ldots. \tag{6.27}$$

   Use arbitrary initial conditions.

2. Although parity (even or oddness upon reflections through the origin) is conserved by this interaction, there is no guarantee that a solution with arbitrary initial conditions will be an eigenstate of parity. To obtain solutions of definite parity, set the initial conditions for the wave function appropriate for it being even or odd. Since even solutions must be finite at the origin with zero slope, while odd solutions must vanish at the origin, with finite slope, try

$$y[0] = 1, \quad y[1] = 0, \quad \text{at } x = 0, \quad \text{even parity}, \tag{6.28}$$

$$y[0] = 0, \quad y[1] = 1, \quad \text{at } x = 0, \quad \text{odd parity}. \tag{6.29}$$

Note that because the normalization of the wave function is arbitrary, the exact value of $y(0)$ is irrelevant.

3. Modify your program so that it outputs and plots the wave function for negative x values. To do that you need only add another loop that reflects the x's to negative values and stores $\pm y[0]$ values depending upon the parity.

4. Use a numerical integration algorithm to normalize your wave function to total probability 1.

5. Examine some cases with large quantum numbers, for instance, $n = 20$–30.

   a. Compute the probability density for these cases.
   b. Do your probabilities show increased values near the classical turning points where $E = V(x)$?
   c. What do your solutions look like for x values that take you beyond the classical turning points?

6. Another approach to this problem is to integrate directly from large negative x's to large positive x's, without any reflections. This is quicker, though less clear as to how to build in states of definite parity.

   a. Try different n values and initial conditions at $x = 5$ in order to investigate if the one step approach also produces states of definite parity.
   b. Compare the solutions obtained with reflection to those obtained with the one-step approach.

7. The analytic solution for the wave function in a 1-D harmonic oscillator is given in terms of Hermite polynomials: $\psi_n(x) = H_n(x)e^{-x^2/2}$. Analytic expressions for the polynomials are deduced from the generating function [Schiff(68)]

$$H_n(x) = (-1)^n e^{x^2} \frac{\partial^n}{\partial x^n} e^{-x^2}. \tag{6.30}$$

Comment upon comparisons of your computed solutions to the analytic solutions.

6.5   Partial Wave Expansions

Much in quantum mechanics is based on the partial wave of the plane wave

$$e^{i\mathbf{k}\cdot\mathbf{r}} = e^{ikr\cos\theta} = e^{ikrz} = \sum_{\ell=0}^{\infty}(2\ell + 1)i^\ell j_\ell(kr)P_\ell(\cos\theta), \tag{6.31}$$

where j_ℓ is the spherical Bessel function and P_ℓ is the Legendre polynomial.

1. For various values of z, calculate and plot the LHS versus RHS of (6.31) as functions of $\cos\theta$.

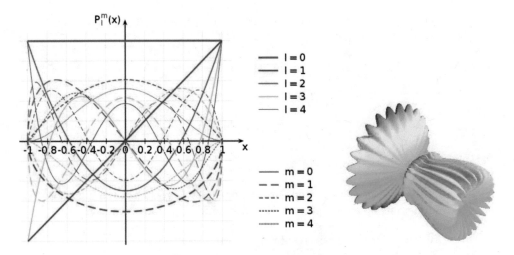

Figure 6.3. *Left:* The associated Legendre polynomials for various ℓ and m values (from Wikipedia). *Right:* A rotatable visualization of a spherical harmonic created with Mayavi in which the radial distance represents the value of the function.

2. For various values of $\cos\theta$, calculate and plot LHS versus the RHS of (6.31) as functions of kr.

6.5.1 Associated Legendre Polynomials

Equation (6.35) in §6.6 expresses the hydrogen wave function in terms of a radial function $\psi(r)$ and the spherical harmonic $Y_\ell^m(\theta, \phi)$. The spherical harmonic, in turn, can be expressed in terms of the associated Legendre polynomial $P_\ell^m(\cos\theta)$ as

$$Y_\ell^m(\theta, \phi) = \sqrt{\frac{2\ell+1}{4\pi}\frac{(l-m)!}{(l+m)!}}\, P_l^m(\cos\theta)e^{im\phi}. \tag{6.32}$$

The associated Legendre polynomials, in turn, satisfy the ODE

$$\frac{d}{dx}\left[(1-x^2)\frac{dP_\ell^m(x)}{dx}\right] + \left[\ell(\ell+1) - \frac{m^2}{1-x^2}\right]P_\ell^m(x) = 0, \quad -\ell \le m \le \ell. \tag{6.33}$$

The behaviors of the $P_\ell^m(x)$ for various values of ℓ and m are shown on the left of Figure 6.3.

Our implementation of the computation of the Legendre polynomials via the solution of an ODE is given in Listing 6.7, with the Matplotlib package used for the graphics, and rk4 for the ODE solution.

1. Use a fourth-order Runge-Kutta algorithm to integrate (6.33) for the associated Legendre polynomials in range $-\ell \leq m \leq \ell$.

2. Normalize the P_ℓ^ms by setting their initial values to:

ℓ Value	m Value	$P_\ell^m(0)$	ℓ Value	m Value	$P_\ell^m(0)$
0, 2	all	1			
$\ell > 2$, even	0	-1	$\ell > 2$, odd	0	1
	> 0	1		> 0	-1
	< 0, $\lvert m \rvert$ even	1		< 0, $\lvert m \rvert$ even	1
	< 0, $\lvert m \rvert$ odd	-1			

3. Verify that the $m = 0$ solutions are the ordinary Legendre polynomials, and are even functions of x for ℓ even, and odd functions of x for ℓ odd.

4. Verify that your numerical results agree with Figure 6.3 left.

5. Verify that your Y_ℓ^m's satisfy the orthonormality condition:

$$\int_0^{2\pi} d\phi \int_{-1}^{+1} d(\cos\theta)\, Y_\ell^m(\theta,\phi)^* \, Y_{\ell'}^{m'}(\theta,\phi) = \delta_{\ell\ell'}\delta_{mm'}. \qquad (6.34)$$

(A double integration is no more than two single integrations, with one of the integration loops within the other.)

6. In Figure 6.3 right we show a 3-D visualization of a $Y_\ell^m(\theta,\phi)$ created with Mayavi, in which the radial distance represents the value of the $Y_\ell^m(\theta,\phi)$. Create several 3-D visualizations of several spherical harmonics. If possible, use color to represent values and make the visualization rotatable.

    a. If you cannot create a 3-D visualization, then make surface plots of $Y_\ell^m(\theta,\phi)$ as separate functions of θ and ϕ.

    b. Another alternative, if you cannot create a 3-D visualization, is to create surface plots of $Y_\ell^m(\theta,\phi)$ as separate functions of x, y, and z.

6.6   Hydrogen Wave Functions

6.6.1   Hydrogen Radial Density

The wave function for a spherically symmetric system can be expressed in spherical coordinates as the product of a radial wave function $R_{n\ell}(r)$ and a spherical harmonic containing the angular dependence [Schiff(68)]:

$$\psi_{n\ell m}(r,\theta,\phi) = R_{n\ell}(r)Y_\ell^m(\theta,\phi), \qquad (6.35)$$

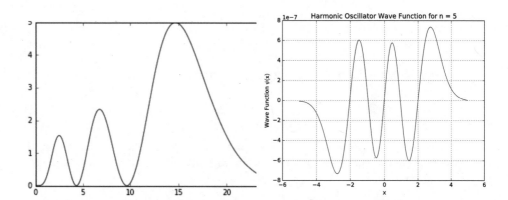

Figure 6.4. *Left:* Hydrogen radial density for $n = 5$, $\ell = 2$ computed by `Hdensity.py`. *Right:* Harmonic oscillator wave function for n = 5 (with odd parity) and n = 8 (with even parity) computed by `HOnumeric.py`.

where some formulations have a $R_{n\ell}(r)/r$ as the radial factor. After a change to dimensionless variables, the radial Schrödinger equation becomes the ODE:

$$\frac{1}{\rho^2}\frac{d}{d\rho}\left[\rho^2\frac{dR_{n\ell}}{d\rho}\right] + \left[\frac{\lambda}{\rho} - \frac{1}{4} - \frac{\ell(\ell+1)}{\rho^2}\right]R_{n\ell} = 0. \qquad (6.36)$$

In order to have a bound state, the solution must be confined, which means there must be an exponential drop-off for large ρ:

$$R_{n\ell}(\rho) = F_{n\ell}(\rho)e^{-\rho/2}. \qquad (6.37)$$

If we substitute this into (6.36) we obtain an ODE for $F_{n\ell}(\rho)$:

$$\frac{d^2F_{n\ell}}{d\rho^2} + \left(\frac{2}{\rho} - 1\right)\frac{dF_{n\ell}}{d\rho} + \left[\frac{\lambda-1}{\rho} - \frac{\ell(\ell+1)}{\rho^2}\right]F_{n\ell} = 0. \qquad (6.38)$$

Analysis indicates that $\lambda = n = 1, 2, 3, ...$ and $\ell(< n) = 0, 1, 2, 3...$.

1. Assume that you do not know the analytic solution for $F_{\ell,n}$, and so need to solve for it numerically.

   a. Show that the standard dynamical and numerical form for this equation is

   $$\frac{dy^{(0)}}{d\rho} = y^{(1)}, \quad \frac{dy^{(1)}}{d\rho} = -\left(\frac{2}{\rho} - 1\right)y^{(1)} - \left[\frac{\lambda-1}{\rho} - \frac{\ell(\ell+1)}{\rho^2}\right]y^{(0)}. \qquad (6.39)$$

   b. Solve for $R_{n\ell}$ for a number of n and ℓ values.

c. The radial density for hydrogen is

$$P(r) = 4\pi r^2 |F_{n\ell}(\rho)e^{-\rho/2}|^2. \tag{6.40}$$

Use a numerical integration algorithm to normalize the radial density so that

$$\int_0^\infty P(r)dr = 1. \tag{6.41}$$

d. Compare your numerical densities with those available in texts or on the Web. One of our solutions for the unnormalized radial density is shown in Figure 6.4 left.

Our program `Hdensity.py` in Listing 6.8 solves the Schrödinger equation numerically using the rk4 algorithm, computes the radial density from the wave function, and plots up the results using Matplotlib. It is straightforward and similar to other programs.

6.6.2 Hydrogen 3-D Wave Functions

In (6.35) we have an expression for the hydrogen wave function in spherical coordinates. In previous sections you have been asked to compute and visualize $R_{n\ell}(r)$ and $Y_\ell^m(\theta,\phi)$. Now it is time to put the pieces together.

1. In Figure 6.3 right we show a 3-D visualization of a $Y_\ell^m(\theta,\phi)$ we created with Mayavi in which the radial distance represents the value of the $Y_\ell^m(\theta,\phi)$. Using whatever software you prefer, create several 3-D visualizations of the hydrogen wave functions $\psi_{n\ell m}(r,\theta,\phi)$. If possible, use color to help represent values and make the visualization rotatable.

   a. If you cannot create a 3-D function visualization like Figure 6.3 left, then try making surface plots of $\psi_{n\ell m}(r,\theta,\phi)$ as separate functions of θ and ϕ.
   b. Another alternative is to create surface plots of $\psi_{n\ell m}(r,\theta,\phi)$ as separate functions of x, y, and z.

6.7 Wave Packets

6.7.1 Harmonic Oscillator Wave Packets

The Schrödinger equation for the time-dependent wave function is:

$$i\hbar\frac{\partial\psi(x,t)}{\partial t} = -\frac{\hbar^2}{2m}\frac{\partial^2\psi(x,t)}{\partial x^2} + \frac{1}{2}kx^2\psi(x,t). \tag{6.42}$$

When considering only stationary states, the time dependence can be factored out of the wave function, leaving an ODE (for the 1-D case) in place of the present PDE.

In §6.7.3 we show how to solve (6.42) as a partial differential equation, while here we just examine some properties of the time-dependent solution.

Consider a wave packet within a harmonic potential whose center of gravity oscillates with the period of the classical motion [Schiff(68)]:

$$\psi(x,t) = \frac{\alpha^{1/2}}{\pi^{1/4}} \exp\left[-\tfrac{1}{2}(x - x_0 \cos \omega t)^2 - i(\tfrac{\omega t}{2} + x x_0 \sin \omega t - \tfrac{1}{4}x_0^2 \sin 2\omega t)\right], \quad (6.43)$$

where $x = \alpha x$, $\alpha = (mk/\hbar^2)^{1/4}$, and $\omega = \sqrt{k/m}$. The corresponding probability density in terms of the original x is

$$|\psi(x,t)|^2 = \frac{\alpha}{\sqrt{\pi}} e^{-\alpha^2[x - a\cos(\omega_c t)]^2}. \quad (6.44)$$

1. Plot this probability density as a function of time and position. Our program HOpacket.py using the Visual package is given in Listing 6.11, with the Matplotlib version, HOpacketMat.py, given in Listing 6.12.

2. Explain why the wave packet does not seem to change its shape with time.

3. Compare the motion of the wave packet to that of a classical oscillator, $x(t) = \cos(t)$, with $k = 1$ and $m = 1$.

4. Is (6.44) an energy or a momentum eigenstate? (Explain.)

6.7.2 Momentum Space Wave Packets

Coordinate and momentum spaces are equally valid places from which to view quantum mechanics, and so we now examine the momentum space version of the wave packet (6.44). While it is straightforward to take the Fourier transform of the density, that would not be correct. Instead, we need to transform the coordinate space wave function into a momentum space wave function, and then form its squared modulus to obtain the probability in momentum space.

1. Evaluate the Fourier transform $\phi(p,t)$ of (6.43) over the x variable to obtain the momentum space wave function. Verify that the inverse transform yields the original coordinate space wave function:

$$\phi(p,t) = \frac{1}{\sqrt{2\pi\hbar}} \int_{-\infty}^{+\infty} \psi(x,t)e^{-ipx/\hbar}, \quad \psi(x,t) = \frac{1}{\sqrt{2\pi\hbar}} \int_{-\infty}^{+\infty} \phi(p,t)e^{+ipx/\hbar}.$$

2. Calculate and plot the momentum space probability

$$\rho(p,t) = |\phi(p,t)|^2, \quad (6.45)$$

and examine its behavior in time.

3. The variance of a probability density $\rho(x)$ is defined as

$$\text{var}(x) = \sigma^2(x) = \int_{-\infty}^{+\infty} (x - \langle x \rangle)^2 \, \rho(x) \, dx, \quad \langle x \rangle = \int_{-\infty}^{+\infty} x \rho(x) \, dx, \qquad (6.46)$$

where $\langle x \rangle$ is the expectation value of x. Compute the variance σ_x of the coordinate space wave packet and the variance σ_p of the momentum space wave packet and see how close they come to satisfying the uncertainty principle:

$$\sigma_p \sigma_n \geq \hbar. \qquad (6.47)$$

Our program for the solution of the partial differential equation describing the harmonic oscillator is given in Listing 6.9. A sample program that produces an animation of this harmonic oscillator wave packet, as well as of the corresponding classical harmonic oscillator, is given in Listing 6.11. In turn, Listing 6.12 has the Matplotlib version of the program.

6.7.3   Solving Time-Dependent Schrödinger Equation

Many quantum mechanics texts start with the time-dependent Schrödinger equation,

$$i \frac{\partial \psi(x,t)}{\partial t} = -\frac{1}{2m} \nabla^2 \psi(x,t) + V(x)\psi(x,t), \qquad (6.48)$$

though rarely solve it ($\hbar = 1$ here). However, (6.48) can be solved via the leapfrog method, much like other PDEs. We express the complex wave function as

$$\psi(x,t) = \text{Re}\,\psi(x,t) + i\,\text{Im}\,\psi(x,t), \qquad (6.49)$$

and so the Schrödinger equation becomes two, coupled PDEs:

$$\frac{\partial \text{Re}\,\psi(x,t)}{\partial t} = -\frac{1}{2m} \frac{\partial^2 \text{Im}\,\psi(x,t)}{\partial x^2} + V(x)\text{Im}\,\psi(x,t), \qquad (6.50)$$

$$\frac{\partial \text{Im}\,\psi(x,t)}{\partial t} = +\frac{1}{2m} \frac{\partial^2 \text{Re}\,\psi(x,t)}{\partial x^2} - V(x)\text{Re}\,\psi(x,t). \qquad (6.51)$$

The solution is based on a finite-difference algorithm in which space is represented along a 1-D grid, while time advances in discrete steps,

$$x = i\Delta x, \quad i = 0, 1, \ldots I_{max}, \qquad t = n\Delta t, \quad n = 0, 1, \ldots N_{max}. \qquad (6.52)$$

The algorithm takes the solution at the present time and advances it one time step ahead. With the discrete values of x, t, and ψ, the derivatives in (6.50) and (6.50) can be expressed as finite differences of the values of ψ on the grid. This converts the differential equations into discrete equations of the form

$$\text{Re}\,\psi(x, t + \tfrac{1}{2}\Delta t) = \text{Re}\,\psi(x, t - \tfrac{1}{2}\Delta t) + [2\beta + V(x)\,\Delta t]\text{Im}\,\psi(x,t)$$

$$- \beta[\text{Im}\,\psi(x + \Delta x, t) + \text{Im}\,\psi(x - \Delta x, t)], \qquad (6.53)$$

where $\beta = \Delta t/\Delta x^2$. There is a similar equation for $\mathrm{Im}\psi(x,t)$.

It is often possible to improve the precision of the solution to wave equations with coupled amplitudes by solving for each amplitude at slightly differing times. In the present case, we improve the degree to which overall probability is conserved by solving for the real and imaginary parts of the wave function at "staggered" times [Askar & Cakmak(78), Visscher(91), Maestri et al.(00)],

$$\mathrm{Re}\,\psi(t=0),\ \mathrm{Re}\,\psi(\Delta t),\ \mathrm{Re}\,\psi(2\Delta t),\ldots,\ \mathrm{Im}\,\psi(\Delta t/2),\ \mathrm{Im}\,\psi(3\Delta t/2)\ldots. \quad (6.54)$$

One then uses an expression for the probability that cancels the leading error terms:

$$\rho(t) = \begin{cases} R^2(t) + I\left(t+\frac{\Delta t}{2}\right)I\left(t-\frac{\Delta t}{2}\right), & \text{for integer } t, \\ I^2(t) + R\left(t+\frac{\Delta t}{2}\right)R\left(t-\frac{\Delta t}{2}\right), & \text{for half-integer } t. \end{cases} \quad (6.55)$$

Finally, we adopt the notation

$$\mathrm{Re}\psi(x,t) \to R_{x=i\Delta x}^{t=n\Delta t}, \qquad \mathrm{Im}\psi(x,t) \to I_{x=i\Delta x}^{t=(1+2n)\Delta t/2}. \quad (6.56)$$

The algorithm now takes the simple form of (future ψ) = (present ψ) + change:[5]

$$R_i^{n+1} = R_i^n - \frac{\Delta t}{(\Delta x)^2}\left(I_{i+1}^n + I_{i-1}^n - 2I_i^n\right) + \Delta t V_i I_i^n \quad (6.57)$$

$$I_i^{n+1} = I_i^n + \frac{\Delta t}{(\Delta x)^2}\left(R_{i+1}^n + R_{i-1}^n - 2R_i^n\right) - \Delta t\, V_i\, R_i^n, \quad (6.58)$$

where the superscript n indicates time and the subscript i position. Note that the algorithm does not require us to store the wave function for all time steps (which can be in the 1000's), but only for the present and future.

Our sample program `HOmovSlow.py` is given in Listing 6.9 and uses the Visual package, with successive plots providing animation. A pseudocode for it is given below. Note that we also have the solution `HOmov.py` in Listing 6.10 that runs faster by using Python's slicing feature (optimized by Bruce Sherwood).

```
# Pseudocode HOmovSlow.py: Solves & animates wavepacket in HO
Import visual
Initialize constants, setup arrays
Loop over i
    Initialize Psi(i, t=0) over all x values
    Store v(i) over all x values
 Begin time propagation
    Loop over i
        Step RePsi forward via algorithm
        Evaluate Rho
        For every 10th step
            Plot values for x & Rho
```

[5]This is a corrected form of the equations appearing in [LPB(08), LPB(15)].

```
Loop over i
    Step ImPsi forward via algorithm
Loop over i
    Set present RePsi & ImPsi  = future RePsi & ImPsi
```

1. Program up the algorithm (6.57)–(6.58) for a wave packet in a 1-D square well potential with infinite walls.

   a. You can use an initial wave function of your choice, with a standard choice being a Gaussian multiplied by a plane wave:

   $$\psi(x, t = 0) = \exp\left[-\frac{1}{2}\left(\frac{x - x_0}{\sigma_0}\right)^2\right] e^{ik_0 x}. \qquad (6.59)$$

   b. Define arrays that store all space values of the wave function, but only the two time values for present and future times.
   c. Boundary conditions are required for a unique solution, and we suggest having the wave function vanish at the edges of a very large box.
   d. When you are sure that the program is running properly, do a production run with at least 5000 time steps.

2. Make surface plots of probability versus position versus time.

3. Check how well the total probability is conserved for early and late times by determining the integral of the probability over all of space, $\int_{-\infty}^{+\infty} dx\, \rho(x)$, and computing its percentage change as a function of time (its specific value doesn't matter because that's just normalization).

4. Investigate a number of different potential wells. In particular check that a harmonic oscillator potential binds a Gaussian wave packet with no dispersion.

5. Extend your solution to a 2-D case such as a parabolic tube

   $$V(x, y) = V_0 x^2, \quad -10 \le x \le 10, \quad 0 \le y \le 20. \qquad (6.60)$$

6.7.4   Time-Dependent Schrödinger with E Field

A particle with charge q is bound in a harmonic oscillator and is described by the initial wave packet

$$\psi(x, t = 0) = e^{-(x/0.5)^2/2} e^{ipx}. \qquad (6.61)$$

The particle is then subjected to an external time-dependent electric field $E(t)$.

1. Write down the time-dependent Schrödinger equation describing this particle.

2. Compute the wave function and probability density for several periods for $E = 70$, $q = 1$, and $k = 50$.

3. Compare the no-field and with-field probabilities.

4. What changes when you change the sign of the charge?

5. Progressively weaken the strength of the harmonic oscillator potential until you are essentially viewing the motion of the wave packet in a pure electric field.

6. Increase the strength of the harmonic oscillator potential and describe the change on the wave packet.

7. Make the electric potential sinusoidal with a strength comparable to that of the harmonic oscillator potential, and view the solution.

8. Vary the frequency of the sinusoidal potential so that it passes through the natural frequency of the harmonic oscillator, and make a plot of the resulting resonance.

Our program `HOchargeMat.py` is in Listings 6.13.

6.8 Scattering

In a scattering experiment, a beam of particles is directed at a target, and the number and energy of particles scattered into different angles are observed. Because the energy of the beam can be varied continuously, scattering occurs in the continuum of positive energies. Because the scattering wave function is not confined, there is no boundary condition leading to an eigenvalue problem, and so all positive energies are allowed. And since the scattering wave function can occupy all of space, scattering wave functions are not normalizeable.

6.8.1 Square Well Scattering

The square well potential is simple enough to permit a partially analytic solution for both scattering and bound states. Thus we save some work here by using an analytic form for the wave function in the inner region. All of the other steps are the same as would be taken in a complete numerical solution, with the only difference being that the complete numerical solution includes integrating the ODE to obtain the inner wave function [Bayley & Townsend(21), Ram & Town(21)].

Consider a projectile of reduced mass μ scattering from a spherically symmetric, 3-D square well,

$$V(r) = \begin{cases} -V_0, & \text{for } 0 \leq r \leq a, \\ 0, & \text{for } r > a. \end{cases} \tag{6.62}$$

We decompose the wave function into radial and angular parts,

$$\psi(r,\theta,\phi) = \sum_{\ell=0}^{\infty} \frac{\chi_l(r)}{r} Y_\ell^m(\theta,\phi), \tag{6.63}$$

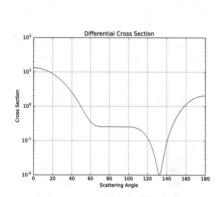
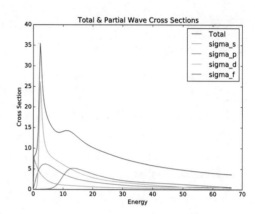

Figure 6.5. *Left:* Differential cross section as a function of scattering angle for scattering from a square-well potential with $V_0 = 15$, and $E = 10$. *Right:* Total and partial cross sections as a function of energy for this same potential.

and substitute it into the Schrödinger equation to obtain:

$$-\frac{\hbar^2}{2\mu}\frac{d^2\chi}{dr^2} + \left[-V_0 + \frac{\ell(\ell+1)\hbar^2}{2\mu r^2}\right]\chi = E\chi, \quad r < a \tag{6.64}$$

$$-\frac{\hbar^2}{2\mu}\frac{d^2\chi}{dr^2} + \left[\frac{\ell(\ell+1)\hbar^2}{2\mu r^2}\right]\chi = E\chi, \quad r > a. \tag{6.65}$$

Because we assume that the potential vanishes for $r > a$, the solution of (6.65) can be expressed as a linear combination of the regular and irregular solutions to the homogeneous Schrödinger equation, that is, as

$$R_\ell(r > a) = e^{i\delta_\ell}[\cos\delta_\ell\, j_\ell(kr) - \sin\delta_\ell\, n_\ell(kr)], \qquad k = \sqrt{2\mu E/\hbar^2}, \tag{6.66}$$

where δ_ℓ is the phase shift, and j_ℓ and n_ℓ are the spherical Bessel and Neumann functions, respectively.

The wave function within the square well is the regular free wave solution though with a wave vector κ that incorporates the strength of the potential:

$$R_\ell(r < a) = j_\ell(\kappa r), \qquad \kappa = \sqrt{2\mu(E - V)}/\hbar. \quad (r < a). \tag{6.67}$$

In order to match the internal wave function (6.67) to the external one (6.65) at $r = a$, we multiply the external wave function by a normalization factor B_ℓ and set the wave functions equal:

$$j_\ell(\kappa a) = B_\ell\, e^{i\delta_\ell}[\cos\delta_\ell\, j_\ell(ka) - \sin\delta_\ell\, n_\ell(ka)] \quad \Rightarrow \quad B_\ell = \frac{j_\ell(\kappa a)\, e^{-i\delta_\ell}}{\cos\delta_\ell\, j_\ell(ka) - \sin\delta_\ell\, n_\ell(ka)}.$$

The phase shifts now follow by matching the wave function and its derivative:

$$\tan \delta_\ell = \frac{kj'_\ell(ka) - \gamma_\ell j_\ell(ka)}{kn'_\ell(ka) - \gamma_\ell n_\ell(ka)}, \qquad \gamma_\ell = \kappa \frac{j'_\ell(\kappa a)}{j_\ell(\kappa a)}. \tag{6.68}$$

Once the phase shifts are known, the differential and total cross section follow:

$$\frac{d\sigma}{d\Omega}(\theta) = \frac{1}{k^2}\left|\sum_{\ell=0}^{\infty}(2\ell+1)e^{i\delta_\ell}\sin\delta_\ell\,P_\ell(\cos\theta)\right|^2, \qquad \sigma_{tot}(E) = \frac{4\pi}{k^2}\sum_{\ell=0}^{\infty}(2\ell+1)\sin\delta_\ell^2. \tag{6.69}$$

Some typical cross sections are shown in Figure 6.5, while a wave function is shown on the left of Figure 6.6.

1. Find a subroutine library such as scipy containing methods for the spherical Bessel and Neumann functions and their derivatives.

2. Calculate the phase shifts for $\ell = 0 - 10$. Use $E = 10$, $V = 15$, $a = 1$, $2\mu = 1$, and $\hbar = 1$.

3. Compute the wave functions for these same ℓ values.

4. Compute the differential and total cross sections.

5. Lower the energy until only S waves make a significant contribution. When the cross section looks flat, then you know only S waves are contributing.

6. In 1920 Ramsauer and Townsend [Ram & Town(21)] studied electron scattering from rare gas atoms and discovered that an extremely low S wave total cross section occurs at certain low energies. See if you can reproduce a reduction in total cross section as you slowly vary the energy. If you look at the wave function while you vary the energy, you should notice that at some energy an extra half cycle of the wave function gets pulled into the well, as compared to the $V = 0$ wave function. Because this is effectively the same as a zero phase shift δ, the S-wave cross section vanishes. To demonstrate this, plot the total cross section versus energy for the a well depth $V_0 = 30.1$, and observe how σ_{tot} gets very small at $E \approx 5.3629$.

Our program for scattering from a square well is given in Listing 6.14. There are a good number of steps required to compute all of the quantities, but the basic step is solving an ODE in an inner and outer region of space for each partial wave, and then matching the solution to each other to extract δ_ℓ. Once matched, we repeat the process for each partial wave. A pseudocode for the program is:

```
# Pseudocode for ScattSqWell.py: Quantum scattering from square well

Import library functions
Initialize constants, plotting arrays
```

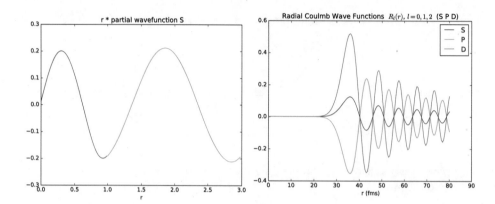

Figure 6.6. *Left:* The S wave radial wave function $rR_\ell(r)$ from square well. *Right:* Real part of regular S, P, and D radial wave functions for the Coulomb scattering of 7 MeV α particles by gold.

```
Gam(): Compute Logarithmic Derivative
    call scipy for spherical Bessels
PhaseShifts(): Compute Phase Shifts
    call Gam()
    compute spherical Bessels & Neumanns
    loop over L values
TotalSig(): Compute Total Cross Sections
PlotSig(): Compute & Plot Partial Cross Sections
DiffSig(): Compute & Plot Differential Cross Sections
WaveFunction(): Compute & Plot Partial Wave Function
    separate internal & external wave functions

Call all functions
```

6.8.2   Coulomb Scattering

A basic assumption of scattering theory is that observations are made at distances far enough from the collision center that the value of the potential at these distances has essentially vanished. This will be possible for potentials with finite or short range, such as a square well or a screened Coulomb potential, but not for the pure Coulomb potential with its $1/r$ falloff. In this latter case the asymptotic ($r \to \infty$) wave function exhibits distortion and so requires a different definition for the phase shifts than (6.66) [Schiff(68)].

In the COM frame the time-independent Schrödinger equation with a Coulomb potential is

$$-\frac{\hbar^2}{2\mu}\nabla^2\psi(\mathbf{r}) + \frac{ZZ'e^2}{r}\psi(\mathbf{r}) = E\psi(\mathbf{r}), \tag{6.70}$$

where μ is the projectile-target reduced mass, Ze is the projectile charge, $Z'e$ is the target charge, and E is the CM energy. The partial wave decomposition of the wave function into angular and radial variables is, as before,

$$\psi(\mathbf{r}) = \sum_{\ell,m} Y_\ell^m(\theta, \phi) \frac{u_\ell(r)}{r}, \tag{6.71}$$

$$\Rightarrow \quad \frac{d^2 u_\ell}{dr^2} + \left[k^2 - \frac{2\eta k}{r} - \frac{\ell(\ell+1)}{r^2} \right] u_\ell = 0. \tag{6.72}$$

$$k = \sqrt{2\mu E/\hbar^2} = \sqrt{2\mu c^2 E/(\hbar c)^2}, \quad \eta = \mu Z Z' e^2/\hbar^2 k. \tag{6.73}$$

We separate out the small r and large r behaviors,

$$u_\ell(r) = e^{ikr}(kr)^{\ell+1} f_\ell(r), \tag{6.74}$$

substitute into the radial equation, and obtain

$$r \frac{d^2 f_\ell(r)}{dr^2} + [2ikr + 2(\ell+1)] \frac{df_\ell(r)}{dr} + [2ik(\ell+1) - 2nk] f_\ell(r) = 0. \tag{6.75}$$

Although maybe not obvious, this equation is equivalent to the often-occurring confluent hypergeometric equation

$$z \frac{d^2 F(z)}{dz^2} + (b - z) \frac{dF(z)}{dz} - aF(z) = 0. \tag{6.76}$$

The solution regular at the origin, sometimes called $1F1$, is

$$f_\ell(r) = C_\ell F(\ell + 1 + in, 2\ell + 2, -2ikr), \tag{6.77}$$

$$C_\ell = \frac{(2ik)^\ell e^{-n\pi/2} \Gamma(\ell + 1 + in)}{v^{1/2}(2\ell)!}, \quad v = \hbar k/\mu, \tag{6.78}$$

where v is the velocity in the COM. These are the analogs of the familiar hydrogenic wave functions, though now for positive energy continuum states and not negative energy bound states.

1. Find a computer subroutine library that computes the confluent hypergeometric function $1F1$ for complex arguments. For example, `mpmath` does this in Python, but not `scipy`.

2. Check that your software package also can calculate the gamma function for complex arguments. `Scipy` does this with Python.

3. Evaluate the equations' parameters so that they describe alpha particles of lab energy 7.07 MeV scattering from a stationary gold nucleus.

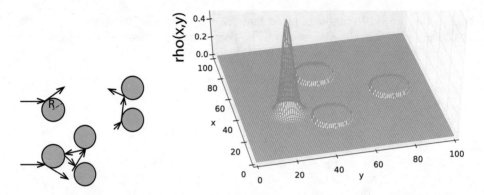

Figure 6.7. *Left:* Classical trajectories for scatterings from one, two, and three stationary disks. *Right:* A wave packet incident on three disks of radius 10 located at the vertices of an equilateral triangle.

4. Evaluate and plot the s, p, and d wave functions for alpha–gold scattering. Our program to do this is given in Listing 6.16, with its output on right of Figure 6.6.

5. Compare the Coulomb wave functions to the free-space (non-Coulomb distorted) wave functions with the same quantum numbers and at the same energy.

6. Investigate the effect on the wave functions of increasing the energy.

6.8.3 Three Disks Scattering; Quantum Chaos

In §6.7.3 we provided the algorithm for and examples of solving the time-dependent Schrödinger equation numerically for the motion of a wave packet. In §3.9 we examined classical scattering from one, two, and three fixed, hard disks (Figure 6.7 left), and investigated how classical chaos can occur when scattering from the three disks. Here we combine the two ideas to examine the scattering of a wave packet from several hard disk configurations (Figure 6.7 right), with quantum chaos possible for the three-disk scattering. Quantum chaos is a new field that has developed only in the last two decades, and deducing and observing its signals can be quite subtle. To avoid computation, much of the research has been with the semiclassical formulation of quantum mechanics. We will examine the solution of the time-dependent Schrödinger equation, where we might be able to see the quantum chaos happening.

To make your work easier, in Listing 6.17 we give our program `3QMdisks.py` for quantum scattering from three fixed, hard disks with $\hbar = 1$. The disks have radius R, a center-to-center separations of a, and are placed at the vertices of an equilateral triangle. As seen in Figure 6.7, a Gaussian wave packet,

$$\psi(x, y, t = 0) = e^{i(k_x x + k_y y)} e^{-A(x-x_0)^2 - A(y-y_0)^2}, \qquad (6.79)$$

is incident upon the disks. *Note:* The program as given has the disks confined within a very small box. This may be appropriate for a bound state billiard problem, but not for scattering. *You will need to enlarge the confining box in order to eliminate the effects of reflections from the boundary.* And while you are free to make surface plots of Re $\psi(x, y)$ and Im $\psi(x, y)$, we find it easier to visualize a surface plot of the probability density $z(x, y) = \rho(x, y)$. Here is a pseudocode version of the program:

```
# 3QMdisks.py: Wavepacket scattering from 3 disks wi MatPlot

Import packages
Assign parameter values
Declare arrays

Function Pot1Disks(x_center,y_center)
    V(x,y) = 5 if distance from center < r; else V  = 0
Function Pot3Disk
    V = Pot1Disk (30,45) + Pot1Disk (70,45) + Pot1Disk (50,80)

Compute initial Psi(x,y), Rho(x,y) for all (x,y)
Evaluate V(x,y) for all (x,y)
Time loop   0 < t < t_max
    Apply leapfrog algorithm for RePsi, ImPsi
    If  in disk, Psi = 0           # Make Psi vanish in disks
Surface plot Z = Rho[X, Y]
```

1. Start with the study of one disk:

    a. Produce surface plots of $\rho(x, y)$ for times from 0 until the packet leaves the scattering region. Note, the edges of the billiard table will reflect the wave packet and ergo should be far from the disks in order not to interfere with the scattering from the disks.

    b. Examine the qualitative effect of varying the size of the disk.

    c. Examine the qualitative effect of varying the momentum of the wave packet.

    d. Vary the initial position of the wave packet so that there are nearly head-on collisions as well as ones at glancing angles.

2. Next repeat your investigation for the two disk system. Try to vary the parameters so that you obtain a high degree of *multiple* scatterings from the disks. In particular see if you can find the analog of the classical case where there is a trapped orbit with unending back-and-forth scatterings. (*Hint:* Try starting the wave packet between the two disks.)

3. Next, extend and repeat the investigation to the three-disk system. Vary the parameters so that you obtain a high degree of multiple scatterings from the disks. In particular see if you can find the analog of the classical case where there are many trapped orbits with unending back-and-forth scatterings.

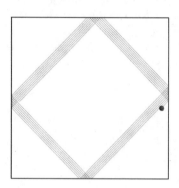
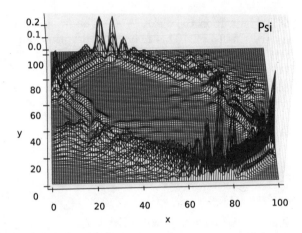

Figure 6.8. *Left:* The path followed by a classical ball bouncing elastically within a square billiard. *Right:* A quantum wave packet, initially Gaussian in shape, after 200 time steps confined within a square billiard table.

a. Develop an algorithm that determines the time delay of the wave packet, that is, the time it takes for most of the initial packet to leave the scattering region.

b. Plot the time delay versus the wave packet momentum and look for indications of chaos, such as sharp peaks or rapid changes. The literature indicates that high degrees of multiple scattering occur when $a/R \simeq 6$.

6.8.4   Chaotic Quantum Billiards

Chaos is an oft-observed behavior in classical nonlinear systems. Observing chaos in quantum systems with their innate statistical properties is more challenging [Gas(14)]. One approach is to examine a quantum system whose classical analog has definite chaotic behavior. As an example, the projectile in classical billiards (§3.9.3 and Figure 3.19) often bounces around endlessly, filling all of the allowed space without ever repeating (ergodic behavior).

In this chapter you should have already solved for the scattering of a wave packet from three hard disks. That same program can be adapted to quantum billiards by removing the disks:

$$I_{i,j}^{n+1} = I_{i,j}^n + \frac{\Delta t}{\Delta x^2}\left(R_{i+1,j}^n + R_{i-1,j}^n - 4R_{i,j}^n + R_{i,j+1}^n + R_{i,j-1}^n\right) \qquad (6.80)$$

$$R_{i,j}^{n+1} = R_{i,j}^n - \frac{\Delta t}{\Delta x^2}\left(I_{i+1,j}^n + I_{i-1,j}^n - 4I_{i,j}^n + I_{i,j+1}^n + I_{i,j-1}^n\right). \qquad (6.81)$$

Imposing the boundary conditions that the wave function vanishes at the table edges leads to reflections. In `3QMdisks.py` in Listing 6.17 we present a simple program that computes the time-dependent motion of a Gaussian wave packet within a square boundary. We impose the boundary condition by requiring $\psi = 0$ at the boundaries. Some results after 200 time steps are shown in Figure 6.8. We see that the quantum system displays a signature of the endless classical reflections from the table's edges.

1. Compute the motion of an initial Gaussian wave packet confined to the top of a square billiards table.

   a. Solve for the motion of the wave packet for initial conditions that lead to periodic orbits of the classical billiard.
   b. Examine the wave function for a series of times ranging from those in which only a small number of reflections has occurred to very large times.
   c. Compute the classical motion of a square billiard for the same range of times and compare your classical results to the corresponding quantum ones.
   d. How many reflections does it take for the wave packet to lose all traces of the classical trajectories?

2. Examine the motion of a quantum wave packet for the various billiards studied classically: a) a circle, b) a stadium, c) a circle with a disk in the middle. In all cases, examine initial conditions that lead to classically periodic orbits.

6.9 Matrix Quantum Mechanics

6.9.1 Momentum Space Bound States (Integral Equations)

Even though the equations of quantum mechanics are usually expressed in coordinate space, they have equally valid representations in momentum space. Often the momentum-space formulations are expressed as integral equations, which is fine since integral equations are straightforward to solve numerically. Here we look at the bound state problem, and refer the reader to [LPB(15)] for a discussion of the more intricate scattering problem.

We start with the coordinate space Schrödinger equation

$$-\frac{1}{2\mu}\frac{d^2\psi(r)}{dr^2} + V(r)\psi(r) = E\psi(r). \tag{6.82}$$

In momentum space, in a partial-wave basis, and for $\ell = 0$, (6.82) becomes

$$\frac{k^2}{2\mu}\psi_n(k) + \frac{2}{\pi}\int_0^\infty dp\, p^2 V(k,p)\psi_n(p) = E_n\psi_n(k), \tag{6.83}$$

where n labels the discrete bound states. The momentum-space wave function $\psi_n(k)$ in (6.83) is the Fourier transform of $\psi_n(r)$:

$$\psi_n(k) = \int_0^\infty dr\, kr\, \psi_n(r)\sin(kr), \tag{6.84}$$

$$
\begin{array}{ccccccc}
\bullet & \bullet & \bullet & \bullet & \bullet & \bullet & \bullet \\
k_1 & k_2 & k_3 & & & & k_N
\end{array}
$$

Figure 6.9. The grid of momentum values on which the integral equation is solved.

and the momentum-space potential $V(k', k)$ is the Fourier transform of the coordinate-space potential,

$$
V(k', k) = \frac{1}{k'k} \int_0^\infty dr \, \sin(k'r) \, V(r) \sin(kr). \tag{6.85}
$$

Equation (6.83) is an integral equation for $\psi_n(k)$, and so differs from an integral representation of $\psi_n(k)$ in that the integral cannot be evaluated until $\psi_n(p)$ is known. However, we can transform this equation into a set of linear equations that can be solved with matrix techniques. We approximate the integral in (6.83) as a weighted sum over N integration points,

$$
\int_0^\infty dp \, p^2 V(k, p) \psi_n(p) = \sum_{j=1}^N w_j k_j^2 V(k, k_j) \psi_n(k_j) = E_n. \tag{6.86}
$$

This converts the integral equation to the algebraic equation

$$
\frac{k^2}{2\mu} \psi_n(k) + \frac{2}{\pi} \sum_{j=1}^N w_j k_j^2 V(k, k_j) \psi_n(k_j) = E_n. \tag{6.87}
$$

Equation (6.87) contains the N unknown function values $\psi_n(k_j)$, the unknown energy E_n, as well as the unknown $\psi_n(k)$. Rather than try to determine $\psi(k)$ for all k values, we restrict k to the same set $k = k_i$ values (Figure 6.9) as used in the approximation of the integral. This leaves N coupled linear equations in $(N + 1)$ unknowns:

$$
\frac{k_i^2}{2\mu} \psi_n(k_i) + \frac{2}{\pi} \sum_{j=1}^N w_j k_j^2 \, V(k_i, k_j) \psi_n(k_j) = E_n \psi_n(k_i), \quad i = 1, N. \tag{6.88}
$$

In matrix form these equations are:

$$
[\mathbf{H}][\psi_n] = E_n[\psi_n], \tag{6.89}
$$

$$
\begin{bmatrix}
\frac{k_1^2}{2\mu} + \frac{2}{\pi} V(k_1, k_1) k_1^2 w_1 & \cdots & \frac{2}{\pi} V(k_1, k_N) k_N^2 w_N \\
\frac{2}{\pi} V(k_2, k_1) k_1^2 w_1 & \cdots & \\
& \ddots & \\
\cdots & \cdots & \frac{k_N^2}{2\mu} + \frac{2}{\pi} V(k_N, k_N) k_N^2 w_N
\end{bmatrix}
\begin{bmatrix}
\psi_n(k_1) \\
\ddots \\
\psi_n(k_N)
\end{bmatrix}
= E_n
\begin{bmatrix}
\psi_n(k_1) \\
\ddots \\
\psi_n(k_N)
\end{bmatrix},
$$

where E_n and the $\psi_n(k)$ vector are unknowns. Since we have N equations in $N + 1$ unknowns, this is an eigenvalue problem for which there are solutions only for certain values of E_N.

6.9.2   k Space Bound States Delta Shell Potential

1. Consider the delta shell potential

$$V(r) = \frac{\lambda}{2\mu}\delta(r - b). \tag{6.90}$$

   This is one of the few potentials for which an analytic solution exists. Show analytically that there is a single bound state and that it satisfies the transcendental equation [Gottfried(66)]

$$e^{-2\kappa b} - 1 = \frac{2\kappa}{\lambda}, \qquad E_n = -\kappa^2/2\mu. \tag{6.91}$$

2. Show that this equation requires $\lambda < 0$ for there to be a bound state.

3. Let $2\mu = 1$, $b = 10$, and λ be large and negative. Find a value of κ that solves (6.91), and determine its corresponding energy.

4. Show that the delta shell potential in momentum-space is

$$V(k', k) = \int_0^\infty \frac{\sin(k'r)}{k'k}\frac{\lambda}{2\mu}\delta(r - b)\sin(kr)dr = \frac{\lambda}{2\mu}\frac{\sin(k'b)\sin(kb)}{k'k}. \tag{6.92}$$

5. Solve numerically in momentum space for the bound states of the delta shell potential.

6. Program up the potential and Hamiltonian matrices $V_{i,j}$ and $H_{i,j}$ for Gaussian quadrature integration with at least $N = 16$ grid points. Use values of b and λ that you know support a bound state.

7. Use a matrix subroutine library to solve the eigenvalue problem (6.89).

8. Perform the matrix multiplication $[H][\psi_n] = E_n[\psi_n]$ to verify that the RHS equals the LHS in the original problem.

9. Compute the bound state energies for a range of λ values and verify that the bound states get deeper as λ increases in magnitude, but that no others appear.

10. *Note:* Your eigenenergy solver may return several eigenenergies. The true bound states will be at negative real energies and should change little as the number of grid points changes. The others eigenenergies are numerical artifacts.

11. Try increasing the number of Gaussian integration points in steps of 8, for example, $N = 16, 24, 32, 64, \ldots$, and see how the energy seems to converge and then develop fluctuations due to round-off error.

12. Extract the best value for the bound-state energy and estimate its precision by observing how it changes with the number of grid points. Compare with (6.91).

13. After you have determined the eigenenergies, determine the momentum-space eigenfunctions $\psi_n(k)$. Does $\psi_n(k)$ falloff at $k \to \infty$? Does $\psi_n(k)$ oscillate? Is $\psi_n(k)$ well behaved at the origin? Is $\psi_n(k)$ normalizeable?

14. Using the same points and weights as used to evaluate the integral in the integral equation, determine the coordinate-space wave function via the Fourier transform

$$\psi_n(r) = \int_0^\infty dk \psi_n(k) \frac{\sin(kr)}{kr} k^2. \qquad (6.93)$$

  a. Does $\psi_n(r)$ falloff as you would expect for a bound state? Does $\psi_n(r)$ oscillate? Is $\psi_n(r)$ well behaved at the origin?

  b. Compare the r dependence of this $\psi_n(r)$ to the analytic wave function:

$$\psi_n(r) \propto \begin{cases} e^{-\kappa r} - e^{\kappa r}, & \text{for } r < b, \\ e^{-\kappa r}, & \text{for } r > b. \end{cases} \qquad (6.94)$$

A sample program that searches for bound states of the momentum space Schrödinger equation is given in Listing 6.5. It imports the `GuassPoints` function to generate the quadrature points.

6.9.3   k Space Bound States Other Potentials

This same method used for solving for bound states of the delta shell potential can be applied to other potentials. For a local potential that is a function of only r, one just has to evaluate the single integral (6.85). For example, here we give two familiar examples, the square well and the Gaussian (harmonic oscillator):

$$V(r) = -V_0\theta(R-r) \Rightarrow V(k',k) = \frac{-V_0}{2k'k}\left[\frac{\sin(k'-k)R}{k'-k} - \frac{\sin(k'+k)R}{k'+k}\right], \qquad (6.95)$$

$$V(r) = -V_0 e^{-\alpha r^2} \Rightarrow V(k',k) = \frac{-V_0\sqrt{\pi}}{4k'k\sqrt{\alpha}}\left[e^{-\frac{(k'-k)^2}{4\alpha}} - e^{-\frac{(k'+k)^2}{4\alpha}}\right]. \qquad (6.96)$$

Repeat the preceding approach to solving the momentum space bound state problem for either or both of these potentials. *Beware:* The sharp edges of a square well cause it to have large amounts of high momentum components. As we see in (6.96), the potential falls off only like $1/k'k$ at large k, as well as being oscillatory there. This slow a falloff may cause the integral in the momentum space Schrödinger equation (6.83) not to converge well, while the subtraction of rapidly oscillating sine functions for the square well leads to numerical imprecision. In contrast, there should be no problem with the momentum space Gaussian potential and its exponential falloff.

6.9.4 Hydrogen Hyperfine Structure

The energy levels of the hydrogen atom exhibit a fine structure splitting arising from the interaction of the electron's spin with its orbital angular momentum. These split levels have an additional hyperfine splitting arising from the interaction of the electron's spin with the proton's spin [Squirtes(02), Brans(91)]. In Gaussian CGS units the magnetic moment of a particle of charge q is related to its angular S by

$$\boldsymbol{\mu} = g\frac{q}{2m}\mathbf{S}, \tag{6.97}$$

where g is the particle's g factor and m its mass. An electron has $q = -e$, $g \simeq -2$, and $\boldsymbol{S} = \hbar\boldsymbol{\sigma}/2$, and so

$$\boldsymbol{\mu}_e \simeq (-2)\frac{-e}{2m_e}\frac{\boldsymbol{\sigma}}{2} = \mu_B\boldsymbol{\sigma}, \qquad \mu_B = \frac{e\hbar}{2m_e} = 5.05082 \times 10^{-27} \text{ joule/Tesla}, \tag{6.98}$$

where μ_B is called the electron's Bohr magneton. Because the proton's mass is so much larger than the electron's, the proton's Bohr magneton is some 2000 times smaller in magnitude than the electron's:

$$\mu_B|_p = \frac{-e\hbar}{2m_p} = -\frac{m_e}{m_p}\mu_B|_e = -\frac{1}{1836.15}\mu_B, \tag{6.99}$$

and consequently the hyperfine structure is some 2000 times smaller than the fine structure.

Even though the electron's and the proton's spins exist in different internal spaces, they are both spin 1/2 and so both can be represented by the Pauli matrices:

$$\boldsymbol{\sigma} = \sigma_x\hat{\epsilon}_x + \sigma_y\hat{\epsilon}_y + \sigma_z\hat{\epsilon}_z \tag{6.100}$$

$$\sigma_x = \begin{bmatrix} 0 & 1 \\ 1 & 0 \end{bmatrix}, \quad \sigma_y = \begin{bmatrix} 0 & -i \\ i & 0 \end{bmatrix}, \quad \sigma_z = \begin{bmatrix} 1 & 0 \\ 0 & -1 \end{bmatrix}. \tag{6.101}$$

In terms of the Pauli matrices, the electron-proton interaction is

$$V = W\boldsymbol{\sigma}_e \cdot \boldsymbol{\sigma}_p = W(\sigma_x^e\sigma_x^p + \sigma_y^e\sigma_y^p + \sigma_z^e\sigma_z^p), \tag{6.102}$$

where we are not specifying the value for W. The spin 1/2 state for either the electron or proton can be either up or down:

$$|\alpha\rangle = \begin{bmatrix} 1 \\ 0 \end{bmatrix}, \quad |\beta\rangle = \begin{bmatrix} 0 \\ 1 \end{bmatrix}. \tag{6.103}$$

For example, the action of the electron–proton interaction (6.102) might be

$$V|\psi\rangle = W\sigma^e \cdot \sigma^p |\alpha^e\alpha^p\rangle = (\sigma_x^e\sigma_x^p + \sigma_y^e\sigma_y^p + \sigma_z^e\sigma_z^p |\alpha^e\alpha^p\rangle \tag{6.104}$$

$$= |\beta^e\beta^p\rangle + i |\beta^e\beta^p\rangle + |\alpha^e\alpha^p\rangle. \tag{6.105}$$

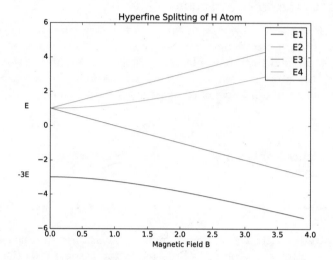

Figure 6.10. The splitting of the hydrogen triplet and singlet ground states induced by an external magnetic field.

1. Show that the Hamiltonian matrix $\langle \alpha^e \alpha^p | H_0 | \alpha^e \alpha^p \rangle$ for the state (6.105) is

$$
H_B = \begin{bmatrix} W & 0 & 0 & 0 \\ 0 & -W & 2W & 0 \\ 0 & 2W & -W & 0 \\ 0 & 0 & 0 & W \end{bmatrix}. \tag{6.106}
$$

2. Use a symbolic manipulation program, such as `simpy` in Python, to verify that the eigenvalues of this Hamiltonian are:

$$
-3W(\text{multiplicity 3, triplet state}), \quad W(\text{multiplicity 1, singlet state}). \tag{6.107}
$$

   The value predicted by [Brans(91)] for the level splitting of the 1S state was

$$
\nu = \hbar \Delta E = 4W/\hbar = 1420 \text{ MHz (predicted)}. \tag{6.108}
$$

   The value measured by [Ram & Town(21)] is:

$$
\nu = 1420.405751800 \pm 0.000000028 \text{ Hz (measured)}. \tag{6.109}
$$

   In addition to being one of the most accurately measured quantities in physics, (6.109) agrees with theory.

3. Determine the effect on the 1S states of hydrogen of introducing an external magnetic field **B** in the z direction.

a. Show that the electron-proton Hamiltonian now includes the term

$$H' = \boldsymbol{\mu}_e \cdot \mathbf{B} + \boldsymbol{\mu}_p \cdot \mathbf{B} = (\mu_e \sigma_z^e + \mu_p \sigma_x^p)B. \qquad (6.110)$$

b. Show, using the same spin states and matrices as before, that the Hamiltonian matrix for this interaction is:

$$H' = \begin{bmatrix} -(\mu_e + \mu_p)B & 0 & 0 & 0 \\ 0 & -(\mu_e - \mu_p)B & 0 & 0 \\ 0 & 0 & -(-\mu_e + \mu_p)B & 0 \\ 0 & 0 & 0 & (\mu_e + \mu_p)B \end{bmatrix}.$$

c. Add this correction to the original Hamiltonian and use a symbolic manipulation program to show that the eigenvalues of this new Hamiltonian are:

$$e_1 = -W - \sqrt{B^2 \mu_e^2 - 2B^2 \mu_e \mu_p + B^2 \mu_p^2 + 4W^2}, \qquad (6.111)$$

$$e_2 = B\mu_e + B\mu_p + W, \qquad e_3 = -B\mu_e - B\mu_p + W, \qquad (6.112)$$

$$e_4 = -W - \sqrt{B^2 \mu_e^2 - 2B^2 \mu_e \mu_p + B^2 \mu_p^2 + 4W^2}. \qquad (6.113)$$

d. Evaluate and plot these four hydrogen 1S energy-level values as a function of the external magnetic field B.
e. Check if your answers change when the proton's contribution is ignored.
f. Set $W = 1$ and plot the eigenvalues as a function of B.

Our program in Python using sympy and Matplotlib is given in Listing 6.20, with typical results shown in Figure 6.10.

6.9.5 SU(3) Symmetry of Quarks

Fundamental theories derive, and experiments confirm, that for each symmetry of nature there is an associated conserved quantity. For example, translational invariance is related to momentum conservation, and rotational invariance is related to angular momentum conservation. Likewise elementary particle interactions are observed to have a number of conserved quantities associated with them, and, presumably, these must be related to symmetries of the strong interactions. For example, Gellman and Zweig independently found that the mass values of the elementary particles can be related to each other if one assumes an internal SU(3) symmetry, the symmetry shown by Special (determinate 1), Unitary matrices of dimension 3.

The SU(3) group has eight generators of transformations that act on complex vectors in an internal 3-D space:

$$\lambda_1 = \begin{bmatrix} 0 & 1 & 0 \\ 1 & 0 & 0 \\ 0 & 0 & 0 \end{bmatrix}, \quad \lambda_2 = \begin{bmatrix} 0 & -i & 0 \\ i & 0 & 0 \\ 0 & 0 & 0 \end{bmatrix}, \quad \lambda_3 = \begin{bmatrix} 1 & 0 & 0 \\ 0 & -1 & 0 \\ 0 & 0 & 0 \end{bmatrix},$$

$$\lambda_4 = \begin{bmatrix} 0 & 0 & 1 \\ 0 & 0 & 0 \\ 1 & 0 & 0 \end{bmatrix}, \quad \lambda_5 = \begin{bmatrix} 0 & 0 & -i \\ 0 & 0 & 0 \\ i & 0 & 0 \end{bmatrix}, \quad \lambda_6 = \begin{bmatrix} 0 & 0 & 0 \\ 0 & 0 & 1 \\ 0 & 1 & 0 \end{bmatrix}, \quad (6.114)$$

$$\lambda_7 = \begin{bmatrix} 0 & 0 & 0 \\ 0 & 0 & -i \\ 0 & 1 & 0 \end{bmatrix}, \quad \lambda_8 = \frac{1}{\sqrt{3}} \begin{bmatrix} 1 & 0 & 0 \\ 0 & 1 & 0 \\ 0 & 0 & -2 \end{bmatrix}.$$

You may think of these generators as an extension of the Pauli matrices to 3-D. In analogy to spin up and down vectors forming a basis for the Pauli matrices, for SU(3) there are three basis vectors:

$$|u\rangle = \begin{bmatrix} 1 \\ 0 \\ 0 \end{bmatrix}, \quad |d\rangle = \begin{bmatrix} 0 \\ 1 \\ 0 \end{bmatrix}, \quad |s\rangle = \begin{bmatrix} 0 \\ 0 \\ 1 \end{bmatrix}. \quad (6.115)$$

These basis vectors are called "up" (p), "down" (n), and "strange" (λ) quarks. While originally imagined as just mathematical objects, we have now come to realized that these quarks represent subparticles within the elementary particles.

Just as the Pauli matrices can be combined into raising and lowering operators, the SU(3) generators can be combined into raising and lowering operators, for instance:

$$I_\pm = \frac{1}{2}(\lambda_1 \pm i\lambda_2), \quad U_\pm = \frac{1}{2}(\lambda_6 \pm i\lambda_7), \quad V_\pm = \frac{1}{2}(\lambda_4 \pm i\lambda_5). \quad (6.116)$$

Use a linear algebra software package such as Scipy to show that:

1. The SU(3) matrices are unitary ($U^\dagger U = 1$).

2. The SU(3) matrices have a determinant of 1.

3. The commutation relations for the Pauli matrices are $[\sigma_i, \sigma_j] = 2i\epsilon_{ijk}$.

4. I_+ raises d to u.

5. V_+ raises s to u.

6. U_+ raises s to d.

7. Verify the actions of the lowering operators $I_-, V_-,$ and U_-.

8. What happens when you apply I_+ to u?

9. What happens when you apply U_- to s?

10. Verify the commutation relations

$$[T_a, T_b] = i \sum_{c=1}^{8} f_{abc} T_c \qquad (6.117)$$

where the f's are the structure constants

$$f_{123} = 1, \qquad f_{458} = f_{678} = \tfrac{\sqrt{3}}{2}, \qquad (6.118)$$
$$f_{147} = -f_{156} = f_{246} = f_{257} = f_{345} = -f_{367} = \tfrac{1}{2}, \qquad (6.119)$$

with other $f_{ijk} = 0$ unless they are related to the above by permutation.

11. Verify the anticommutation relations

$$\{T_a, T_b\} = \tfrac{1}{3}\delta_{ab} + \sum_{c=1}^{8} d_{abc} T_c, \qquad (6.120)$$

where the d's are symmetric and take the values

$$d_{118} = d_{228} = d_{338} = -d_{888} = \tfrac{1}{\sqrt{3}}, \qquad (6.121)$$
$$d_{448} = d_{558} = d_{668} = d_{778} = \tfrac{-1}{2\sqrt{3}}, \qquad (6.122)$$
$$d_{146} = d_{157} = -d_{247} = d_{256} = d_{344} = d_{355} = -d_{366} = -d_{377} = \tfrac{1}{2}. \qquad (6.123)$$

12. Verify that there is an invariant quadratic Casimir operator

$$\sum_{i=1}^{8} \lambda_i \lambda_i = 16/3. \qquad (6.124)$$

The program SU3.py in Listing 6.22 performs some of these operations.

6.10 Coherent States and Entanglement

6.10.1 Glauber Coherent States

A particle within a harmonic oscillator (HO) potential is described by the Hamiltonian

$$H = \tfrac{1}{2}p^2 + \tfrac{1}{2}\omega_0^2 q^2, \qquad (6.125)$$

where we have set $m = 1$. This Hamiltonian can also be written as

$$H = \hbar\omega_0(aa^\dagger + \tfrac{1}{2}), \quad a = \frac{1}{\sqrt{2\hbar\omega_0}}(\omega_0 q + ip), \quad a^\dagger = \frac{-1}{\sqrt{2\hbar\omega_0}}(\omega_0 q - ip), \qquad (6.126)$$

where a and $a^\dagger$ are operators that annihilate and create a quanta, respectively. Because these operators acting on the HO basis $|n\rangle$ change the number n of quanta present, the HO basis vectors are not eigenstates of the annihilation operator:

$$a\,|n\rangle = \sqrt{n}\,|n-1\rangle\,, \quad a^\dagger\,|n\rangle = \sqrt{n+1}\,|n+1\rangle\,. \tag{6.127}$$

However, if we construct a "coherent state" as a linear superposition of HO basis state,

$$|\alpha\rangle = e^{-\alpha^2/2} \sum_{n=0}^{\infty} \frac{\alpha^n}{\sqrt{(n!)}}\,|n\rangle\,, \tag{6.128}$$

then it can be verified by substitution that this state is an eigenstate of both the annihilation and creation operators [Hartley(82), Stephen(87), Carruthers & Nieto(65)]:

$$a\,|\alpha\rangle = \alpha\,|\alpha\rangle\,, \qquad H\,|\alpha\rangle = (\alpha^2 + \tfrac{1}{2})\,|\alpha\rangle\,, \tag{6.129}$$

as well as the Hamiltonian. Such states have some classical-like properties and are useful in describing the quantum state of lasers. In fact, as we shall see, a coherent state behaves much like a wavepacket in which its position and momentum change with time, though not its shape.

To investigate the spatial dependence of these coherent states, recall the coordinate space solution of the time-independent Schrödinger equation for the HO is:

$$\langle x\,|n\rangle = H_n(\beta x)\,e^{-\beta^2 x^2/2}\,, \qquad \beta^2 = \sqrt{k}/\hbar, \tag{6.130}$$

where k is the elastic constant and H_n is the Hermite polynomial of order n. Since the coherent state can be expanded in the HO basis, it is straightforward to include the time dependence:

$$|\alpha, t\rangle = e^{-\alpha^2/2} \sum_{n=0}^{\infty} \frac{\alpha^n}{(n!)^{1/2}} e^{-i\omega_0(n+\frac{1}{2})t}\,|n\rangle\,, \tag{6.131}$$

where $\omega_0 = \sqrt{k/m}$ is the classical frequency.

1. Verify by substitution that the Glauber state $|\alpha\rangle$ is an eigenstate of the annihilation operator as well as the Hamiltonian.

2. Write a program that constructs a time-dependent coherent state (6.131) by summing up to $n = n_{max}$.

3. Use $E_\alpha = \alpha^2 + 1/2$, to find α for a given energy, and $m = k = \omega_0 = \hbar = 1$, and $n_{max} = 5$ to start.

4. Plot the coherent state as time progresses.

5. Increase the value of n_{max} until a stable wavepacket results.

6. What is the normalization of the coherent state?

7. How does the normalization change with time?

Our program GlauberState.py is given in Listing 6.19.

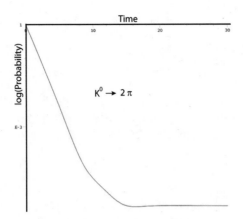

Figure 6.11. A semilog plot of the probability of a neutral kaon decaying into two pions as function of time. Note the change in slopes and the deviation from a straight line.

6.10.2   Neutral Kaons as Superpositions of States

One way physicists have tried to introduce some order into the zoo of elementary particles is by observing what appears to be a conserved quantity, and then relating that quantity to an underlying symmetry in the particles' interactions. An example of this is the pair of neutral kaons, K^0 and $\overline{K^0}$. These two particles are antiparticles of each other, are spinless, have the same mass, though with the K^0 having strangeness $+1$ while the $\overline{K^0}$ has strangeness -1. The differing strangeness, being conserved by the strong and electromagnetic interactions, means that these interactions cannot connect one particle to the other. Consequently, the K^0 and $\overline{K^0}$ states are eigenstates of the strong and electromagnetic interaction, and so retain their identities when only these two interactions dominate. However, since strangeness is not conserved by the weak interaction, it can lead to a slow and indirect conversion between the particles [Fraunfelder & Henley(91)]:

$$K^0 \rightleftharpoons 2\pi \rightleftharpoons \overline{K^0}. \tag{6.132}$$

Even though the weak interaction does not conserve strangeness, experiments indicated, and theoretical considerations supported, the belief that the weak interaction does conserve the CP, the combined operation of charge conjugation and parity reflection. However, the strongly interacting version of these particles, K^0 and $\overline{K^0}$, get converted into each other by the CP operation:

$$CP|K^0\rangle = -|\overline{K^0}\rangle, \qquad CP|\overline{K^0}\rangle = -|K^0\rangle. \tag{6.133}$$

Consequently, an eigenstate of CP, and , accordingly, the type of state that decays via the weak interaction, can be constructed from (6.133) as:

$$|K_1\rangle = |K^0\rangle + |\overline{K^0}\rangle \qquad\qquad |K_2\rangle = |K^0\rangle - |\overline{K^0}\rangle, \qquad (6.134)$$
$$\Rightarrow \quad CP|K_1\rangle = +|K_1\rangle, \qquad\qquad CP|K_2\rangle = -|K_2\rangle. \qquad (6.135)$$

The important point to remember here is that K^0 and $\overline{K^0}$ are created by the strong interaction, but that they decay via the weak interaction. Accordingly, the particles that decay are actually the CP eigenstates K_1 and K_2 that conserve CP in their decays:

$$K_1 \to 2\pi, \qquad K_2 \to 3\pi, \qquad (6.136)$$

where the pion is an eigenstate of CP with negative eigenvalue, $CP|\pi\rangle = -|\pi\rangle$. Since there is a more limited phase space (energy-momentum states) available for decay into three pions than into two pions, the 2π decay should be faster, and indeed [Patrignani et al.l(16)]:

$$\tau(K_1) \simeq 0.90 \times 10^{-10}s, \qquad \tau(K_2) \simeq 5.2 \times 10^{-8}s. \qquad (6.137)$$

So now we can work backwards. An experiment starts with a strong interaction that creates K^0 and $\overline{K^0}$. These initial states can be expressed in terms of K_1 and K_2 as:

$$|K^0\rangle = |K_1\rangle + K_2\rangle, \qquad |\overline{K^0}\rangle = |K_1\rangle - |K_2\rangle. \qquad (6.138)$$

Yet since K_1 and K_2 decay with different lifetimes, we expect these states to have the mixed time dependence

$$|K^0(t)\rangle = |K_1\rangle e^{-\frac{t}{2\tau_1}} + |K_2\rangle e^{-\frac{t}{2\tau_2}}, \qquad |\overline{K^0}\rangle = |K_1\rangle e^{-\frac{t}{2\tau_1}} - |K_2\rangle e^{-\frac{t}{2\tau_2}}, \qquad (6.139)$$

where we are using units in which $\hbar = c = 1$. This mixture of lifetimes means that a beam of neutral kaons will first decay into two pions, and then, after moving along a significant distance, decay into three pions.

1. If a neutral kaon beam is moving with $v = c/2$, how far will it travel, on the average, before exhibiting two pion decays?

2. If a neutral kaon beam is moving with $v = c/2$, how far will it travel, on the average, before exhibiting three pion decays?

3. Plot the probability (unnormalized) of K^0 decay as a function of time. Is this purely exponential?

4. Plot the probability (unnormalized) of $\overline{K^0}$ decay as a function of time. Is this purely exponential?

The peculiarities of neutral kaons never seem to cease. Continuing experiments have revealed that, though rare, a K_1 sometimes decays into 3π's, and, though rare, sometimes a K_2 decays into 2π's. This means that CP symmetry is not completely conserved by the weak interaction, and that the physical states which decay are combinations of CP eigenstates:

$$|K_S\rangle = |K_1\rangle + \epsilon|K_2\rangle, \qquad |K_L\rangle = |K_2\rangle - \epsilon|K_1\rangle, \tag{6.140}$$

where S and L refer to the short- and long-lifetime states. Experiment indicates that ϵ is small, $\epsilon \simeq 0.0023$. In addition, this violation of CP means that K_S and K_L have slightly different masses,

$$\frac{m_S - m_L}{\langle m \rangle} = \frac{\Delta m}{498 MeV} \simeq 6 \times 10^{-19}. \tag{6.141}$$

While it may appear that this small a detail may be of much importance, it is a matter-antimatter asymmetry that may play a part in big bang nucleosynthesis, and for this reason be related to why our universe is mainly matter, with little antimatter. In addition, this small mass difference means that the rest energies of the particle and antiparticle differ ever so slightly, which in turn means that their state vectors, in addition to having different exponential time decays, also have different oscillatory time dependences:

$$|K^0(t)\rangle = |K_S\rangle e^{-(im_S + 1/\tau_S)t/2} + |K_L\rangle e^{-(im_L + 1/\tau_L)t/2}, \tag{6.142}$$

$$|\overline{K^0}(t)\rangle = |K_S\rangle e^{-(im_S + 1/\tau_S)t/2} - |K_L\rangle e^{-(im_L + 1/\tau_L)t/2}. \tag{6.143}$$

1. By what percentage are K_S and K_L not true eigenstates of CP?

2. Calculate an analytic expression for the probability (unnormalized) of the decay of a K_S into two pions. (*Hint:* It should contain both decaying exponentials and a periodic term.)

3. Plot the probability of the decay of a K^0 into two pions as a function of time. You may want to use the value $\Delta m = 0.53 \times 10^{10}/s$ so you can measure time in seconds. Our result is shown in Figure 6.11. Is this purely exponential?

6.10.3   Double Well Transitions

Given two identical potential wells with a barrier of infinite height and finite width between them, as seen on the left of Figure 6.12, a particle placed in one well will remain there forever. However, if there is a perturbation ΔE that lowers the barrier between them, as seen on the right of Figure 6.12, then one can expect that the bound particle will make a transition from the left side to the right side with a probability [Fraunfelder & Henley(91)]:

$$\text{Probability}(L \rightarrow R) = \sin^2(\Delta E t/\hbar). \tag{6.144}$$

This is the essence of the oscillations, examined in §6.10.2, between one kind of neutral kaon and the other.

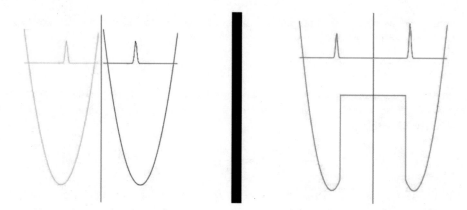

Figure 6.12. Output from the code `TwoWells.py` that solves the time-dependent Schrödinger equation. On the left we have one wavepacket confined to the left well, and another confined to the right well. On the right we have a single well with a barrier in the middle, and two independent wave packets in that well.

1. Write a simulation that creates an animation of a wavepacket moving between two wells. We have created such simulations using the code `TwoWells.py` in Listing 6.18, with Figure 6.12 showing a single frame from the animation.

   a. In realistic situations, such as kaon oscillations, the bound particle would effectively bounce around inside its potential well a large number of times before a transition occurs. This is a consequence of the perturbation connecting the two wells being small relative to the energy of the states. As you are likely to see in the simulations, the wavepacket tends to break up into component pieces after just a few collisions. This is due in part to a break-down of the numerics. Accordingly, you may need to make the time step Δt used in the leapfrog algorithm very small, which in turn means that your code will have to run for a proportionally longer time. And, of course, when you change the size of Δt, you may also need to adjust the space step size Δx so as still to respect the Courant condition (4.6).

   b. Explore how the transition from one well to another changes as the energy of the particle changes. Barrier penetration tends to be highly sensitive to the value of the particle's energy.

   c. Explore how the transition from one well to another changes as the width of the potential barrier changes. Barrier penetration tends to be highly sensitive to the potential barrier's width.

   d. Explore how the transition from one well to another changes as the height of the potential barrier changes. Barrier penetration tends to be highly sensitive to the potential barrier's height.

6.10.4 Qubits

In the classical model of a computer's memory, the bit 0 is stored in an "up" magnetic core, and the bit 1 in a "down" core. In quantum computing, information is stored in a two state system known as a quantum bit or *qubit*, for example, two spin states of a particle [McMahon(08), Kaye et al.(07), Benenti et al.(04)]. Because the quantum system can be in one state, or in the other state, or in a superposition of the two states, this is clearly different from the classical model.

Furthermore, qubits can interact with each other, which leads to qubits being *entangled*. When we say that two quantum systems A and B are *entangled*, we mean that there is a correlation between the values of some properties of A with the corresponding properties of B. Consider a composite state of two qubits. If the qubits are prepared independently of each other, and each one is isolated from the other, then each one forms a closed system. If the two qubits can interact, though are otherwise closed systems, then the qubits are *entangled*. If two systems are not entangled, then they are *separable*. A formal criterion is:

> If state $|\alpha\rangle$ belongs to a Hilbert space H_1, and state $|\beta\rangle$ belongs to a Hilbert space H_2, then the states are entangled if the state H in the Hilbert space of both states cannot be expressed as a tensor (direct) product of the two states.

For example, if the two states are spin up and spin down, the up state can be physically far from the down state, yet still be entangled. This means that the up state cannot be described by just a single particle state, but requires the entire state vector. So if the total state has spin zero and one particle is spin up, the other, far away, must be correlated and have spin down. Just how the two particles communicate with each other is unclear. And so, the state

$$|\psi_2\rangle = \frac{1}{\sqrt{2}}\left(|01\rangle + |10\rangle\right) \tag{6.145}$$

is *separable* because it can be expressed as the tensor product:

$$|\psi_2\rangle = \frac{1}{\sqrt{2}}\left(|0\rangle + |1\rangle\right) \otimes |1\rangle. \tag{6.146}$$

Nevertheless, the state

$$|\psi_1\rangle = \frac{1}{\sqrt{2}}\left(|00\rangle + |11\rangle\right), \tag{6.147}$$

cannot be so expressed, and is , accordingly, *entangled*. More formally, given two arbitrary $\mathbb{C}^2$ states,

$$|\phi\rangle = \begin{pmatrix} a \\ b \end{pmatrix}, \qquad |\chi\rangle = \begin{pmatrix} c \\ d \end{pmatrix}, \tag{6.148}$$

the tensor or direct product of these states is

$$|\Psi\rangle \equiv |\phi\rangle \otimes |\phi\rangle = \begin{pmatrix} a \\ b \end{pmatrix} \otimes \begin{pmatrix} c \\ d \end{pmatrix} = \begin{pmatrix} ac \\ ad \\ bc \\ bd \end{pmatrix}. \qquad (6.149)$$

This direct product belongs to the 4-D Hilbert space of complex numbers $\mathbb{C}^4$. A way of determining if a state is entangled here is to consider a composite system in $\mathbb{C}^4$:

$$|\psi\rangle = \begin{pmatrix} w \\ x \\ y \\ z \end{pmatrix}. \qquad (6.150)$$

The state is *separable* if and only if $wz = xy$, which for this state means $acbd = adbc$, which is the case here.

   The problems to follow are similar to those in §6.9.4 dealing with hyperfine structure of hydrogen, except that here we follow the notation of quantum computing in which states get identified with a zero and 1 bit.

   Two interacting magnetic dipoles A and B are separated a distance r. The dipole-dipole interaction Hamiltonian has the form

$$H = \frac{\mu^2}{r^3}(\vec{\sigma}_A \cdot \vec{\sigma}_B - 3Z_A Z_B), \qquad Z_A = \vec{\sigma}_A \cdot \hat{r}, \quad Z_B = \vec{\sigma}_B \cdot \hat{r}. \qquad (6.151)$$

The states on which this Hamiltonian acts are:

$$|0_A 0_B\rangle = |0_A\rangle|0_B\rangle, \quad |0_A 1_B\rangle = |0_A\rangle|1_B\rangle, \quad |1_A 0_B\rangle = |1_A\rangle|0_B\rangle, \quad |1_A 1_B\rangle = |1_A\rangle|1_B\rangle.$$

Here $\vec{\sigma}_A = X_A \hat{\mathbf{i}} + Y_A \hat{\mathbf{j}} + Z_A \hat{\mathbf{k}}$, and, in the notation of quantum computing, X, Y, and Z are the Pauli matrices:

$$X \equiv \sigma_x = \begin{pmatrix} 0 & 1 \\ 1 & 0 \end{pmatrix}, \quad Y \equiv \sigma_y = \begin{pmatrix} 0 & -i \\ i & 0 \end{pmatrix}, \quad Z \equiv \sigma_z = \begin{pmatrix} 1 & 0 \\ 0 & -1 \end{pmatrix}. \qquad (6.152)$$

These Pauli matrices act in the internal spin space on the up and down states (again in quantum computing notation on the 0 and 1 bits):

$$|0\rangle \equiv |\uparrow\rangle = \begin{bmatrix} 1 \\ 0 \end{bmatrix}, \quad |1\rangle \equiv |\downarrow\rangle = \begin{bmatrix} 0 \\ 1 \end{bmatrix}. \qquad (6.153)$$

Accordingly, the actions of the Pauli matrices on these qubit states are:

$$X|0\rangle = |1\rangle, \quad X|1\rangle = +|0\rangle, \quad Y|0\rangle = i|1\rangle, \quad Y|1\rangle = -i|0\rangle, \quad Z|0\rangle = |0\rangle, \quad Z|1\rangle = -|1\rangle.$$

Note that states $|0_A\rangle$ and $|1_A\rangle$ form a basis in a Hilbert space $\mathbb{C}^2$ of 2-D complex numbers, while states $|0_B\rangle$ and $|1_B\rangle$ form a basis in separate $\mathbb{C}^2$ Hilbert space.

1. Show that these direct products form a basis for $\mathbb{C}^4$:

$$|00\rangle = \begin{pmatrix} 1 \\ 0 \end{pmatrix} \otimes \begin{pmatrix} 1 \\ 0 \end{pmatrix}, \qquad |01\rangle = \begin{pmatrix} 1 \\ 0 \end{pmatrix} \otimes \begin{pmatrix} 0 \\ 1 \end{pmatrix}, \qquad (6.154)$$

$$|10\rangle = \begin{pmatrix} 0 \\ 1 \end{pmatrix} \otimes \begin{pmatrix} 1 \\ 0 \end{pmatrix}, \qquad |11\rangle = \begin{pmatrix} 0 \\ 1 \end{pmatrix} \otimes \begin{pmatrix} 0 \\ 1 \end{pmatrix}. \qquad (6.155)$$

Hint: For example, $|11\rangle = \begin{pmatrix} 0 \\ 0 \\ 0 \\ 1 \end{pmatrix}$.

2. If P and Q are the operators

$$P = \begin{pmatrix} p_{11} & p_{12} \\ p_{21} & p_{22} \end{pmatrix}, \qquad Q = \begin{pmatrix} q_{11} & q_{12} \\ q_{21} & q_{22} \end{pmatrix}, \qquad (6.156)$$

in separate Hilbert spaces, show that their tensor product is

$$P \otimes Q = \begin{pmatrix} p_{11}q_{11} & p_{11}q_{12} & p_{12}q_{11} & p_{12}q_{12} \\ p_{11}q_{21} & p_{11}q_{22} & p_{12}q_{21} & p_{12}q_{22} \\ p_{21}q_{11} & p_{21}q_{12} & p_{22}q_{11} & p_{22}q_{12} \\ p_{21}q_{21} & p_{21}q_{22} & p_{22}q_{21} & p_{22}q_{22} \end{pmatrix}. \qquad (6.157)$$

3. Show that the Hamiltonian (6.151), when represented in the direct product space, takes the form

$$H = \frac{\mu^2}{r^3}(X_A \otimes X_B Y_A \otimes Y_B + Z_A \otimes Z_B - 3Z_A \otimes Z_B). \qquad (6.158)$$

4. Show that the tensor product:

$$X_A \otimes X_B = \begin{pmatrix} 0 & 0 & 0 & 1 \\ 0 & 0 & 1 & 0 \\ 0 & 1 & 0 & 0 \\ 1 & 0 & 0 & 0 \end{pmatrix}. \qquad (6.159)$$

5. Evaluate the tensor products $Y_A \otimes Y_B$ and $Z_A \otimes Z_B$ as 4×4 matrices.

6. Show that the Hamiltonian in the direct product space is:

$$H = \frac{\mu^2}{r^3} \begin{pmatrix} -2 & 0 & 0 & 0 \\ 0 & 2 & 2 & 0 \\ 0 & 2 & 2 & 0 \\ 0 & 0 & 0 & -2 \end{pmatrix}. \qquad (6.160)$$

7. Use a linear algebra package to show that the eigenvalues of H are 4, 0, -2, and -2, and that the corresponding eigenvectors are:

$$\phi_1 = \frac{1}{\sqrt{2}} \begin{pmatrix} 0 \\ 1 \\ 1 \\ 0 \end{pmatrix} = \frac{|01\rangle + |10\rangle}{\sqrt{2}}, \qquad \phi_2 = \begin{pmatrix} 0 \\ 0 \\ 0 \\ 1 \end{pmatrix} = |11\rangle, \qquad (6.161)$$

$$\phi_4 = \frac{1}{\sqrt{2}} \begin{pmatrix} 0 \\ 1 \\ -1 \\ 0 \end{pmatrix} = \frac{|01\rangle - |10\rangle}{\sqrt{2}}, \qquad \phi_3 = \begin{pmatrix} 1 \\ 0 \\ 0 \\ 0 \end{pmatrix} = |00\rangle. \qquad (6.162)$$

8. Recall the discussion of entanglement. Of the four eigenstates just obtained, determine which ones are separable and which ones are entangled.

9. Using these eigenvectors states as new basis states, express the Hamiltonian matrix H in these basis states, that is, evaluate the matrix elements in

$$H = \begin{pmatrix} \langle\phi_1|H|\phi_1\rangle & \langle\phi_1|H|\phi_2\rangle & \langle\phi_1|H|\phi_3\rangle & \langle\phi_1|H|\phi_4\rangle \\ \langle\phi_2|H|\phi_1\rangle & \langle\phi_2|H|\phi_2\rangle & \langle\phi_2|H|\phi_3\rangle & \langle\phi_2|H|\phi_4\rangle \\ \langle\phi_3|H|\phi_1\rangle & \langle\phi_3|H|\phi_2\rangle & \langle\phi_3|H|\phi_3\rangle & \langle\phi_3|H|\phi_4\rangle \\ \langle\phi_4|H|\phi_1\rangle & \langle\phi_4|H|\phi_2\rangle & \langle\phi_4|H|\phi_3\rangle & \langle\phi_4|H|\phi_4\rangle \end{pmatrix}. \qquad (6.163)$$

If you have done this correctly, the Hamiltonian should now be diagonal with the eigenvalues as the diagonal elements.

In Listing 6.21 we present the program `Entangle.py` that performs the necessary linear algebra using the `numpy` package.

6.11   Feynman Path Integral Quantum Mechanics ⊙

Feynman was looking for a formulation of quantum mechanics different from the Schrödinger picture in which there would be a more direct connection to classical mechanics. It is said that Dirac suggested Hamilton's principle of least action as a possible $\hbar \to 0$ limit of a quantum least-action principle [Sakurai(67)]. Seeing that Hamilton's principle deals with the paths of particles through space-time, Feynman postulated that the quantum-mechanical wave function describing the propagation of a free particle from the space-time point $a = (x_a, t_a)$ to the point $b = (x_b, t_b)$ (Figure 6.13 left) can expressed as [Feynman & Hibbs(65), Mannheim(83)]:

$$\psi(x_b, t_b) = \int dx_a G(x_b, t_b; x_a, t_a)\psi(x_a, t_a), \qquad (6.164)$$

where G is the *Green's function* or *propagator*

$$G(x_b, t_b; x_a, t_a) \equiv G(b, a) = \sqrt{\frac{m}{2\pi i(t_b - t_a)}} \exp\left[i\frac{m(x_b - x_a)^2}{2(t_b - t_a)}\right]. \qquad (6.165)$$

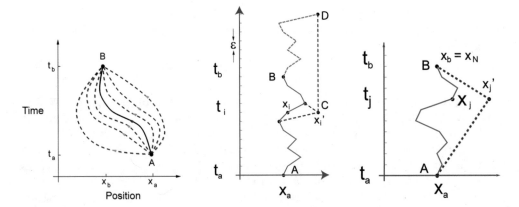

Figure 6.13. *Left:* A collection of paths connecting the initial space-time point A to the final point B. The solid line is the classical trajectory that minimizes the action S. The dashed lines are additional paths sampled by a quantum particle. *Middle:* A path through a space-time lattice that starts and ends at $x = x_a = x_b$. The dotted path BD is a transposed replica of path AC. *Right:* The dashed path joins the initial and final times in two equal time steps; the solid curve uses N steps each of size ε. The position of the curve at time t_j defines the position x_j.

Equation (6.164) is a form of Huygens's wavelet principle in which the new wave $\psi(x_b, t_b)$ is created by summing over a spherical wavelet $G(b; a)$ emitted by each point on the wavefront $\psi(x_a, t_a)$.

Hamilton's principle deals with the classical *action* S taken as the line integral of the Lagrangian along a particle's path:

$$S[\bar{x}(t)] = \int_{t_a}^{t_b} dt\, L\,[x(t), \dot{x}(t)]\,, \quad L = T\,[x, \dot{x}] - V[x]. \tag{6.166}$$

Here T is the kinetic energy, V is the potential energy, $\dot{x} = dx/dt$, and square brackets indicate a *functional*. Feynman imagined that another way of interpreting (6.164) is as a form of Hamilton's principle in which the wave function ψ for a particle to be at B equals the sum over all *paths* through space-time originating at time A and ending at B (Figure 6.13). This view incorporates the statistical nature of quantum mechanics by having different probabilities for travel along different paths, with the most likely path being the classical trajectory because it minimizes the action S. The probability for a path is determined by its Green's function, which is related to the action:

$$G(b, a) = \sqrt{\frac{m}{2\pi i(t_b - t_a)}}\, e^{iS[b,a]/\hbar}, \tag{6.167}$$

$$S[b, a] = \frac{m}{2}\,(\dot{x})^2\,(t_b - t_a) = \frac{m}{2}\frac{(x_b - x_a)^2}{t_b - t_a}. \tag{6.168}$$

Feynman extended these ideas as a new view of quantum mechanics in which its statistical nature arises from varying probabilities for a particle to take different *paths* from a to b,

$$G(b, a) = \sum_{\text{paths}} e^{iS[b,a]/\hbar} \quad \text{(path integral)}. \tag{6.169}$$

This expression is called a *path integral* because it sums over all paths in Figure 6.13 left, with the classical action (6.166) for each path computed as an integral along the path. The connection between classical and quantum mechanics is the realization that since $\hbar \simeq 10^{-34}$Js, the action is a very large number, $S/\hbar \geq 10^{20}$, and so even though all paths enter into the sum (6.169), the main contributions come from those paths adjacent to the classical trajectory $\bar{x}$, where the action is a minimum.

As the references derive, there is connection between the bound state wave function and the Green's function in this theory:

$$|\psi_0(x)|^2 = \lim_{\tau \to \infty} e^{E_0\tau} G(x, -i\tau; x, 0), \tag{6.170}$$

where the limit is for large imaginary times. The use of imaginary times changes the Lagrangian function into a Hamiltonian function in the expression for the action, which in turn leads to there being an energy in the exponentials:

$$|\psi_0(x)|^2 = \frac{1}{Z} \lim_{\tau \to \infty} \int dx_1 \cdots dx_{N-1} e^{-\varepsilon\mathcal{E}}, \qquad Z = \lim_{\tau \to \infty} \int dx dx_1 \cdots dx_{N-1} e^{-\varepsilon\mathcal{E}}. \tag{6.171}$$

This is what we want to compute. However, because there is still more theory and implementation needed, which we will not discuss, for this problem we will give you a code with which to explore the path integral formulation of quantum mechanics. Basically, the program sets up a lattice in space and time so that positions and times are discrete, with integrals evaluated as sums over values at the lattice points, and derivatives evaluated as the differences in values at successive lattice points. A Metropolis algorithm (§7.4) is used to vary the trajectory and search for the state with the lowest energy. The classical trajectory remains most likely, with nearby paths being less likely. The program `QMC.py` is given in Listing 6.23 and outlined in the pseudocode below.

```
# Pseudocode for QMC.py: Quantum MonteCarlo (path integration)

Import packages
Set constants, initialize array path to zeros & prob to zeros
Define function  PlotAxes                          # Axis path plot
Define function WaveFunctionAxes            # Axis Psi plot
Define function Energy                              # Sum link energies
    E = PE + KE = (Delta x)^2 + x^2
Define function PlotPath:                       #  Paths values for plot
Define function PlotWF                          # Psi values for Plot
Set up plots
Eold = Energy(path)                                # E initial
```

```
while true                              # Set random path elements
    pick random element
    pick random change for element
    change path
    calculate new energy
    if  Enew < Eold & e^(Eold - Enew) < random number
            reject change              # Metropolis
            plot path
            accumulate wave function values
            plot wave function
```

1. Plot some of the actual space-time paths used in the simulation along with the classical trajectory.

2. Make the x lattice spacing smaller and see if this leads to a more continuous picture of the wave function.

3. Sample more points (run longer) with smaller time step ε, and see if this leads to a more precise value for the wave function.

4. Because there are no sign changes in a ground-state wave function, you can ignore the phase, assume $\psi(x) = \sqrt{\psi^2(x)}$, and then estimate the energy via

$$E = \frac{\langle\psi|\,H\,|\psi\rangle}{\langle\psi|\psi\rangle} = \frac{\omega}{2\langle\psi|\psi\rangle} \int_{-\infty}^{+\infty} \psi^*(x)\left(-\frac{d^2}{dx^2} + x^2\right)\psi(x)dx, \qquad (6.172)$$

where the space derivative is evaluated numerically.

5. Explore the effect of making $\hbar$ larger and hence permitting greater fluctuations around the classical trajectory. Do this by decreasing the value of the exponent in the Boltzmann factor. Determine if this makes the calculation more or less robust in its ability to find the classical trajectory.

6.12   Code Listings

```
# HOnumeric.py: 1-D HO wave functions via rk4

import numpy as np, matplotlib.pylab as plt
from rk4Algor import rk4Algor

rVec = np.zeros((1000),float)              # x values for plot
psiVec = np.zeros((1000),float)          # Wave function values
fVec = np.zeros(2)
y = np.zeros((2))
n = 6                                            # n = npr L+1

def f(x,y):                                        # ODE RHS
    fVec[0] = y[1]
```

```
      fVec[1] = -(2*n+1-x**2)*y[0]
      return fVec

if(n%2==0):    y[0]=1e-8                                      # Set parity
else:  y[0]=-1e-8
y[1] = 1.;  i = 0
f(0.0,y)                                                      # RHS at r = 0
dr = 0.01
for r in np.arange(-5,5,dr):              # Compute WF steps of dr
    rVec[i] = r
    y = rk4Algor(r, dr, 2, y, f)
    psiVec[i] = y[0]
    i = i+1                                                   # Advance i & r
plt.figure()
plt.plot(rVec,psiVec)
plt.grid()
plt.title('Harmonic Oscillator Wave Function n = 6')
plt.xlabel('x')
plt.ylabel('$\psi(x)$')
plt.show()
```

Listing 6.1. **HOnumeric.py** solves the Schrödinger equation numerically for the 1-D harmonic oscillator wave function using the rk4 algorithm and arbitrary initial conditions.

```
# HOanal.py: symbolic soltn of HO ODE using sympy

from sympy import *

f, g = symbols('f g',cls=Function)          # makes f a function
t, kap, w0 = symbols('t kap w0')
f(t)
f(t).diff(t)
# The ODE
diffeq = Eq(f(t).diff(t,t) + kap*(f(t).diff(t)) + (w0*w0)*f(t))
print ("\n ODE to be solved:")
print( diffeq)
print( "\n Solution of ODE:")
ff = dsolve(diffeq,f(t))      # Solves ODE
F = ff.subs(t,0)
print( ff)
```

Listing 6.2. **HOanal.py** solves the Schrödinger equation for the 1-D harmonic oscillator wave function using symbolic manipulation package Sympy.

```
# QuantumEigenCall.py: Finds E & psi via rk4 + bisection

from numpy import *; from rk4Algor import rk4Algor
import numpy as np, matplotlib.pyplot as plt

# m/(hbar*c)**2 = 940MeV/(197.33MeV-fm)**2 = 0.4829
eps = 1e-1; Nsteps = 501;  h=0.04; Nmax = 100 # Params
E = -17.; Emax = 1.1*E;   Emin = E/1.1
```

```
def f(x, y):                                # RHS for ODE
    global E
    F = zeros((2), float)
    F[0] = y[1]
    F[1] = -(0.4829)*(E-V(x))*y[0]
    return F

def V(x):                                   # Potential
    if (abs(x) < 10.):  return (-16.0)
    else:               return (0.)

def diff(h,E):                              # Change in log deriv
    #global E
    y = zeros((2),float)
    i_match = Nsteps//3                     # Matching radius
    nL = i_match + 1
    #print('nL',nL)
    y[0] = 1.E-15;                          # Initial left wf
    y[1] = y[0]*sqrt(-E*0.4829)
    for ix in range(0,nL + 1):
        x = h * (ix  -Nsteps/2)
        y = rk4Algor(x, h, 2, y, f)
    left = y[1]/y[0]                        # Log  derivative
    y[0] = 1.E-15;              # Slope for even; reverse if odd
    y[1] = -y[0]*sqrt(-E*0.4829)            # Initialize R wf
    for ix in range( Nsteps,nL+1,-1):
        x = h*(ix+1-Nsteps/2)
        y = rk4Algor(x, -h, 2, y, f)
    right = y[1]/y[0]                       # Log derivative
    return( (left - right)/(left + right) )

def plot(h):                    # Repeat integrations for plot
    global xL,xR,Rwf,Lwf
    x = 0.                                  # Matching radius
    Lwf=[]                                  # Left wave function
    Rwf=[]                                  # Right wave function
    xR=[]                                   # x for right wf
    xL=[]
    Nsteps = 1501                           # Integration steps
    y = zeros((2),float)
    yL = zeros((2,505),float)
    i_match = 500                           # Matching radius
    nL = i_match + 1;
    print('nL',nL)
    y[0] = 1.E-40                           # Initial left wf
    y[1] = -sqrt(-E*0.4829) *y[0]
    for ix in range(0,nL+1):
        yL[0][ix] = y[0]
        yL[1][ix] = y[1]
        x = h * (ix -Nsteps/2)
        y = rk4Algor(x, h, 2, y, f)
    y[0] = -1.E-15              # - slope: even; reverse for odd
    y[1] = -sqrt(-E*0.4829)*y[0]
    for ix in range(Nsteps -1,nL + 2,-1):       # Right WF
        x = h * (ix + 1 -Nsteps/2)              # Integrate in
        y = rk4Algor(x, -h, 2, y, f)
        xR.append(x)
        Rwf.append (y[0] )
    x = x-h
    normL = y[0]/yL[0][nL]
    for ix in range(0,nL+1):    # Normalize L wf & derivative
        x = h * (ix-Nsteps/2 + 1)
        y[0] = yL[0][ix]*normL
```

```
            y [ 1 ]  =  yL [ 1 ] [ ix ] * normL
            xL . append ( x )
            Lwf . append ( y [ 0 ] )            # Factor for scale

    fig = plt . figure ()
    ax = fig . add_subplot (111)
    ax . grid ()            #j +=1

    for count in range (0 , Nmax ) :            # Main program
        E = ( Emax + Emin ) / 2 .            # Bisec E range
        Diff = diff ( h , E )
        Etemp = E
        E = Emax
        diffMax = diff ( h , E )
        E = Etemp
        if ( diffMax * Diff > 0 ) : Emax = E    # Bisection algor
        else :                      Emin = E
        print ("Iteration , E =" , count , E )
        if ( abs ( Diff ) < eps ) :     break
        if count >3:
            fig . clear ()            # Erase previous figure
            plot ( h )
            plt . plot ( xL , Lwf )
            plt . plot ( xR , Rwf )
            plt . text (3 , -200 , 'Energy= %10.4f '%( E ) , fontsize =14)
            plt . pause (0.8)  # Pause to delay figures

        plt . xlabel ('x')
        plt . ylabel ('$\psi(x) $' , fontsize =18)
        plt . title ('R & L Wavefunctions Matched at x = 0')

    print ("Final eigenvalue E =" , E )
    print ("Iterations = " , count ," , max = " , Nmax )
    plt . show ()
```

Listing 6.3. QuantumEigenCall.py solves the Schrödinger equation for arbitrary potential by matching inner and outer wave functions.

```
# QuantumNumerov.py: quantum bound state via Numerov algorithm
# General method, but here for HO V(x)= 9.4*x*x/2
# hbarc* omega=hbarc*sqrt(k/m)=19.733,  r mc**2=940 MeV, k=9.4
# E =(N+1/2)hbarc*omega = (N+1/2)19.733, N even, change N odd

from numpy import *
import numpy as np, matplotlib.pyplot as plt

n = 1000; m = 2;  imax = 100;   Xleft0 = -10; Xright0 = 10
h  = 0.02; amin= 81.; amax= 92.;  e = amin;  de = 0.01
eps= 1e-4; im = 500; nl = im + 2;  nr = n - im + 2
xmax = 5.0
print("nl, nr",nl, nr)
print(h)

xLeft = arange(-10,0.02,0.02); xRight = arange(10,0.02,-0.02)
xp = arange(-10,10,0.02)        # Bisection interval
uL =  zeros((503),float);  uR =  zeros([503],float)
k2L = zeros([1000],float); k2R = zeros([1000],float)
uL[0] = 0; uL[1] =0.00001;  uR[0] = 0; uR[1] = 0.00001
```

```
def V(x):                            # Potential  harmonic  oscillator
    v = 4.7*x*x
    return v

def setk2(e):                        # Sets k2L=(sqrt(e-V))^2 and k2R
    for i in range(0,n):
        xLeft = Xleft0 + i*h
        xr = Xright0 - i*h
        fact=0.04829  # 2 m*c**2/hbarc**2=2*940/(197.33)**2
        k2L[i] = fact*(e-V(xLeft))
        k2R[i] = fact*(e-V(xr))

def Numerov (n,h,k2,u,e):
    setk2(e)
    b=(h**2)/12.0                                # L & R Psi
    for i in range( 1,n):
        u[i+1]=(2*u[i]*(1-5.*b*k2[i])-(1+b*k2[i-1])
        *u[i-1])/(1+b*k2[i+1])

def diff(e):
    Numerov (nl,h,k2L,uL,e)              # Left wf
    Numerov (nr,h,k2R,uR,e)             # Right wf
    f0 = (uR[nr-1]+uL[nl-1]-uR[nr-3]-uL[nl-3])/(h*uR[nr-2])
    return f0

istep = 0
x1 = arange(-10,.02,0.02);    x2 = arange(10,-0.02,-0.02)
fig = plt.figure()
ax = fig.add_subplot(111)
ax.grid()

while abs(diff(e)) > eps :          # Bisection algorithm
    e =(amin + amax)/2
    print(e,istep)
    if diff(e)*diff(amax) > 0: amax = e
    else: amin = e
    ax.clear()
    plt.text(3,-200,'Energy= %10.4f'%(e),fontsize=14)
    plt.plot(x1,uL[:-2])
    plt.plot(x2,uR[:-2])
    plt.xlabel('x')
    plt.ylabel('$\psi(x) $',fontsize=18)
    plt.title('R & L Wavefunctions Matched at x = 0')
    istep = istep+1
    plt.pause(0.8)  # Pause to delay figures
plt.show()
```

Listing 6.4. QuantumNumerov.py solves for bound states of the Schrödinger equation numerically using the Numerov algorithm.

```
# BoundCall.py: p space bound state; imports GaussPoints, matrix

from numpy import*
from numpy.linalg import*
from GaussPoints import GaussPoints

min1 =0.;      max1 =200.;   u =0.5;    b =10.
eps = 3.e-10                         # Precision for Gauss points
```

```
N = 16; Lambda = -1024
H = zeros((N,N), float)                                    # Hamiltonian
WR = zeros((N), float)                             # Eigenvalues, potential
k = zeros((N), float); w = zeros((N),float);          # Pts & wts
GaussPoints(N, min1, max1, k, w, eps)          # Call gauss points

for i in range(0,N):
    for j in range(0,N):
        VR = (Lambda/2/u)*sin(k[i]*b)/k[i]*sin(k[j]*b)/k[j]
        H[i,j] = 2./math.pi*VR*k[j]*k[j]*w[j]     # Hamiltonian
        if (i == j):  H[i,j] += k[i]*k[i]/2/u
Es, evectors = eig(H)
ReE = real(Es) ;    ImE = imag(Es)                    # Eigenvalues

for j in range(0,N):
    print(" Npoints =",N, "Lambda =",Lambda," ReE =",ReE[j])
    print(" ImE = ", ImE)
    break
```

Listing 6.5. **BoundCall.py** solves for bound state of delta shell potential in momentum space; imports `GaussPoints` and matrix subs.

```
# DecaySound.py spontaneous decay simulation

from vpython import *     # Was: from visual.graph import *
import random, winsound

lambda1 = 0.005                                        # Decay constant
max = 80.;   time_max = 500;    seed = 68111
number = nloop = max                              # Initial value
graph1 = graph(title ='Spontaneous Decay',xtitle='Time',
               ytitle = 'Number',xmin=0,xmax=500,ymin=0,ymax=90)
decayfunc = gcurve(color = color.green)

for time in arange(0, time_max + 1):                  # Time loop
    for atom in arange(1, number + 1 ):            # Decay loop
        decay = random.random()
        if (decay < lambda1):
            nloop = nloop - 1                      # A decay
            winsound.Beep(600, 100)                # Sound beep
    number = nloop
    decayfunc.plot( pos = (time, number) )
    rate(30)
```

Listing 6.6. **DecaySound.py** simulates spontaneous decay by comparing a random number to the decay rate constant. The *winsound* package plays a beep each time there is a decay, and this leads to the sound of a Geiger counter.

```
# Plm.py: Associated Legendre Polynomials via Integration

import numpy as np, matplotlib.pylab as plt
```

```
from rk4Algor import rk4Algor

CosTheta = np.zeros((1999),float)
Plm = np.zeros((1999),float)
y = [0]*(2);    dCos = 0.001
el = 4;  m = 2    #m intger  m<=el,   m = 1,2,3,...
if el == 0 or el == 2: y[0] = 1
if (el>2 and (el)%2 == 0):
    if m  ==  0: y[0] = -1
    elif( m>0 ): y[0] = 1
    elif m<0 and abs(m)%2 == 0: y[0] = 1
    elif m<0 and abs(m)%2 == 1: y[0] = -1
if (el>2 and el%2 == 1) :
    if m == 0: y[0] = 1
    elif m>0:  y[0] = -1
    elif m<0:  y[0] = 1
y[1] = 1

def f(Cos, y):                             # RHS of equation
    rhs = [0]*(2)                  # Declare array dimension
    rhs[0] = y[1]
    rhs[1] = 2*Cos*y[1]/(1-Cos**2)-(el*(el+1)
        -m**2/(1-Cos**2))*y[0]/(1-Cos**2)
    return rhs

f(0,y)       # Call function for xi = 0 with init conds.
i = -1
for Cos in np.arange(-0.999999, 1-dCos, dCos):
    i = i+1
    CosTheta[i] = Cos
    y = rk4Algor(Cos, dCos, 2, y, f)  # call runge kutt
    Plm[i] =  y[0]   #

plt.figure()
plt.plot(CosTheta,Plm)
plt.grid()
plt.title('Unormalized $\mathbf{P_l^m(Cos)}$')
plt.xlabel('cos(theta)')
plt.ylabel('$\mathbf{P_l^m(Cos)}$')
plt.show()
```

Listing 6.7. Plm.py solves an ODE to determine the associated Legendre Polynomials $P_\ell^m(x)$. The Matplotlib package is used for the graphics, and rk4 for the ODE solution.

```
# Hdensity.py:  Hydrogen Radial density calling rk4Algor

import numpy as np, matplotlib.pylab as plt
from rk4Algor import rk4Algor

n = 5; el = 2; dr = 0.01                    # n = npr+el+1
rVec = np.zeros((2500),float)           # array for plot
RhoVec = np.zeros((2500),float)         # Density array
fvector = [0]*(2)
y = [0]*(2);  y[0] = 1e-8;  y[1] = 0

def f(r,y):                             # RHS of ODE
    fvector[0] = y[1]
    fvector[1] = -(2/r-1)*y[1]-((n-1)/r-el*(el+1)/r**2)*y[0]
```

```
      return fvector

f(0.001,y)                                          # f(t= 0)
i = 0
for r in np.arange(0.001,25,dr):
    rVec[i] = r
    y = rk4Algor(r, dr, 2, y, f)    # call rk4 algorithm
    RhoVec[i] = 4*3.141593*(y[0]*np.exp(-0.5*r) )**2 *r**2
    i = i+1
plt.figure()
plt.plot(rVec,RhoVec)
plt.title('Hydrogen Radial Density n = 5')
plt.xlabel('Radius r')
plt.ylabel('Density')
plt.show()
```

Listing 6.8. **Hdensity.py** solves the Schrödinger equation numerically for hydrogen using the rk4 algorithm, and then computes the radial density from the wave function. The Matplotlib package is used for the graphics.

```
# HOcharge.py: Quantum particle in HO & E field with Visual

from visual import *

dx = 0.04;  dx2 = dx*dx;  k0 = 5.5*pi;  dt = dx2/20.;  xmax = 6.
xs = arange(-xmax,xmax+dx/2,dx)            # Array x positions
E = 210                                    # Magnitude E field
g = display(width=500,height=250,
    title='Wave packet, harmonic well plus E field',range=10)
PlotObj= curve(x=xs, color=color.yellow, radius=0.1)
g.center = (0,2,0)
psr = exp(-0.5*(xs/0.5)**2) * cos(k0*xs)   # Re part Initial psi
psi = exp(-0.5*(xs/0.5)**2) * sin(k0*xs)   # Im part Initial psi
V = 25.0*xs**2-E*xs                        # Electric potential

while True:   # Solution as time transpires
    rate(500)
    psr[1:-1] = psr[1:-1] - (dt/dx2)*(psi[2:]+psi[:-2]
        -2*psi[1:-1]) +dt*V[1:-1]*psi[1:-1]
    psi[1:-1] = psi[1:-1] + (dt/dx2)*(psr[2:]+psr[:-2]
        -2*psr[1:-1]) -dt*V[1:-1]*psr[1:-1]
    PlotObj.y = 4*(psr**2 + psi**2)
```

Listing 6.9. **HOmovSlow.py** solves the time-dependent Schrödinger equation via a finite difference algorithm for the harmonic oscillator, using the Visual package for animation.

```
# HOmov: Solve t-dependent Sch Eqt for HO wi animation

from visual import *

# Initialize wave function, probability, potential
```

```
dx = 0.04;        dx2 = dx*dx;    k0 = 5.5*pi;    dt = dx2/20.0;
xmax = 6.0;   beta = dt/dx2
xs = arange(-xmax,xmax+dx/2,dx)                    # Array x values

g = display(width=500, height=250, title='Wave Packet in HO Well')
PlotObj = curve(x=xs, color=color.yellow, radius=0.1)
g.center = (0,2,0)                                 # Center of scene

R = exp(-0.5*(xs/0.5)**2) * cos(k0*xs)    # Re, Im Initial packet
I = exp(-0.5*(xs/0.5)**2) * sin(k0*xs)
V   = 15.0*xs**2

while True:
   rate(500)
   R[1:-1] = ←
       R[1:-1]-beta*(I[2:]+I[:-2]-2*I[1:-1])+dt*V[1:-1]*I[1:-1]
   I[1:-1] = ←
       I[1:-1]+beta*(R[2:]+R[:-2]-2*R[1:-1])-dt*V[1:-1]*R[1:-1]
   PlotObj.y = 4*(R**2 + I**2)
```

Listing 6.10. **HOmov.py** solves the time-dependent Schrödinger equation via a finite difference algorithm for the harmonic oscillator, using slicing for speed.

```
# HOpacket.py: HO wave packet in motion via Visual

from numpy import *
from visual.graph import *

oneoverpi = 1/math.sqrt(math.pi); a = 1.          # m=1=hbar
xx = 5; tt = 0                    # Initial x & t (classical)
wavef = display(x=0, y=0, width=600, height=600, range=8)
plotob = curve(color=color.yellow, radius=0.1)
spring = helix(pos=(-4,-3,0), radius=0.4, color=color.white,
                     coils=10.4, axis=(10,0,0))       # Spring
mass = box(pos=(1,-3,0), length=1, width=1, height=1,
   color=color.yellow)

for t in arange(0,20,0.1):                         # Time loop
   xx = cos(tt)
   x  =  arange(-5.0, 5.0, 0.001)
   rate(3)
   y = oneoverpi*exp(-(x-a*math.cos(t))**2)
   plotob.x = x                                    # x coord
   plotob.y = 4*y                                  # y coord
   spring.axis = vector(4+xx,0,0)       # Classical oscillator
   mass.pos = (xx,-3,0)                 # Position oscillator
   tt = tt+0.1
```

Listing 6.11. **HOpacket.py** computes the motion of a wave packet within a harmonic oscillator well using Visual package for animation.

```
# HOpacketMat.py: HO Wave Packet wi Matplotlib Animation
```

```
from numpy import *
import numpy as np, matplotlib.pyplot as plt
from matplotlib import animation

a = 1; oneoverpi = 1.0/(sqrt(np.pi))
fig = plt.figure()
ax = fig.add_subplot(111,autoscale_on=False,xlim=(-5,5),
    ylim=(0,1.5))
ax.grid()                                   # Plot a grid
plt.title("Wave Packet in H. O. potential")
plt.xlabel("x")
plt.ylabel (" $|\psi(x,t)|^2$")
line, = ax.plot([],[], lw=2)

def init():                                 # base frame
    line.set_data([],[])
    return line,

def animate(t):                    # Called repeatedly
    y=oneoverpi*np.exp(-(x-a*np.cos(0.01*t))**2) # Plot ea 0.01*t
    line.set_data(x,y)
    return line,

x = np.arange(-5,5,0.01)                     # range for x values
ani = animation.FuncAnimation(fig, animate,init_func=init,
    frames=10000, interval=10,blit=True)
plt.show()
```

Listing 6.12. HOpacketMat.py computes the motion of a wave packet within a harmonic oscillator well using Matplotlib package for animation.

```
# HOchargeMat.py charge in HO plus E field wi Matplotlib

from numpy import *
import numpy as np, matplotlib.pyplot as plt
import matplotlib.animation as animation

dx = 0.06;   dx2 = dx*dx;   k0 = 5.5*pi;   dt = dx2/8.;   xmax = 6.
xs = np.arange(-xmax,xmax+dx/2,dx)              # x array
psr = exp(-0.5*(xs/0.5)**2) * cos(k0*xs)        # Re Psi
psi = exp(-0.5*(xs/0.5)**2) * sin(k0*xs)        # Im Psi
E = 70                                          # E field
v = 25.0*xs**2 -E*xs                            # V HO + E
fig = plt.figure()
ax = fig.add_subplot(111, autoscale_on=False,
    xlim=(-xmax,xmax), ylim=(0, 1.5))
ax.grid()                                       # Plot grid
plt.title("Charged Harmonic Oscillator in E Field")
line, = ax.plot(xs, psr*psr+psi*psi, lw=2)

def animate(dum):
    psr[1:-1] = psr[1:-1] - (dt/dx2)*(psi[2:]+psi[:-2]
        -2*psi[1:-1]) + dt*v[1:-1]*psi[1:-1]
    psi[1:-1] = psi[1:-1] + (dt/dx2)*(psr[2:]+psr[:-2]
        -2*psr[1:-1]) - dt*v[1:-1]*psr[1:-1]
    line.set_data(xs,psr**2+psi**2)
    return line,
```

```
ani = animation.FuncAnimation(fig, animate, 1, blit=True)
plt.show()
```

Listing 6.13. HOchargeMat.py computes the wave packet for a charged particle within a harmonic oscillator well also exposed to an electric field using Matplotlib package for animation.

```python
# ScattSqWell.py: Quantum scattering from square well

import scipy.special, matplotlib.pyplot as plt,  numpy as np
from math import *

a = 1;  V = 15;  E = 10; nLs = 10;  Nin = 100;  Nout = 100;n=10;
alpha = np.sqrt(V+E);   beta = np.sqrt(E)
delta  = np.zeros((nLs),float);  SigL = np.zeros((nLs,200),float)
partot=np.zeros((n,200),float)
nexpts=100
def Gam(n,xx): # Spherical Bessel ratio
    gamma = np.zeros((n),float)
    for nn in range(0,n):  jn,jpr = scipy.special.sph_jn(nn,xx)
    gamma = alpha*jpr/jn   # gamma match psi outside-inside
    return gamma

def phaseshifts(n,alpha,beta):
    gamm = Gam(n,alpha)
    num = np.zeros((n),float); den = np.zeros((n),float)
    jnb,jnpr = scipy.special.sph_jn(n,beta)
    ynb,yprb = scipy.special.sph_yn(n,beta)
    for i in range(0,n):
        num1 = gamm[i]*jnb[i]
        den1 = gamm[i]*ynb[i]
        num[i] = beta*jnpr[i]-num1
        den[i] = beta*yprb[i]-den1
        td = atan2(num[i],den[i])
        delta[i] = td
    return delta

def totalcrossect(n,alpha,beta):
    delta = phaseshifts(n,alpha,beta)
    suma  = 0
    for i in range (0,n): suma = suma+(2*i+1)*(sin(delta[i]))**2
    return 4*np.pi*suma/beta**2

def plotcross(alpha,beta):
    e = 0.
    cross = np.zeros((200),float)              # total crossection
    delta = phaseshifts(n,alpha,beta)
    en = np.zeros((200),float)                          # energies
    for i in range(1,200):
        e = e + 100/300.
        en[i] = e
        alpha = np.sqrt(V+e)
        beta = np.sqrt(e)
        cross[i] = totalcrossect(n,alpha,beta)
        for m in range(0,n):
            partot[m,i] = 4*pi*(2*m+1)*(sin(delta[m]))**2/beta**2
    f2 = plt.figure()
```

```
        ax2 = f2.add_subplot(111)
        plt.plot(en,cross,label = "Total")
        plt.plot(en,partot[0,:],label = "S ")
        plt.plot(en,partot[1,:], label ="P ")
        plt.plot(en,partot[2,:],label = "D")
        plt.plot(en,partot[3,:],label = "E")
        plt.title("Total & Partial Cross Sections")
        plt.legend()
        plt.xlabel("Energy")

def diffcrossection():
        zz2 = np.zeros((n),complex)
        dcr = np.zeros((180),float)
        delta = phaseshifts(n,alpha,beta)       # phaseshifts
        for i in range(0,n):                     # n partial waves
            cosd = cos(delta[i])
            sind = sin(delta[i])
            zz = complex(cosd,sind)
            zz2[i] = zz*sind
        for ang in range(0,180):
            summ = 0.
            radi = cos(ang*pi/180.)
            for i in range(0,n):                 #  partial wave loop
                poL = scipy.special.eval_legendre(i,radi)
                summ+= (2*i+1)*zz2[i]*poL
            dcr[ang] = (summ.real**2 +summ.imag**2)/beta**2
        angu = np.arange(0,180)
        f1 = plt.figure()                        # plot separate figure
        ax1 = f1.add_subplot(111)
        plt.semilogy(angu,dcr)                   # Semilog dsig/dw plot
        plt.xlabel("Scattering Angle")
        plt.title ("Differential Cross Section")
        plt.grid()

def wavefunction():                    # Compute Psi(<1) & Psi(>1)
        delta = phaseshifts(n,alpha,beta)
        BL = np.zeros((n),complex)
        Rin = np.zeros((n,Nin),float)            # Psi(r<1), nLs
        Rex = np.zeros((n,nexpts),float)
        for i in range (0,10):                   # BL for matching
            jnb,jnpr = scipy.special.sph_jn(n,alpha) # SphBes
            jnf,jnfr = scipy.special.sph_jn(n,beta)
            ynb,yprb = r = scipy.special.sph_yn(n,beta)
            cosd = cos(delta[i])
            sind = sin(delta[i])
            zz = complex(cosd,-sind)
            num = jnb[i]*zz
            den = cosd*jnf[i]-sind*ynb[i]
            BL[i] = num/den          # For wavefunction match
        intr = 1.0/Nin                           # Points increment
        for i in range(0,n):                     # Internal Psi
            rin = intr
            for ri in range(0,Nin):              # PsiIn plot
                alpr = alpha*rin
                jnint,jnintpr = scipy.special.sph_jn(n,alpr)
                Rin[i,ri] = rin*jnint[i]
                rin = rin+intr
        extr = 2./nexpts
        for i in range(0,n):
            rex = 1.0
            for rx in range(0,nexpts):           # PsiIn plot
                argu = beta*rex
                jnxt,jnintpr = scipy.special.sph_jn(n,argu)
                nxt,jnintpr = scipy.special.sph_yn(n,argu)
```

```
                    factr = jnxt[i]*cos(delta[i])-nxt[i]*sin(delta[i])
                    fsin = sin(delta[i])*factr
                    fcos = cos(delta[i])*factr
                    Rex[i,rx] = rex*(fcos*BL.real[i]-fsin*BL.imag[i] )
                    rex = rex+extr
     ai = np.arange(0,1,intr)
     nwaf = 0        # PsiL to plot, CHANGE FOR OTHER WAVES
     f3 = plt.figure()
     ax3 = f3.add_subplot(111)
     plt.plot(ai,Rin[nwaf, :])
     ae = np.arange(1,3,extr)
     plt.title("$\Psi(r<1), \ \ \Psi(r>1), \ \ \ell = 0$")
     plt.xlabel ("$r$")
     plt.plot(ae,Rex[nwaf, :])

   diffcrossection()              # Diff crossection
   plotcross(alpha,beta)          # Total crossections
   wavefunction()                 # Psi
```

Listing 6.14. **ScattSqWell.py** computes the scattering from a square well calling a subroutine library to evaluate spherical Bessel and Neumann functions.

```
# SqBillardQM.py: Solve t-dependent Schroedinger eq on square table

import matplotlib as mpl, numpy as np, matplotlib.pylab as plt
from mpl_toolkits.mplot3d import Axes3D

Nmax = 101; dt = 0.01;  dx = 0.2;  dx2 = dx*dx;   fc = dt/dx2
Tmax = 200; Kx = 10.;   Ky = 15.;  xin = 50.;     yin = 50.

I = np.zeros((Nmax,Nmax),float);   R = np.zeros((Nmax,Nmax),float)
for i in range(1,Nmax-1):                           # Initial psi
        for j in range(1,Nmax-1):
            Gauss = np.exp(-.05*(i-yin)**2-.05*(j-xin)**2)
            R[i,j] = Gauss*np.cos(Kx*j + Ky*i)
            I[i,j] = Gauss*np.sin(Kx*j + Ky*i)
for t in range(0,Tmax):                             # Step thru time
    R[1:-1,1:-1] = R[1:-1,1:-1] - fc*(I[2:,1:-1] + I[0:-2,1:-1]
                    -4*I[1:-1,1:-1] + I[1:-1, 2:] + I[1:-1,0:-2])
    I[1:-1,1:-1] = I[1:-1,1:-1]  + fc*(R[2:,1:-1] + R[0:-2,1:-1]
                    -4*R[1:-1,1:-1] + R[1:-1,2:  ] + R[1:-1,0:-2])
x = y = np.arange(0, Nmax)
X, Y = plt.meshgrid(x, y)
Z = (I**2 + R**2)
fig = plt.figure()
ax = Axes3D(fig)
ax.plot_wireframe(X ,Y ,Z)
ax.set_xlabel('x');   ax.set_ylabel('y');   ax.set_zlabel('Psi')
ax.set_title('Psi (x,y,t) at 200 Times on Square Billiard Table')
plt.show()
```

Listing 6.15. SqBilliardQM.py solves the time-dependent Shrödinger equation for motion of a wave packet on a square billiard table.

```
# CoulWF.py:    Regular Coulomb scattering wave function

from scipy import special; from mpmath import * # hypergeometric
import matplotlib.pyplot as plt, numpy as np
from math import *

f1 = np.zeros((10),complex); Rea = np.zeros((10,161),float)
zi = complex(0,1.0)
mAu =   196.966569*931.494;  mAlpha =   4.002602*931.494
Zau = 79;   Zalph = 2
mu = mAlpha*mAu/(mAlpha + mAu)
hbarc = 197.33                           # MeV-fm, E in MeV, r in fm
Elab =   7.
Ecom = Elab*mAu/(mAlpha + mAu)
vel = sqrt(Ecom*2/mu)
ka = sqrt(2.0*mu*Ecom)/hbarc
etaco = Zalph*Zau*mu/(hbarc*ka*137.)              # Coulomb parameter
expi = exp(-0.5*etaco*pi)

i = 0                         # Main loop over r and i
for r in np.arange(0.1,80.5,0.5):
    rho = complex(0,-2*ka*r)                        # -2ikr
    expo = complex(cos(ka*r),sin(ka*r))          # exp(ikr)
    for L in range(0,10):
        a = L + 1.0 + etaco * zi    # Arg gamma function
        sol = hyp1f1(a, 2*L+2., rho)        # Hypergeometric
        rhoL = (-rho)**L
        gam = special.gamma(a)                 # Gamma(l+1+in)
        upar = rhoL*expo *sol*gam*expi/factorial(2*L)
        f1[L] = upar/sqrt(vel)
        Rea[L,i] = f1.real[L]                  # Real psi
    i += 1

rr = np.arange(0.1,80.5,0.5)
plt.plot(rr,Rea[0,:],label = 'S')
plt.plot(rr,Rea[1,:],label = 'P',linewidth=2)
plt.plot(rr,Rea[2,:],label = 'D',linewidth=3)
plt.legend()
plt.xlabel("r (fermis)")
plt.title ("Radial Coulmb Wave Functions $y_l(r)$ for $l = 0,1,2$")
plt.show()
```

Listing 6.16. CoulWF.py computes the regular Coulomb scattering wave functions by evaluating confluent hypergeometric function with a complex argument.

```
# 3QMdisks.py: Wavepacket scattering from 3 disks wi MatPlot

import matplotlib.pylab as p,   numpy as np
from mpl_toolkits.mplot3d import Axes3D

r = 10;          N = 101;       x1 = 51; # 51 = 90.*sqrt 3/2-30
dx = 0.1;        dx2 = dx*dx;   k0  = 20.;    k1 = 0.
dt  =  0.002;    fc = dt/dx2; Xo = 40;    Yo =  25
    # Declare arrays
V      = np.zeros((N,N),float);   Rho   = np.zeros((N,N),float)
RePsi = np.zeros((N,N),float);    ImPsi = np.zeros((N,N),float)
ix = np.arange(0, 101);           iy = np.arange(0,101)
```

```
X, Y = np.meshgrid(ix, iy)
fig = p.figure();  ax = Axes3D(fig)                    # Create figure

def Pot1Disk(xa,ya):                         # Potential single disk
    for i in range (ya-r,ya+r+1):
        for j in range(xa-r,xa+r+1):
            if np.sqrt((i-ya)**2+(j-xa)**2)<=r:  V[i,j] = 5.

def Pot3Disks():                             # Potential three disk
    Pot1Disk(30,45);   Pot1Disk(70,45);  Pot1Disk(50,80)

def Psi_0(Xo,Yo):                                       # Initial Psi
    for i in np.arange(0,N):
        for j in np.arange(0, N):
            Gaussian = np.exp(-0.03*(i-Yo)**2-0.03*(j-Xo)**2)
            RePsi[i,j] = Gaussian*np.cos(k0*i+k1*j)
            ImPsi[i,j] = Gaussian*np.sin(k0*i+k1*j)
            Rho[i,j] = RePsi[i,j]**2 + ImPsi[i,j]**2 + 0.01
Psi_0(Xo,Yo)  # Psi and Rho initial
Pot3Disks()                                             # Initial Psi

for t in range(0, 150):  # 120->30          # Compute Psi t < 120
    if t%5 == 0:  print( 't =', t)                   # Print ea 5th t
    ImPsi[1:-1,1:-1] = ImPsi[1:-1,1:-1] + fc*(RePsi[2: ,1:-1]
        + RePsi[:-2 ,1:-1] - 4*RePsi[1:-1,1:-1] + RePsi[1:-1,2: ]
        + RePsi[1:-1, :-2]) + V[1:-1,1:-1]*dt*RePsi[1:-1,1:-1]
    RePsi[1:-1,1:-1] = RePsi[1:-1,1:-1] - fc*(ImPsi[2: ,1:-1]
        +ImPsi[ :-2,1:-1] - 4*ImPsi[1:-1,1:-1] + ImPsi[1:-1,2: ]
        +ImPsi[1:-1, :-2]) + V[1:-1,1:-1]*dt*ImPsi[1:-1,1:-1]
    for i in range(1, N-1):                         # Compute Rho
        for j in range(1,N-1):     # Hard Disk, psi = 0
            if V[i,j] !=0: RePsi[i,j] = 0; ImPsi[i,j] = 0
            Rho[i,j] = 0.1*(RePsi[i,j]**2
                + ImPsi[i,j]**2) + 0.0002*V[i,j]
X, Y = np.meshgrid(ix, iy)
Z = Rho[X,Y]
ax.set_xlabel('y')
ax.set_ylabel('x')
ax.set_zlabel('Rho(x,y)')
ax.plot_wireframe(X, Y, Z, color = 'g')
print("finito")
p.show()
```

Listing 6.17. **3QMdisks.py** solves the time-dependent Shrödinger equation describing the scattering of a wave packet from three hard disks with visualization by Matplotlib. 3QMdisksVis.py solves same problem using the Visual package.

```
# TwoWells.py:  Time-dependent Schroedinger packets in two wells

from visual.graph import *

dx = 0.08; dx2 = dx*dx; k0 = 5.; dt = dx2/8; Nmax = 200; addi=250
V_L = zeros((Nmax),float);  V_R = zeros((Nmax),float)
V2 = zeros((Nmax+addi),float)
RePsiL = zeros((Nmax+1),float);  ImPsiL = zeros((Nmax+1),float)
Rho = zeros((Nmax+1),float);  RhoR = zeros((Nmax+1),float)
RePsiR = zeros((Nmax+1),float);  ImPsiR = zeros((Nmax+1),float)
```

```
RePsi2L = zeros((Nmax+addi),float)
ImPsi2L = zeros((Nmax+addi),float)
RhoAL = zeros((Nmax+addi),float)
Rho2R =    zeros((Nmax+addi),float)
RePsi2R = zeros((Nmax+addi),float)
Psi2R =    zeros((Nmax+addi),float)
Xleft = arange(-18.,-2.,0.08)
Xright = arange(2.0,18.,0.08)
Xall = arange(-18,18,0.08)

g = display(width=500,height=500);  g.center = (0,0,20)
cL = curve(color=color.red,x=Xleft)
cR = curve(color=color.yellow,x=Xright)
curve(pos=[(0,250),(0,-250)])                 # Vert line tru x=0
PlotObj =  curve(x=Xleft, color=color.red,      radius=0.8)
PlotObjR = curve(x=Xleft, color=color.yellow, radius=0.8)
escena2=display(width=500,height=500,x=500);
allc = curve(color=color.green,x=Xall)
curve(pos=[(0,250),(0,-250)])                 # Vertical line tru x=0
PlotAllR = curve(x=Xall,color=color.cyan,radius=0.8,
  display=escena2)

def potentials():
    for i in range(0,Nmax):
        xL = -18.0 + i*dx                 # left well, left figure
        V_L[i] = 10*(xL+10)**2/2
        xR = 2.0 + i*dx
        V_R[i] = 10*(xR-10)**2/2           # right well left figure
    for j in range(0,Nmax+addi):
        xL= -18+j*dx
        if j<=125: V2[j] = 10.*(xL+10)**2/2   # LHS
        if j>125 and j<325:  V2[j] = V2[125]  # Pert lowers
        if j>=325:  V2[j] = 10.0*(xL-10)**2/2 # RHS right side

potentials()

def plotpotentials(i=0):
    cL.x = 10*Xleft+15; cR.x = 10*Xright-15      # Widen
    cL.y = 10*(Xleft+10)**2/2-100; cR.y = 10*(Xright-10)**2/2-100
    allc.x = 8 * Xall
    allc.y = V2 - 100
    i = i+1

plotpotentials()
RePsiL = exp(-5*((Xleft+10))**2) * cos(k0*Xleft)    # Initial psi
ImPsiL = exp(-5*((Xleft+10))**2) * sin(k0*Xleft)
Rho  = RePsiL*RePsiL + ImPsiL*ImPsiL
RePsiR = exp(-5*((Xright-10))**2)*cos(-k0*Xright)   # Just On side
ImPsiR = exp(-5*((Xright-10))**2)*sin(-k0*Xright)
RhoR  = RePsiR**2 + ImPsiR**2
for i in range(0,450):                        # initial conditions
    x = -18+i*dx                              # gives -18 <=x <=18
    if i<=225:
        RePsi2L[i] = 0*exp(-5*(x+10)**2)*cos(k0*x)  # to middle
        ImPsi2L[i] = 0*exp(-5*(x+10)**2)*sin(k0*x)
    else:                                     # too small set=0
        RePsi2L[i] = 0.
        ImPsi2L[i] = 0.
    RhoAL[i] = 50.*(RePsi2L[i]**2 + ImPsi2L[i]**2) # Right psi
for j in range(0,450):
    x = -18+j*dx
    if j<=225:
        RePsi2R[j] = 0.  # too small, make it 0
        Psi2R[j] = 0.
```

```
        else:
            RePsi2R[j] = exp(-5*(x-10)**2)*cos(-k0*x)  # Left psi
            Psi2R[j]  =  exp(-5*(x-10)**2)*sin(-k0*x)
        Rho2R[j] = 50.*(RePsi2R[j]**2 + Psi2R[j]**2)
for t in range(0,2900):
    rate(100)
    RePsiL[1:-1] = RePsiL[1:-1] - (dt/dx2)*(ImPsiL[2:]
        +ImPsiL[:-2]-2*ImPsiL[1:-1]) +dt*V_L[1:-1]*ImPsiL[1:-1]
    ImPsiL[1:-1] = ImPsiL[1:-1] + (dt/dx2)*(RePsiL[2:]
        +RePsiL[:-2]-2*RePsiL[1:-1]) -dt*V_L[1:-1]*RePsiL[1:-1]
    PlotObj.x = 10*(Xleft)+15  # RHS left figure
    PlotObj.y = 50*(RePsiL**2 + ImPsiL**2) +150
    RePsiR[1:-1] = RePsiR[1:-1] - (dt/dx2)*(ImPsiR[2:]+ImPsiR[:-2]
        -2*ImPsiR[1:-1]) +dt*V_R[1:-1]*ImPsiR[1:-1]
    ImPsiR[1:-1] = ImPsiR[1:-1] + (dt/dx2)*(RePsiR[2:]+RePsiR[:-2]
        -2*RePsiR[1:-1]) -dt*V_R[1:-1]*RePsiR[1:-1]
    PlotObjR.x = 10*(Xright)-15  # LHS left figure
    PlotObjR.y = 50*(RePsiR**2 + ImPsiR**2) +150
    RePsi2L[1:-1] = RePsi2L[1:-1] - ↵
        (dt/dx2)*(ImPsi2L[2:]+ImPsi2L[:-2]
        -2*ImPsi2L[1:-1]) +dt*V2[1:-1]*ImPsi2L[1:-1]
    ImPsi2L[1:-1] = ImPsi2L[1:-1] + ↵
        (dt/dx2)*(RePsi2L[2:]+RePsi2L[:-2]
        -2*RePsi2L[1:-1]) -dt*V2[1:-1]*RePsi2L[1:-1]
    RePsi2R[1:-1] = RePsi2R[1:-1] - (dt/dx2)*(Psi2R[2:]
        +Psi2R[:-2]-2*Psi2R[1:-1]) +dt*V2[1:-1]*ImPsiR[1:-1]
    Psi2R[1:-1] = Psi2R[1:-1] + (dt/dx2)*(RePsi2R[2:]+RePsi2R[:-2]
        -2*RePsi2R[1:-1]) -dt*V2[1:-1]*RePsi2R[1:-1]
    PlotAllR.x = 8*(Xall)
    PlotAllR.y = 70*(RePsi2R**2 + Psi2R**2)+150
    + 50*(RePsi2L**2 + ImPsi2L**2)
```

Listing 6.18. **TwoWells.py** Wave packet mixing between two connected potential wells.

```
# GlauberState.py:  Glauber's Coherent Quantum State

from numpy import *
import numpy as np, matplotlib.pyplot as plt, math

sqpi = np.sqrt(np.pi)
E = 3. ;    alpha = np.sqrt(E-0.5)          # E, Coherent Eigenvalue
factr = np.exp(-0.5*alpha*alpha);   nmax = 20

def Hermite(x, n):                          # Hermite polynomial
    if(n == 0): p = 1.0
    elif(n == 1): p = 2*x
    else:
        p0 = 1
        p1 = 2*x
        for i in range(1,n):
            p2 = 2*x*p1-2*i*p0
            p0 = p1
            p1 = p2
            p = p2
    return p

def glauber(x,t,nmax):                      # Coherent state
    Reterm = 0.0
```

```python
        Imterm = 0.0
        factr = np.exp(-0.5*alpha*alpha)
        for n in range (0, nmax):
            fact = np.sqrt(1.0/(math.factorial(n)*sqpi*(2**n)))
            psin = fact*Hermite(x,n)*np.exp(-0.5*x*x)
            den =np.sqrt(math.factorial(n))
            num = factr*(alpha**n)*psin
            Reterm += num*(np.cos((n+0.5)*t))/den
            Imterm += num*(np.sin((n+0.5)*t))/den
        phi = np.sqrt(Reterm*Reterm+Imterm*Imterm)
        return phi

def animate(t):
    y = glauber(xx,t,nmax)  # Find coherent stat
    s=str(t)
    plt.plot(xx,y,label=s)
    leg=plt.legend(loc='best',ncol=4,mode="expand",shadow=True)

fig = plt.figure()
ax = fig.add_subplot(111,autoscale_on=False, xlim=(-6,6),
    ylim=(0,1.5))
ax.grid()                                 # Plot a grid
plt.title("Glauber states at different times")
plt.xlabel("x")
plt.ylabel (" $|\psi(x,t)|^2$")
xx = np.arange(-6.0,6.0,0.2)                  # Range for x values
for t in np.arange(0,3.6,0.5):
    animate(t)
plt.show()
```

Listing 6.19. GlauberState.py computes the coherent Glauber state for the harmonic oscillator and plots the result as an animation over time.

```python
# Hyperfine.py: Symbolic Hydrogen hyperfine structure using Sympy

from sympy import *
import numpy as np, matplotlib.pyplot as plt

W, mue, mup, B = symbols('W mu_e mu_p B')       # Symbols & ↵
    Hamiltonian
H  = Matrix([[W,0,0,0],[0,-W,2*W,0],[0,2*W,-W,0],[0,0,0,W]])
Hmag = Matrix([[-(mue+mup)*B,0,0,0],[0,-(mue-mup)*B,0,0],
    [0,0,-(-mue+mup)*B,0],[0,0,0,(mue+mup)*B]])
print ("\n Hyperfine Hamiltonian H =", H )
print ("\n Eigenvalues and multiplicities of H =",H.eigenvals() )
print( "\n Hmag =", Hmag)
Htot = H + Hmag                           # Hamiltonian + perturbation
print("\n Htot = H + Hmag =", Htot)
print("\n Eigenvalues of matrix HB" )
e1, e2, e3, e4 = Htot.eigenvals()                  # 4 eigenvalues
print(" e1 = ", e1, "\n e2 = ", e2, "\n e3 = ",e3, "\n e4 = ",e4)
print ("\n After substitute mu_e = 1, and mu_p = 0 in eigenvalues")
print( " e1 = ",e1.subs([(mue,1),(mup,0)]),"\n e2 = ",
e2.subs([(mue,1),(mup,0)]))
print( " e3 = ",e3.subs([(mue,1),(mup,0)]),"\n e4 = ",
e4.subs([(mue,1),(mup,0)]))
b = np.arange(0,4,0.1)
E = 1
```

```
E4 = -E + np.sqrt(b**2 +4*E**2)
E3 = E - b
E2 = E + b
E1 = -E - np.sqrt(b**2 +4*E**2)
plt.figure()
plt.plot(b,E1, label='E1');   plt.plot(b,E2, label='E2')
plt.plot(b,E3, label='E3');   plt.plot(b,E4, label='E4')
plt.legend();  plt.text(-0.4,1, 'E')
plt.xlabel(' Magnetic Field B')
plt.title('Hyperfine Splitting of H Atom 1S Level')
plt.show()
```

Listing 6.20. **Hyperfine.py** calculates the hyperfine splitting in hydrogen using the Python symbolic package Sympy.

```
# Entangle.py: Calculate quantum entangled states

from numpy import * ; from numpy.linalg import *

nmax = 4
H = zeros((nmax,nmax), float)
XAXB = array([[0, 0, 0, 1],[0,0,1,0],[0,1,0,0],[1,0,0,0]])
YAYB = array([[0,0,0,-1],[0,0,1,0],[0,1,0,0],[-1,0,0,0]])
ZAZB = array([[1,0,0,0],[0,-1,0,0],[0,0,-1,0],[0,0,0,1]])
SASB = XAXB + YAYB + ZAZB - 3*ZAZB        # Hamiltonian
print( '\nHamiltonian without mu^2/r^3 factor \n', SASB, '\n')

es,ev = eig(SASB)                 # Eigenvalues and eigenvectors
print ('Eigenvalues\n', es,  '\n')
print( "Eigenvectors (in columns)\n", ev, "\n")
phi1 = (ev[0,0],ev[1,0],ev[2,0],ev[3,0])   # Eigenvectors
phi4 = (ev[0,1],ev[1,1],ev[2,1],ev[3,1])
phi3 = (ev[0,2],ev[1,2],ev[2,2],ev[3,2])
phi2 = (ev[0,3],ev[1,3],ev[2,3],ev[3,3])
basis=[phi1, phi2, phi3,phi4]          # List eigenvectors

for i in range(0,nmax):            # Hamiltonian in new basis
    for j in range(0,nmax):
        term  = dot(SASB,basis[i])
        H[i,j] = dot(basis[j],term)
print( "Hamiltonian in Eigenvector Basis\n", H      )
```

Listing 6.21. **Entangle.py** computes Hamiltonian, eigenvalues, and eigenvectors for entangled quantum states using numpy.

```
# SU3.py: SU3 matrix manipulations

from numpy import *
from numpy.linalg import *

L1 = array([[0,1,0],[1,0,0],[0,0,0]])          # Eight generators
L2 = array([[0,-1j,0],[1j,0,0],[0,0,0]])
```

```
L3 = array([[1,0,0],[0,-1,0],[0,0,0]])
L4 = array([[0,0,1],[0,0,0],[1,0,0]])
L5 = array([[0,0,-1j],[0,0,0],[1j,0,0]])
L6 = array([[0,0,0],[0,0,1],[0,1,0]])
L7 = array([[0,0,0],[0,0,-1j],[0,1j,0]])
L8 = array([[1,0,0],[0,1,0],[0,0,-2]])*1/sqrt(3)
u = array([1,0,0])                              # Up quark
d = array([0,1,0])                              # Down quark
s = array([0,0,1])                              # Strange quark
Ip = 0.5*(L1+1j*L2)                          # Raising operators
Up = 0.5*(L6+1j*L7)
Vp = 0.5*(L4+1j*L5)
Im = 0.5*(L1-1j*L2)                       # Lowering operators
Um = 0.5*(L6-1j*L7)
Vm = 0.5*(L4-1j*L5)
Ipxd = dot(Ip,d)                                # Raise d to u
print("Ipxd", Ipxd)
Vpxs = dot(Vp, s)                               # Raise s to u
print("Vpxs", Vpxs)
Upxs=dot(Up,s)                                  # Raise s to d
print("Upxs ", Upxs)
```

Listing 6.22. SU3.py uses the numpy linear algebra package to construct the matrices representing the operators of the symmetry group SU(3) describing constituent quarks.

```
# QMC.py: VPthon, Quantum MonteCarlo, Feynman path integration

import random
from vpython import *
import numpy as np

N = 101;            M = 101;            xscale = 10. # Initialize
path = np.zeros((M), float); prob =np.zeros([M], float)
trajec = canvas(width=300, height=500, title='Spacetime ↵
    Trajectories')
trplot = curve(color=color.magenta, display=trajec, radius=0.8)

def trjaxs():                       # Plot axis for trajectories
    trax = curve(pos=[(- 97,-100,0), (100,-100,0)],
        color = color.cyan, canvas = trajec)
    curve(pos = [(0,  - 100,0), (0, 100,0)],
        color = color.cyan, canvas = trajec)
    label(pos=vector (0,-110,0), text='0', box=0, canvas = trajec)
    label(pos=vector (60,-110,0), text='x', box=0, canvas = trajec)

wvgraph = canvas(x = 340, y = 150, width = 500, height = 300,
    title = 'Ground State Probability')
wvplot = curve(x = range(0, 100), canvas = wvgraph)    # Probability
wvfax = curve(color = color.cyan)

def wvfaxs():
    wvfax = curve(pos = [(-600,-155,0), (800,-155,0)],
        canvas=wvgraph, color=color.cyan)
    curve(pos = [(0,-150,0), (0,400,0)], display=wvgraph, ↵
        color=color.cyan)
    label(pos = vector(-80,450,0), text='Probability', box=0, ↵
        canvas=wvgraph)
    label(pos = vector(600,-220,0), text='x', box=0, canvas=wvgraph)
```

```
    label(pos = vector(0,-220,0), text='0', box=0, canvas=wvgraph)

trjaxs()
wvfaxs()                                                 # Plot axes

def energy(path):                                        # HO energy
    sums = 0.
    for i in range(0, N-2):
        sums += (path[i+1]-path[i])*(path[i+1]-path[i])
    sums += path[i+1]*path[i+1];
    return sums

def plotpath(path):                          # Plot trajectory in xy scale
    for j in range (0, N):
        trplot.append(pos=vector(20*path[j], 2*j - 100,0))

def plotwvf(prob):                                       # Plot prob
    for i in range (0, 100):
        wvplot.color = color.yellow
        wvplot.append(pos=vector(8*i-400, 4.0*prob[i]-150,0))

oldE = energy(path)                              # find E of path

#while True:                                    # pick random element
for i in range (0,1500):
    rate(50)                                           # slows paintings
    element = int(N*random() )                # Metropolis algorithm
    change = 2.0*(random() - 0.5)
    path[element]   += change                      # Change path
    newE = energy(path);                           # Find new E
    if  newE > oldE and np.exp( - newE + oldE)<= random():
        path[element]   -= change                      # Reject
        trplot.clear()          # Erase previous trajctory
        plotpath(path)
        trplot.visible=True    # Make visible new trajectory
    elem = int(path[element]*16 + 50)        # if path = 0, elem = 50
    if elem < 0:
        elem = 0                      # negative case not allowed
    if elem > 100:
        elem = 100                         # if exceed max
    prob[elem] += 1              # increase probability for that x
    plotwvf(prob)                            # plot prob
    oldE = newE
```

Listing 6.23. **QMC.py** determines the ground-state probability via a Feynman path integration using the Metropolis algorithm to simulate variations about the classical trajectory.

7

Thermodynamics & Statistical Physics

7.1 Chapter Overview

This chapter starts with a numerical solution of heat equation using a leapfrog (time-stepping) algorithm, as done with waves in Chapter 4. Section 7.3 develops simulations for three stochastic processes: random walks, diffusion-limited aggregation, and deposition of particles onto a surface. As we did in §2.9, the output from these simulations may be analyzed for fractal characteristics. A related spontaneous decay simulation is covered in §6.3, with similar stochastic simulations applicable to biological systems discussed in Chapter 8. In §7.4 we present some problems on the thermal behavior of magnetic materials, starting with a simple search for the roots of the equation relating magnetization to temperature. Then there's another simple problem related to the counting of accessible spin states. In §7.5 we present the Ising model and the Metropolis algorithm that simulates thermal equilibrium. This is one of the most important algorithms of Computational Physics, with output that students oft find revealing. The chapter ends with molecular dynamics (MD), a big subject with a variety of research-level codes often used in physics and chemistry. We keep things simple, while covering the basics.

7.2 The Heat Equation

The heat equation describes the variation of temperature T throughout a region of space $\mathbf{x}$ as a function of time t:

$$\frac{\partial T(\mathbf{x},t)}{\partial t} = \kappa \nabla^2 T(\mathbf{x},t), \qquad \kappa = \frac{K}{C\rho}. \qquad (7.1)$$

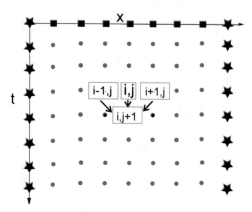

Figure 7.1. The algorithm for the heat equation in which the temperature at the location $x = i\Delta x$ and time $t = (j+1)\Delta t$ is computed from the temperature values at three points of an earlier time. The solid boxed nodes on top correspond to the known initial condition, while the starred nodes correspond to the unchanging boundary conditions.

Here K is the thermal conductivity of the region, C is its specific heat, and ρ is its density. If there is a temperature variation in only the x direction, then (7.1) becomes:

$$\frac{\partial T(x,t)}{\partial t} = \kappa \frac{\partial^2 T(x,t)}{\partial x^2}. \tag{7.2}$$

A unique solution to this second-order PDE requires knowledge of initial conditions and boundary conditions; namely, the initial temperature distribution $T(\mathbf{x}, 0)$, and the temperature along the region's boundaries $T(\mathbf{x}, t)$.

7.2.1 Algorithm for Heat Equation

As is true for most PDEs, the simplest path to a numerical solution is obtained by converting the differential equation to a finite-difference equation. The finite difference technique discretizes space and time on a lattice (Figure 7.1), and solves the PDE only on the lattice sites. The boxed sites along the top row correspond to the known, initial values of the temperature, while the starred sites on the right and left edges sides correspond to the fixed temperature along the boundaries. Since we know the spatial dependence of the solution at $t = 0$, we use a forward-difference approximation for the time derivative incorporating the initial solution:

$$\frac{\partial T(x,t)}{\partial t} \simeq \frac{T(x, t + \Delta t) - T(x,t)}{\Delta t}. \tag{7.3}$$

Because the boundary conditions provide the spatial variation of the temperature along the entire top row and the left and right sides, we can use the more accurate central-difference approximation for the space derivative.

1. Show that the application of the central-difference approximation for the space derivative leads to

$$\frac{\partial^2 T(x,t)}{\partial x^2} \simeq \frac{T(x+\Delta x,t) + T(x-\Delta x,t) - 2T(x,t)}{(\Delta x)^2}. \tag{7.4}$$

2. Show that substitution of the approximate space and time derivatives into (7.2) yields the difference equation:

$$\frac{T(x,t+\Delta t) - T(x,t)}{\Delta t} = \frac{K}{C\rho} \frac{T(x+\Delta x,t) + T(x-\Delta x,t) - 2T(x,t)}{\Delta x^2}. \tag{7.5}$$

3. Reorder (7.5) into the form of an algorithm in which T can be stepped forward in time $t = j\Delta t$ for all of space $x = i\Delta x$ starting with the initial conditions:

$$T_{i,j+1} = T_{i,j} + \eta \left[T_{i+1,j} + T_{i-1,j} - 2T_{i,j} \right], \qquad \eta = \frac{K\Delta t}{C\rho\Delta x^2}. \tag{7.6}$$

4. Start with difference equation and show that the solution will be stable (not grow in time) if

$$\eta = \frac{K\,\Delta t}{C\rho\,\Delta x^2} < \frac{1}{2}. \tag{7.7}$$

This is known as the von Neumann or Courant stability condition. It tells us that if we make the time step Δt smaller, we will always improve the stability, as to be expected. On the other hand if we decrease the space step Δx, without a (nonproportional) *increase* in the time step, the solution becomes unstable.

7.2.2   Solutions for Various Geometries

1. You are given an aluminum bar of length $L = 1\,\text{m}$ and width w aligned along the x axis (Figure 7.2 left). It is insulated along its length though not at its ends. Initially the entire bar is at a uniform temperature of $T_0 = 100\,^\circ\text{C}$, and then both ends are placed in contact with ice water at $0\,^\circ\text{C}$. Heat flows through the noninsulated ends only. Determine how the temperature varies along the length of the bar as a function of time.

   a. Derive the series solution to this problem,

$$T(x,t) = \sum_{n=1,3,\ldots}^{\infty} \frac{4T_0}{n\pi} \sin k_n x \, e^{-k_n^2 \kappa t}, \qquad k_n = \frac{n\pi}{L}. \tag{7.8}$$

   b. What form does the solution (7.8) take after imposing the initial and boundary conditions?

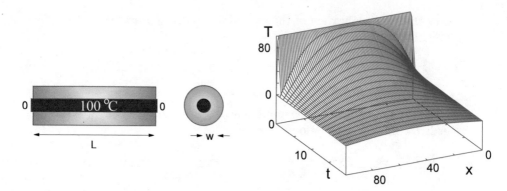

Figure 7.2. *Left:* A metallic bar insulated along its length with its ends in contact with ice. The bar is colored solid red and the insulation is of lighter color. *Right:* Results of numerical calculation of the temperature versus position and versus time, with isotherm contours projected onto the horizontal plane.

c. Program up the finite difference solution of the heat equation. Our programs are given in `EqHeat.py` and `Eq.HeatMov.py`, and here's a pseudocode version of the former:

```
# EqHeat.py: solves heat equation via finite differences, 3-D plot
Import packages
Define constants, declare arrays
Initialize T[Nx, 2] for Nx x values
Set boundary conditions
for 1 < t < Nt
    for 1 < x < Nx-1
        Tnew = Told + change (the algorithm)
    Plot T[all x,1] after every 300 time steps
    set T[all x,0] = T(all x,1) (present = previous future)
Finish plot
```

d. Have your program create a rotatable 3-D plot of $T(x,t)$, as shown in Figure 7.2 right. This helps greatly in debugging and tuning.

e. Vary the time and space steps in your calculation so that you obtain solutions that are stable in time and vary smoothly in both space and time.

f. Test what happens when the von Neumann/Courant stability condition (7.7) is not satisfied.

g. Check that your output always agrees with the boundary conditions.

h. Check that your simulation reaches equilibrium, and estimate the precision of your answer.

i. Compare the analytic and numeric solutions. If the solutions differ, suspect the one that does not appear smooth and continuous.

j. Make a 3-D (surface) plot of temperature versus position versus time, including the isotherms (contours of constant temperature).

2. Repeat the calculation for a poor conductor, such as wood.

3. Start your simulation for an initial temperature distribution $T(x, t = 0) = \sin(\pi x/L)$. Compare to the analytic solution, $T(x,t) = \sin(\pi x/L)e^{-\pi^2 K t/(L^2 C \rho)}$.

4. Two identical bars of lengths $0.25\,\mathrm{m}$ and temperatures $0^\circ C$ are placed in contact with each other along one of their ends, with their other ends kept at $0^\circ C$. One bar is in a heat bath at $100^\circ C$, and the other in a bath of $50^\circ C$. Determine the temperature variation as a function of time within both bars.

5. A bar is in contact with an environment at a temperature T_e. **Newton's law of cooling** tells us that the rate of temperature change of the bar as a result of radiation is

$$\frac{\partial T}{\partial t} = -h(T - T_e), \tag{7.9}$$

with h a positive constant. Show that inclusion of radiation leads to the modified heat equation

$$\frac{\partial T(x,t)}{\partial t} = \frac{K}{C\rho}\frac{\partial^2 T}{\partial^2 x} - hT(x,t). \tag{7.10}$$

   a. Modify the algorithm used to solve the heat equation to now include Newton's cooling.
   b. Compare the cooling of a radiating bar with that of the insulated bar.

6. Solve for conductive and radiative heat flow within a 2-D rectangular iron plate which is initially at a temperature of $100^\circ C$. Three sides are maintained at 100 $^\circ C$, while the top is placed in contact with ice water at $0^\circ C$.

   a. Make separate surface plots of $T(x, y = \text{fixed}, t)$ and $T(x = \text{fixed}, y, t)$, or a series of plots of $T(x, y, t = \text{fixed})$ for various t values.
   b. After equilibrium is reached, make a plot of the isotherms $T(x, y, t = \infty)$.

7. Sometimes an equilibrium is reached in which the temperature no longer changes as a function of time. In this case the heat equation (7.1) takes the form

$$\nabla^2 T(\mathbf{x}, t) = 0. \tag{7.11}$$

   This is the same as Laplace's equation, which is studied in §5.2.1, and can be solved using the same *relaxation* algorithm developed there. And so, solve for isotherms within a 2-D rectangular iron plate in which three sides are maintained at $100^\circ C$ while the fourth side remaining in contact with ice water at $0^\circ C$.

8. Compute the rate of heat flow from the center to the surface of a sphere of constant thermal conductivity. The center of the sphere is kept at $0^\circ C$ and the surface is kept at $100^\circ C$.

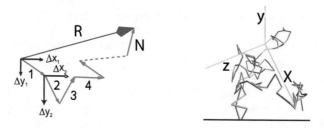

Figure 7.3. *Left:* A schematic of the N steps in a random walk simulation that end up a distance R from the origin. Notice how the Δx's for each step add vectorially. *Right:* A simulated walk in 3-D from *Walk3D.py*.

9. A sphere with constant thermal conductivity is initially at 0°C and is then placed in a heat bath of 100°C. Compute the temperature profile of the sphere as a function of time.

10. A composite sphere is composed of material of high thermal conductivity up to its middle, and low conductivity from its middle to its outside. Compute the rate of heat flow from the center to the surface of a sphere if the center is kept at 0°C and the surface is kept at 100°C.

7.3 Random Processes

7.3.1 Random Walks

Normal diffusion can be modeled as a *random walk* in which a digital walker starts at the origin and takes N steps in random directions. In our simple 2-D random walk model (Figure 7.3 left), the x and y components of each step add algebraically, and so the radial distance R from the starting point after N steps is

$$R^2 = (\Delta x_1 + \Delta x_2 + \cdots + \Delta x_N)^2 + (\Delta y_1 + \Delta y_2 + \cdots + \Delta y_N)^2 + (x \to y). \quad (7.12)$$

If we take the average of a large number of steps in which all directions are equally likely, the cross terms in (7.12) will average out to zero and we will be left with

$$R^2_{\text{rms}} \simeq \langle \Delta x_1^2 + \Delta y_1^2 \rangle + \langle \Delta x_2^2 + \Delta y_2^2 \rangle + \cdots = N \langle r^2 \rangle = N r^2_{\text{rms}},$$
$$\Rightarrow \quad R_{\text{rms}} \simeq \sqrt{N} r_{\text{rms}}, \quad (7.13)$$

where $r_{\text{rms}} = \sqrt{\langle r^2 \rangle}$ is the *root-mean-square* individual step size.

1. Construct a random walk simulation to determine how many collisions, on the average, will a perfume molecule make before it reaches the back of a room, some 20 m away. You are given the fact that the mean free path in air is approximately 68 nm?

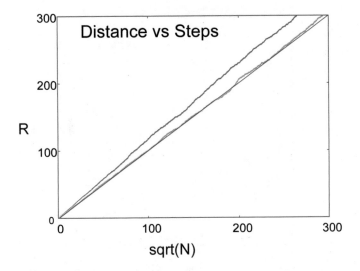

Figure 7.4. The distance covered in walks of N steps using two different schemes for including randomness. The theoretical prediction (7.13) is the straight line.

a. For each step, choose separate random values for $\Delta x'$ and $\Delta y'$ in the range $-1 \to 1$, and then normalize them so that each step is of unit length:

$$\Delta x = \frac{1}{L}\Delta x', \quad \Delta y = \frac{1}{L}\Delta y', \quad L = \sqrt{\Delta x'^2 + \Delta y'^2}. \tag{7.14}$$

Our program Walk.py in Listing 7.4 uses two random numbers to determine the x and y components, which can be positive or negative, of each step:

```
x  +=  (random.random() - 0.5)*2.    #  -1  =<  x  =<  1
y  +=  (random.random() - 0.5)*2.    #  -1  =<  y  =<  1
```

b. Plot six 2-D random walks, each of 1000 steps and each with a different seed.

c. You should expect agreement with (7.13) only as the average over many, independent walks ("trials"). Calculate the mean square distance R^2 for each trial, and then take the average of R^2 for all your $K \simeq \sqrt{N}$ trials,

$$\langle\, R^2(N)\,\rangle = \frac{1}{K}\sum_{k=1}^{K} R^2_{(k)}(N). \tag{7.15}$$

d. Verify numerically the validity of the assumptions made in deriving the theoretical result (7.13), and obtain an estimate of the expected precision of your simulation, by checking how well

$$\frac{\langle \Delta x_i \Delta x_{j\neq i}\rangle}{R^2} \simeq \frac{\langle \Delta x_i \Delta y_j\rangle}{R^2} \simeq 0. \tag{7.16}$$

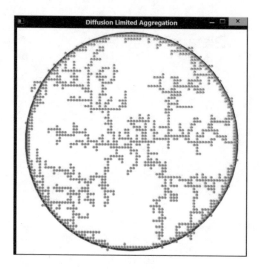

Figure 7.5. A globular cluster of particles created by the diffusion-limited aggregation simulation `DLA.py`.

    e. Plot the root-mean-square distance $R_{\mathrm{rms}} = \sqrt{\langle R^2(N)\rangle}$ as a function of $\sqrt{N}$. Values of N should start with a small number, where $R \simeq \sqrt{N}$ is not expected to be accurate, and end at a quite large value, where several places of accuracy should be expected.

2. Now answer the original question: how many collisions, on the average, will a perfume molecule make before it reaches the back of a room, some 20 m away, given the fact that the mean free path in air is approximately 68 nm.

3. Repeat the preceding and following analysis for a 3-D walk, including the R_{rms} versus N plot.

4. Repeat the preceding and following analysis for a 1-D walk, including the R_{rms} versus N plot.

5. Repeat the calculation of R_{rms} versus $\sqrt{N}$ using a random angle for each step, as opposed to random values for Δx and Δy. Because this method includes less randomness in the simulation, we would expect a lower level of agreement with (7.13).

7.3.2   Diffusion-Limited Aggregation, a Fractal Walk

The shapes of clusters of grapes, dendrites, colloids, and other natural objects appear to arise from an aggregation process. The process is a type of *Brownian motion* related to diffusion, with a characteristic relation between a cluster's perimeter and its mass [Witten & Sander(83)].

1. Deduce the relation between an aggregate's size and its mass via the simulation outlined below.

   (a) Start with an empty 2-D lattice grid[400,400] = 0.

   (b) Place a permanent seed particle in the center by setting grid[199,199]=1.

   (c) Draw a circle around the seed.

   (d) Start a particle off on a walk from random θ on the circle.
      a. Restrict the walk to vertical or horizontal jumps between lattice sites.
      b. Generate a Gaussian-weighted random number in the interval $[-\infty, \infty]$, and use this for the size of the step, with the sign indicating direction. (We indicate how to create weighted distributions in §1.6.2.)

   (e) Imagine a circle of radius 180 lattice spacings centered at grid[199,199]. This is the circle from which we release particles.

   (f) Compute the x and y positions of the new particle on the circle.

   (g) Determine whether the particle moves horizontally or vertically by generating a uniform random number $0 < r_{xy} < 1$, and then having the particle move vertically if $r_{xy} < 0.5$, or horizontally if $r_{xy} > 0.5$.

   (h) Before a jump, check whether a nearest-neighbor site is occupied:
      a. If occupied, the particles stick together, the walker remains in place, and the walk ends.
      b. If unoccupied, jumps one lattice spacing.
      c. Continue jumping until the particle sticks, or leaves the circle, in which case our poor little walker is lost forever.

   (i) Once a random walk is over, another particle is released, and the process repeats for as long as desired. Because many particles are lost, you may need to generate hundreds of thousands of particles to form a cluster of several hundred particles.

2. What relation does your simulation predict between an aggregate's mean size and its mass? Might this relation indicate a fractal?

7.3.3 Surface Deposition

A number of natural and artificial processes involve the deposition of particles onto a surface to form a film. Because the particles are typically evaporated from a hot filament, there is randomness in the emission process; however, the produced films turn out to have well-defined structures (Figure 7.6).

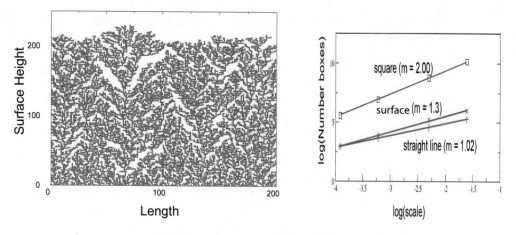

Figure 7.6. *Left:* A simulation of the ballistic deposition of 20,000 particles onto a substrate of length 200. The vertical height increases in proportion to the length of deposition time, with the top being the final surface. *Right:* Fractal dimensions of a line, box, and surface (coastline) determined by box counting. The slope at vanishingly small scale determines the dimension.

1. Create a Monte Carlo simulation of particles falling onto and sticking to a horizontal line of length L composed of 200 deposition sites.

   a. Start all particles from the same height.
   b. Simulate the different emission velocities by having the particles land at random distances along the line.
   c. Have a particle stick to the site on which it lands.
   d. Because there will be columns of deposed particles of various heights, a particle's trajectory may be blocked before it gets down to the line, or it may bounce among columns and then fall into a hole. Therefore assume that if the column height at which the particle lands is greater than that of both its neighbors, it will add to that height.
   e. If the particle lands in a hole, or if there is an adjacent hole, it fills the hole.
   f. Make a continuous plot as the surface is being deposited.

2. Determine the relation between the average height of a column and deposition time. The original research found a linear relation.

3. Extend the simulation of random deposition to two dimensions, so rather than making a line of particles we now deposit an entire surface. Make a continuous plot as the surface is being deposited.

7.4 Thermal Behavior of Magnetic Materials

7.4.1   Roots of a Magnetization vs. Temperature Equation

1. N spin-1/2 particles each with magnetic moment μ are in equilibrium at temperature T in the presence of internal magnetic field B. What is the magnetization in the system as a function of temperature?

   a. Prove that the Boltzmann distribution law [Kittel(05)] predicts the number of particles in the lower energy level (spin up) and in the upper energy level (spin down) to be:

   $$N_L = N\frac{e^{\mu B/(kT)}}{e^{\mu B/(kT)} + e^{-\mu B/(kT)}}, \qquad N_U = N\frac{e^{-\mu B/(kT)}}{e^{\mu B/(kT)} + e^{-\mu B/(kT)}}. \quad (7.17)$$

   b. The *magnetization* M within the system is $\mu \times (N_L - N_U)$, and the molecular magnetic field $B = \lambda M$. Show that the *magnetization* $M(T)$ is proportional to the net number of up spins,

   $$M(T) = N\mu \tanh\left(\frac{\lambda \mu M(T)}{kT}\right). \quad (7.18)$$

   c. Is it possible to find an analytic solution for $M(T)$?

   d. Show that (7.18) can be expressed in terms of the reduced magnetization m, the reduced temperature t, and the Curie temperature T_c:

   $$m(t) = \tanh\left[\frac{m(t)}{t}\right], \qquad m(T) = \frac{M(T)}{N\mu}, \qquad t = \frac{T}{T_c}, \qquad T_c = \frac{N\mu^2\lambda}{k}. \quad (7.19)$$

   e. Show that a solution to this problem is equivalent to finding a root (zero) of

   $$f(m,t) = m - \tanh\left[\frac{m(t)}{t}\right]. \quad (7.20)$$

   f. Create several plots of $f(m,t)$ as a function of the reduced magnetization m.

   g. How many solutions to (7.20) do your plots indicate?

   h. Use the bisection algorithm to find the roots of (7.20) to six significant figures for $t = 0.5$.

   i. Use the Newton-Raphson algorithm to find the roots of (7.20) to six significant figures for $t = 0.5$.

   j. Compare the time it takes the bisection and Newton-Raphson algorithm to find a root. Python's timeit function might help here.

   k. Construct a plot of the reduced magnetization $m(t)$ as a function of the reduced temperature t. This is a numerical inversion of (7.20).

7.4.2   Counting Spin States

1. You are given N indistinguishable spins in a magnetic field **B**, with K particles having spin up and $N - K$ having spin down. You are told that the number

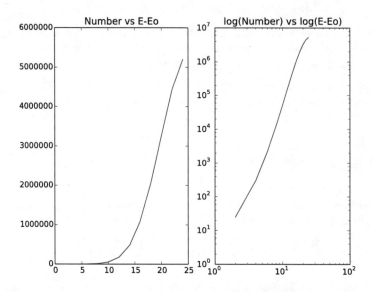

Figure 7.7. *Left*: The number of accessible states versus energy. *Right*: The number of accessible states versus energy.

of accessible states $\Omega(E)$ as a function of energy is proportional to $(E - E_0)^f$, where E_0 is the minimum of energy. Compute the number of states available and the energy of the system for $K = 1, 2, \ldots, 25$. (Our program `Permute.py` is in Listing 7.8.)

a. Show that the number of possible combinations of the particles is:

$$C(N, K) = \frac{N!}{(N - K)!K!}. \qquad (7.21)$$

b. Show that the energy of the system $E = -N\mu B + K\mu B$, where μ is the magnetic moment of each particle.
c. For $\mu = 1$ and $B = 1$, find the minimum energy E_0 for 25 spins.
d. Find the number of accessible states $\Omega(E)$ for $K = 1, 2, \ldots N$.
e. Plot $\Omega(E)$ versus $E - E_0$.
f. Plot $\log\Omega(E)$ versus $\log E - E_0$.
g. Estimate the value of f.

Our program `Permutations.py` produced the plots in Figure 7.7.

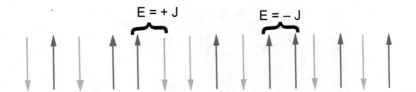

Figure 7.8. The 1-D lattice of N spins used in the Ising model of magnetism. The interaction energy between nearest-neighbor pairs $E = \pm J$ is shown for aligned and opposing spins.

7.5 Ising Model

Figure 7.8 shows N magnetic dipoles fixed in place on an infinite linear chain. Because the particles are fixed, their spins $s_i = \pm 1/2$ are the only dynamical variables. The configuration is described by the state vector

$$|\alpha_j\rangle = |s_1, s_2, \ldots, s_N\rangle = \left\{\pm \frac{1}{2}, \pm \frac{1}{2}, \ldots\right\}, \quad j = 1, \ldots, 2^N. \tag{7.22}$$

We assume that the spins interact with each nearest neighbor via a dipole-dipole interaction, as well as with an external magnetic field $\mathbf{B}$:

$$V_i = -J\mathbf{s}_i \cdot \mathbf{s}_{i+1} - g\mu_b \, \mathbf{s}_i \cdot \mathbf{B}. \tag{7.23}$$

The parameter J is the *exchange energy*, g is the gyromagnetic ratio, and $\mu_b = e\hbar/(2m_e c)$ is the Bohr magneton. The energy and magnetization of this system in state α_k are:

$$E_{\alpha_k} = \left\langle \alpha_k \left| \sum_i V_i \right| \alpha_k \right\rangle = -J \sum_{i=1}^{N-1} s_i s_{i+1} - B\mu_b \sum_{i=1}^{N} s_i, \tag{7.24}$$

$$\mathcal{M}_j = \sum_{i=1}^{N} s_i. \tag{7.25}$$

With two possible values for the spin of any of the N particles, there are 2^N different possible states. Since 2^N gets to be very large for realistic values of N, a statistical approach is usually followed.

The equilibrium alignment of the spins depends critically on the sign of the exchange energy J. If $J > 0$, the lowest energy state will tend to have neighboring spins aligned, as in a *ferromagnet*. If $J < 0$, the lowest energy state will tend to have neighbors with opposite spins, as in an *antiferromagnet*. For simplicity we assume $B = 0$; however, since this removes a preferred direction in space, there may occur instabilities associated with the reversal of all the spins. We also assume a *canonical ensemble* in which the temperature, volume, and number of particles remain fixed.

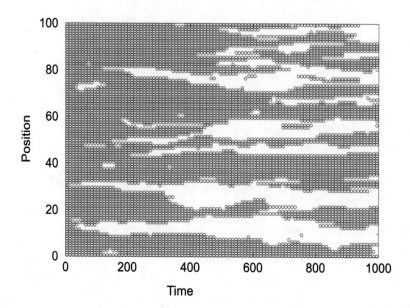

Figure 7.9. An Ising model simulation on a 1-D lattice of 100 initially aligned spins (on the left). Up spins are indicated by circles, and down spins by blank spaces. Although the system starts with all up spins (a "cold" start), the system is seen to form separate domains of up and down spins as time progresses.

Fixed temperature implies that the dipoles are in thermodynamic equilibrium with an average kinetic energy proportional to T. Although this may be an equilibrium state, it is also a dynamic one in which the object's energy fluctuates as it exchanges energy with its environment. The probability $P(\alpha_j)$ of the state α having energy E_α is given by the Boltzmann distribution:

$$\mathcal{P}(E_{\alpha_j}, T) = \frac{e^{-E_{\alpha_j}/k_B T}}{Z(T)}, \qquad Z(T) = \sum_{\alpha_j} e^{-E_{\alpha_j}/k_B T}, \tag{7.26}$$

where k is Boltzmann's constant, T is the temperature, and $Z(T)$ is the partition function.

7.5.1 Metropolis Algorithm

The *Metropolis algorithm* is a technique for computing the Monte Carlo calculation of averages that accurately simulates the fluctuations occurring during thermal equilibrium [Metropolis et al.(53)]. It changes individual spins randomly, but weights the changes such that on the average a Boltzmann distribution results. At any one temperature, the algorithm is repeated until thermal equilibrium is reached. Then, continued application of the algorithm generates the statistical fluctuations about equilibrium from which are deduced thermodynamic quantities. Because the 2^N possible configurations can be very large in number, the amount of computer time needed can be very long; however, equilibration is often obtained after only $\simeq 10N$ iterations.

The explicit steps of the Metropolis algorithm are:

1. Start with an arbitrary spin configuration $\alpha_k = \{s_1, s_2, \ldots, s_N\}$.

2. Generate a *trial* configuration α_{k+1}:

   a. Pick a particle i randomly and flip its spin.

   b. Calculate the energy $E_{\alpha_{\mathrm{tr}}}$ of the trial configuration.

   c. If $E_{\alpha_{\mathrm{tr}}} \leq E_{\alpha_k}$, accept the trial by setting $\alpha_{k+1} = \alpha_{\mathrm{tr}}$.

   d. If $E_{\alpha_{\mathrm{tr}}} > E_{\alpha_k}$, accept with relative probability $\mathcal{R} = \exp(-\Delta E/k_B T)$:

      a. Choose a uniform random number $0 \leq r_i \leq 1$.

      b. Set $\alpha_{k+1} = \begin{cases} \alpha_{\mathrm{tr}}, & \text{if } \mathcal{R} \geq r_j \ (\text{accept}), \\ \alpha_k, & \text{if } \mathcal{R} < r_j \ (\text{reject}). \end{cases}$

The heart of the algorithm is the generation of a random spin configuration α_j (7.22) with probability

$$\mathcal{P}(E_{\alpha_j}, T) \propto e^{-E_{\alpha_j}/k_B T}. \tag{7.27}$$

It does that by setting the ratio of probabilities for a trial configuration of energy E_t to that of an initial configuration of energy E_i as

$$\mathcal{R} = \frac{\mathcal{P}_{\mathrm{tr}}}{\mathcal{P}_i} = e^{-\Delta E/k_B T}, \quad \Delta E = E_{\alpha_{\mathrm{tr}}} - E_{\alpha_i}. \tag{7.28}$$

If the trial configuration has a lower energy ($\Delta E \leq 0$), the relative probability will be greater than 1 and we will accept the trial configuration as the new initial configuration without further ado. However, if the trial configuration has a higher energy ($\Delta E > 0$), we will not reject it out of hand, though instead accept it with relative probability $\mathcal{R} = \exp(-\Delta E/k_B T) < 1$. To accept a configuration with a probability, we pick a uniform random number between 0 and 1, and if the probability is greater than this number, we accept the trial configuration; if the probability is smaller than the chosen random number, we reject it. (You can remember which way this goes by letting $E_{\alpha_{\mathrm{tr}}} \to \infty$, in which case $\mathcal{P} \to 0$ and nothing is accepted.) When the trial configuration is rejected, the next configuration is identical to the preceding one.

1. Three fixed spin-$\frac{1}{2}$ particles interact with each other at temperature $T = 1/k_b$ such that the energy of the system is

$$E = -(s_1 s_2 + s_2 s_3). \tag{7.29}$$

To better understand the algorithm, try doing a simulation by hand. Start with a configuration ↑↓↑. Use the Metropolis algorithm and the series of random numbers 0.25, 0.2, 0.5, 0.3 to determine the results of just three thermal fluctuations of these three spins.

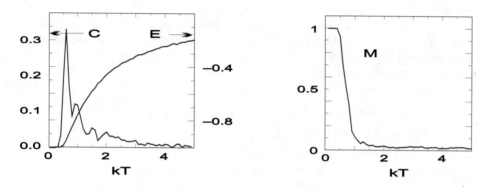

Figure 7.10. Simulation results from a 1-D Ising model of 100 spins. *Left:* energy and specific heat as functions of temperature; *Right:* magnetization as a function of temperature.

Listing `IsingVP.py` is our implementation of the Metropolis algorithm. The key element in the program is the loop

```
for  j in range (1,500):
    test = state
    r = int(N*random.random());   # Flip spin randomly
    test[r] *= -1
    ET = energy(test)
    p = math.exp((ES-ET)/(k*T))   #  Boltzmann test
    enplot.plot(pos=(j,ES))       # Adds segment to curve
    if p >= random.random():
        state = test
        spstate(state)
        ES = ET
```

1. Write a program that implements the Metropolis algorithm, or use the program `IsingVP.py` given in Listing 7.9.

2. The key data structure is an array `S[N]` containing the spin values.

3. Make regular printouts of `S[N]` to examine the formation of spin patterns as the simulation progresses.

4. Set the exchange energy $J = 1$ (ferromagnetic) to fix the energy scale.

5. Start with $k_B T = 1$.

6. Use periodic boundary conditions, `S[0]=S[N-1]`, to minimize end effects.

7. Debug with $N \simeq 20$, but use much larger values for production runs.

8. Verify that the system equilibrates to similar looking states for
   a. an ordered initial configuration (cold start)
   b. a random initial configuration (hot start).

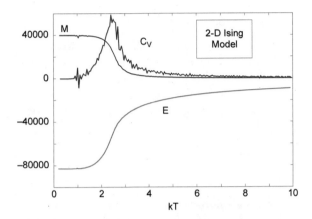

Figure 7.11. The energy, specific heat, and magnetization as a function of temperature from a 2-D Ising model simulation with 40,000 spins. Evidence of a phase transition at the Curie temperature $kT = \simeq 2.5$ is seen in all three functions. The values of C and E have been scaled to fit on the same plot as M. (Courtesy of J. Wetzel.)

7.5.2   Domain Formation

1. Produce figures showing the thermal equilibrium state for a *small* number of atoms at a variety of temperatures.

2. Produce figures showing the thermal equilibrium state for a *large* number of atoms at a variety of temperatures.

3. Verify that there are large fluctuations at high temperatures, or for small numbers of atoms, but smaller fluctuations at lower temperatures, or for large number of atoms.

4. Watch your simulation equilibrate for some large value of $k_B T$ (small values take longer to equilibrate.) There should be a lot of spin flipping even at equilibrium. Look for the occurrence of instabilities in which there is a flipping of a large number of spins. Even a small value for the external B field eliminates this instability.

5. Observe the formation of all up or all down spin domains. Explain why larger and fewer domains should lower the energy of the system.

6. Verify that there are larger and fewer domains at lower temperatures, and, accordingly, that the system has lower energy.

7. Make a graph of average domain size *versus* temperature.

7.5.3 Thermodynamic Properties

Equations (7.24)–(7.25) are used to compute the energy and magnetization of a specific spin configuration. At high temperatures we expect a random assortment of spins and so a vanishing small magnetization. At low temperatures when all the spins are aligned, we expect $\mathcal{M}$ to approach $N/2$, and thus a large magnetization. The internal energy $U(T)$ is just the average value of the energy, and the specific heat is just the derivative of the internal energy:

$$U(T) = \langle E \rangle, \qquad C = \frac{1}{N}\frac{dU}{dT}. \tag{7.30}$$

1. After equilibration, calculate the internal energy U and the magnetization $\mathcal{M}$ for the chain as a function of $k_B T$. Your results should resemble Figure 7.10.

2. Reduce statistical fluctuations by running the simulation a number of times with different initial conditions and taking the average of the results.

3. For large N, the analytic results from statistical mechanics should be valid. Compare the results of your simulation to the analytic expressions.

4. Check that the simulated thermodynamic quantities are independent of initial conditions (within statistical uncertainties). In practice, your cold and hot start results should agree.

7.5.4 Extensions

1. Extend the model to 2-D.

2. Extend the model to 3-D.

3. Extend the model to also include next-nearest neighbor interactions.

4. Extend the model to higher-multiplicity spin states.

5. By examining the heat capacity and magnetization as functions of temperature, search for a phase transition from ordered to unordered configurations in a 2-D or 3-D system. The former should diverge, while the latter should vanish at the phase transition (Figure 7.11).

7.6 Molecular Dynamics

Although quantum mechanics is the proper theory for molecular interactions, it is hard to solve for many-body systems, and so approximate techniques are often employed. Molecular dynamics (MD) solves Newton's laws from first principles for a large number of particles, and has proven to be accurate for properties of materials that do not

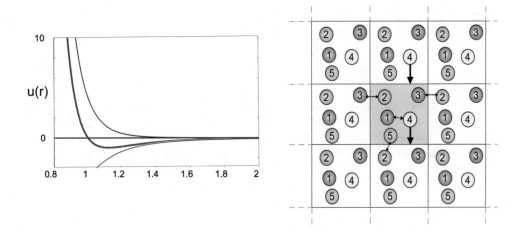

Figure 7.12. *Left:* The Lennard-Jones effective potential used in many MD simulations. Note that the infinitely high central repulsion is not shown. *Right:* The imposition of periodic boundary conditions. The two-headed arrows indicate an interaction with the nearest version of another particle. The vertical arrows indicate how the image of particle 4 enters when the actual particle 4 exits.

depend much on small-r behaviors [Gould et al.(06), Thijssen(99), Fosdick et al.(96)]. MD simulations are similar to thermal Monte Carlo simulations, but with the energy E and volume V of the N particles fixed (*microcanonical ensemble*).

We apply Newton's laws with the assumption that the net force on each particle is the sum of the two-body forces with all other $(N-1)$ particles, with the forces deriving from the sum of the particle-particle potentials $u(r_{ij})$:

$$m\frac{d^2\mathbf{r}_i}{dt^2} = \sum_{i<j=0}^{N-1} \mathbf{f}_{ij}, \quad i = 0, \dots, (N-1), \tag{7.31}$$

$$\mathbf{f}_{ij} = -\frac{du(r_{ij})}{dr_{ij}} \left(\frac{x_i - x_j}{r_{ij}}\hat{\mathbf{e}}_x + \frac{y_i - y_j}{r_{ij}}\hat{\mathbf{e}}_y + \frac{z_i - z_j}{r_{ij}}\hat{\mathbf{e}}_z \right). \tag{7.32}$$

Here $r_{ij} = |\mathbf{r}_i - \mathbf{r}_j| = r_{ji}$ is the distance between the centers of molecules i and j, and the limits on the sums are such that no interaction is counted twice. A common choice is the Lennard-Jones potential (Figure 7.12 left):

$$u(r) = 4\epsilon \left[\left(\frac{\sigma}{r}\right)^{12} - \left(\frac{\sigma}{r}\right)^{6} \right], \qquad \mathbf{f}(r) = -\frac{du}{dr}\frac{\mathbf{r}}{r} = \frac{48\epsilon}{r^2}\left[\left(\frac{\sigma}{r}\right)^{12} - \frac{1}{2}\left(\frac{\sigma}{r}\right)^{6} \right]\mathbf{r}. \tag{7.33}$$

Here the parameter $\epsilon = 1.65^{-21}$ J governs the strength of the interaction, $\sigma = 3.4 \times 10^{-10}$m determines the length scale, and $m = 6.7 \times 10^{-26}$ kg is the particle mass. In order to make the program simpler and to avoid under- and overflows, it is helpful to measure all variables in the natural units formed by these constants. The interparticle potential and force then take the forms

$$u(r) = 4 \left[\frac{1}{r^{12}} - \frac{1}{r^6} \right], \qquad f(r) = \frac{48}{r} \left[\frac{1}{r^{12}} - \frac{1}{2r^6} \right]. \qquad (7.34)$$

To be practical, MD simulations must limit the number of particles and the size of the integration region. These shortcomings are reduced somewhat by the imposition of *periodic boundary conditions* (Figure 7.12 right). If a particle leaves the simulation region, then we bring an *image* particle back through the opposite boundary:

$$x \;\Rightarrow\; \begin{cases} x + L_x, & \text{if } x \leq 0, \\ x - L_x, & \text{if } x > L_x. \end{cases} \qquad (7.35)$$

In principle, a particle interacts with all other particles and their images. Yet since the potential falls off very rapidly, far-off particles do not contribute significantly, and the effect of the potential is ignored for distances greater than a reasonably chosen cutoff radius:

$$u(r) = \begin{cases} 4\left(r^{-12} - r^{-6}\right), & \text{for } r < r_{\text{cut}}, \\ 0, & \text{for } r > r_{\text{cut}}. \end{cases} \qquad (7.36)$$

Accordingly, if the simulation region is large enough for $u(r > L_i/2) \simeq 0$, an atom interacts with only the *nearest image* of another atom. In the interests of speed and simplicity, the equations of motion are usually integrated with the velocity-Verlet algorithm instead of rk4:

$$\mathbf{r}_i(t+h) \simeq \mathbf{r}_i(t) + h\mathbf{v}_i(t) + \frac{h^2}{2}\mathbf{F}_i(t) + \mathrm{O}(h^3), \qquad (7.37)$$

$$\mathbf{v}_i(t+h) \simeq \mathbf{v}_i(t) + h\,\overline{\mathbf{a}(t)} + \mathrm{O}(h^2) \qquad (7.38)$$

$$\simeq \mathbf{v}_i(t) + h\left[\frac{\mathbf{F}_i(t+h) + \mathbf{F}_i(t)}{2}\right] + \mathrm{O}(h^2)\mathbf{v}_i(t) + \frac{h}{2}[\mathbf{F}_i(t+h) + \mathbf{F}_i(t)].$$

Our MD simulation programs, MD1D.py and MD2d.py, are given in Listings 7.10 and 7.11. Although you can write your own, there are lots of items to include and it may be best to modify our codes.

1. Start with a 1-D simulation at zero temperature. The particles should remain in place.

2. Increase the temperature and note how the particles begin to move.

3. At a low, though nonzero, temperature, start off all your particles at the minima in the Lennard-Jones potential. The particles should remain bound in a lattice formation until you raise the temperature.

4. Repeat the simulations for a 2-D system. The trajectories should resemble billiard ball-like collisions.

5. Create an animation of the time-dependent locations of several particles.

6. Calculate and plot the root-mean-square displacement of molecules as a function of time and temperature:

$$R_{\text{rms}} = \sqrt{\left\langle |\mathbf{r}(t + \Delta t) - \mathbf{r}(t)|^2 \right\rangle}, \qquad (7.39)$$

where the average is over all the particles in the box.

7. Determine the approximate time dependence of R_{rms}.

8. Test your system for time-reversal invariance. Stop it at a fixed time, reverse all velocities, and see if the system retraces its trajectories back to the initial configuration after this same fixed time. This would be extremely unlikely to occur for a thermal system, but likely for an MD simulation.

9. It is well known that light molecules diffuse more quickly than heavier ones. Change your MD simulation to handle particles of differing masses [Satoh(11)].

    a. Generalize the velocity-Verlet algorithm so that it can be used for molecules of different masses.
    b. Modify the simulation code so that it can be used for five heavy molecules of mass $M = 10$ and five light molecules of mass $m = 1$.
    c. Start with the molecules placed randomly near the center of the square simulation region.
    d. Assign random initial velocities to the molecules.
    e. Run the simulation several times and verify visually that the lighter molecules tend to diffuse more quickly than the heavier ones.
    f. For each ensemble of molecules, calculate the rms velocity at regular instances of time, and then plot the rms velocities as functions of time. Do the lighter particles have a greater rms velocity?

7.6.1   16 Particles in a Box

A small number of particles are placed in a box. The forces between the particles derive from the Lennard-Jones potential,

$$\mathbf{f}(r) = \frac{48}{r^2} \left[\left(\frac{1}{r} \right)^{12} - \frac{1}{2} \left(\frac{1}{r} \right)^6 \right] \mathbf{r}. \qquad (7.40)$$

A number of independent snapshots are made of the particles in the box, and the number N_{rhs} of particles on the RHS of the box is recorded for each. If n is the

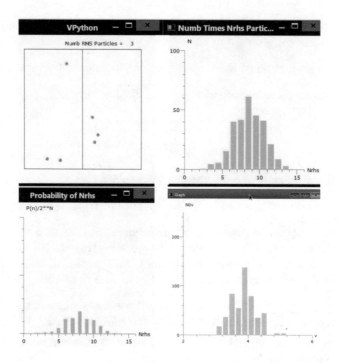

Figure 7.13. *Top Left:* Positions of particles at a single time. *Top Right:* Distribution showing the number of times N_{rhs} particles are present in the RHS of the box. *Lower Left:* The probability distribution for finding N_{rhs} particles in the RHS of the box. *Lower Right:* The velocity distribution for 16 particles in the box.

number of frames that show N_{rhs} particles in the right hand side, then the probability of finding N_{rhs} particles on the RHS is:

$$\mathcal{P}(n) = \frac{C(n)}{2^{N_{rhs}}}. \tag{7.41}$$

1. Modify the previously developed MD program so that it runs for 16 particles inside a 2-D box of side $L = 1$. Assume periodic boundary conditions, and compute the positions and velocities of the particles using the velocity-Verlet algorithm (7.37).

   a. Extend the program so that at the end of each time step it counts the number of particles N_{rhs} on the RHS of the box.

   b. Create, plot, and update continually a histogram containing the distribution of the number of times n that a N_{rhs} occurs, as a function of N_{rhs}.

   c. Make a histogram showing the probability (7.41) of finding N_{rhs} particles on the RHS, as a function of N_{rhs}.

   d. Compare your plots to those in Figure 7.13, created by MDpBC.py.

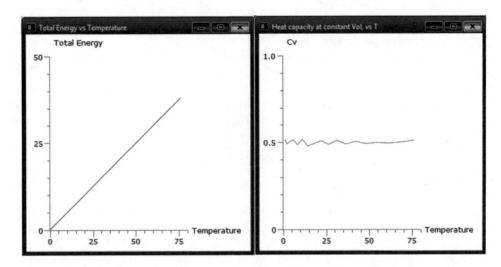

Figure 7.14. *Left:* The total energy versus temperature for 16 particles in a box. *Right:* The heat capacity at constant volume versus temperature for 16 particles in a box.

2. Even though an MD simulation is deterministic, the particles do tend to equilibrate after a rather small number of collisions, in which case the system resembles a thermal one. This is consistent with the result from ergodic theory that after a long time a dynamical system tends to forgets its initial state. Test this hypothesis by randomly assigning several different sets of initial positions and velocities to the 16 particles, and then determining the distributions for each initial condition. If the theorem is valid, the distributions should be much the same.

3. Use your simulation to determine the velocity distribution of the particles.

   a. Create a histogram by plotting the number of particles with velocity in the range v to $v + \Delta v$ versus v.
   b. Start with random values for the initial positions of the particles.
   c. Start all particles off with the same speed v_0, though with random values for directions.
   d. Update the histogram after each step and continue until it looks like a normal distribution.

4. Compute and plot the heat capacity at constant volume, $C_V = \partial E / \partial T$, as a function of temperature for the 16 particles in a box.

   a. As before, start with random values for the initial positions of the particles.
   b. Start all particles off with the same speed v_0, though with random values for directions.

c. Take an average of the temperature for 10 initial conditions, all with the same v_0. Relate the temperature to the total energy.

d. Repeat the computations for increasing values of v_0. (We suggest $0.5 \leq v_0 \leq 20$ in steps of 1.)

e. Plot total energy as a function of the average temperature.

f. Use numerical differentiation to evaluate the heat capacity at constant volume, $C_V = \partial E / \partial T$. Unless you change the parameters of the system, you should expect C_V to be constant, within statistical fluctuations.

7.7   Code Listings

```
# EqHeat.py: heat equation via finite differences, 3-D plot

from numpy import *; import matplotlib.pylab as p
from mpl_toolkits.mplot3d import Axes3D

Nx = 101;          Nt = 3000;       Dx = 0.03;       Dt = 0.9
kappa = 210.; C = 900.; rho = 2700.
T = zeros((Nx,2), float);   Tpl = zeros((Nx, 31), float)

print("Working, wait for figure after count to 10")

for ix in range (1, Nx-1):
   T[ix, 0] = 100.0;
T[0,0] = 0.0 ;   T[0,1] = 0.                 # Initial T
T[Nx-1,0] = 0. ; T[Nx-1,1] = 0.0             # 1st & last T = 0
cons = kappa/(C*rho)*Dt/(Dx*Dx);
m = 1                                        # Counter

for t in range (1, Nt):
    for ix in range (1, Nx - 1):
       T[ix,1] = T[ix,0] + cons*(T[ix+1,0]+ T[ix-1,0]-2.*T[ix,0])
    if t%300 == 0 or t == 1:
        for ix in range (1, Nx - 1, 2): Tpl[ix, m] = T[ix, 1]
        print(m)
        m = m + 1
    for ix in range (1, Nx - 1):  T[ix, 0] = T[ix, 1]
x = list(range(1, Nx - 1, 2))
y = list(range(1, 30))
X, Y = p.meshgrid(x, y)

def functz(Tpl):
    z = Tpl[X, Y]
    return z

Z = functz(Tpl)
fig = p.figure()                              # Create figure
ax = Axes3D(fig)
ax.plot_wireframe(X, Y, Z, color = 'r')
ax.set_xlabel('Position')
ax.set_ylabel('time')
ax.set_zlabel('Temperature')
p.show()
print("finished")
```

Listing 7.1. **EqHeat.py** solves the heat difference equation with a leapfrog algorithm.

```
# EqHeatMov.py: Matplotlib, animated heat eqn via finite diffs

from numpy import *
import numpy as np, matplotlib.pyplot as plt
import matplotlib.animation as animation

Nx = 101;    Dx = 0.01414;    Dt = 0.6
kappa = 210.;    sph = 900.;   rho = 2700.
cons = kappa/(sph*rho)*Dt/(Dx*Dx)    # For algorthim
T = np.zeros( (Nx, 2), float)         # Temp @ first 2 times

def init():
    for ix in range (1, Nx - 1):    T[ix , 0] = 100.0 # Initial T
    T[0,  0] = 0.0;  T[0, 1] = 0.              # Bar ends, t = 0
    T[Nx - 1,  0] = 0.;   T[Nx - 1,  1] = 0.0

init()
k = range(0,Nx)
fig = plt.figure()
# select axis; 111: only one plot, x,y, scales given
ax = fig.add_subplot(111, autoscale_on=False, xlim=(-5, 105),
    ylim=(-5, 110.0))
ax.grid()                              # Plot a grid
plt.ylabel("Temperature")          # Temperature of  bar
plt.title("Cooling of a bar")
line, = ax.plot(k, T[k,0],"r", lw=2)
plt.plot([1,99],[0,0],"r",lw=10)        # Bar
plt.text(45,5,'bar',fontsize=20)

def animate(dum):                      # Dummy num for animation
    for ix in range (1, Nx-1):
        T[ix ,  1] = T[ix ,  0] + cons*(T[ix+1,0]+T[ix -1,0]-2*T[ix ,  0])
    line.set_data(k,T[k,1] )    # (x, y)=(k, temperature)
    for ix in range (1, Nx-1):  T[ix ,0] = T[ix ,1]    # 100 positions
    return line,
ani = animation.FuncAnimation(fig, animate,1)          # 1=dummy
plt.show()
```

Listing 7.2. **EqHeatMov.py** solves the heat difference equation with a leapfrog algorithm and produces an animation of the results.

```
# DecaySound.py spontaneous decay simulation

from vpython import *    # Was: from visual.graph import *
import random, winsound

lambda1 = 0.005                         # Decay constant
max = 80.;  time_max = 500;    seed = 68111
```

```
number = nloop = max                               # Initial  value
graph1 = graph( title ='Spontaneous Decay', xtitle='Time',
                 ytitle = 'Number', xmin=0,xmax=500,ymin=0,ymax=90)
decayfunc = gcurve(color = color.green)

for time in arange(0, time_max + 1):               # Time loop
    for atom in arange(1, number + 1 ):            # Decay loop
        decay = random.random()
        if (decay  <  lambda1):
            nloop = nloop  - 1                         # A decay
            winsound.Beep(600, 100)                 # Sound beep
    number = nloop
    decayfunc.plot( pos = (time, number) )
    rate(30)
```

Listing 7.3. **DecaySound.py** simulates spontaneous decay and beeps at each decay.

```
# Walk.py  Random walk with graph

from vpython import *
import random

random.seed(None)                       # None => system clock
jmax = 20
x    = 0.;              y = 0.          # Start at origin

graph1 = graph(width=300, height=300, title='Random Walk',
    xtitle='x', ytitle='y')
pts = gcurve(color = color.blue)

for i in range(0, jmax + 1):
    pts.plot(pos = (x, y) )                 # Plot points
    x += (random.random() - 0.5)*2.       # -1 =< x =< 1
    y += (random.random() - 0.5)*2.       # -1 =< y =< 1
    pts.plot(pos = (x, y))
    rate(100)
```

Listing 7.4. **Walk.py** simulates a random walk. It calls the random-number generator from the random package. A different seed is needed to obtain a different sequence.

```
# Walk3D.py  3-D Random walk with 3-D graph

from vpython import *
import random

random.seed(None)                       # None => system clock
jmax = 1000
xx =yy = zz =0.0                        # Start at origin

graph1 = canvas(x=0,y=0,width = 500, height = 500,
    title = '3D Random Walk', forward=vector(-0.6,-0.5,-1))
pts   = curve(x=list(range(0, 100)), radius=10.0,
```

```
      color=color.yellow)
xax    = curve(x=list(range(0,1500)), color=color.red,
   pos=[vector(0,0,0),vector(1500,0,0)], radius=10.)
yax    = curve(x=list(range(0,1500)), color=color.red,
   pos=[vector(0,0,0),vector(0,1500,0)], radius=10.)
zax    = curve(x=list(range(0,1500)), color=color.red,
   pos=[vector(0,0,0),vector(0,0,1500)], radius=10.)
xname = label( text = "X", pos = vector(1000, 150,0), box=0)
yname = label( text = "Y", pos = vector(-100,1000,0), box=0)
zname = label( text = "Z", pos = vector(100, 0,1000), box=0)

for i in range(1, 100):
    xx += (random.random() - 0.5)*2.  # -1 =< x =< 1
    yy += (random.random() - 0.5)*2.  # -1 =< y =< 1
    zz += (random.random() - 0.5)*2.  # -1 =< z =< 1
    pts.append(pos=vector ( 200*xx-100, 200*yy-100, 200*zz-100))
    rate(100)
print("Walk's distance R =", sqrt(xx*xx + yy*yy+zz*zz))
```

Listing 7.5. **Walk3D.py** simulated a 3-D random walk. Note, a different seed is needed to obtain a different sequence.

```
# DLA.py:     Diffusion Limited aggregation

from vpython import *
import numpy as np, random
Maxx=500;     Maxy=500                    #canvas width, height
escene = canvas(width=Maxx, height=Maxy,
   title='Diffusion Limited Aggregation', range=40)
escene.center=vector(0,0,15)

def gauss_ran():
    r1 = random.random()
    r2 = random.random()
    fac = sqrt(-2*log(r1))
    mem = int(fac*20000*cos(2*pi*r2))
    return mem

rad = 40.;    step = 0; trav = 0;    size = 60;    max = 500
grid =np.zeros((size,size))# Particle locations, 1=occupied
ring(pos=vector(0,0,0),axis=vector(0,0,1),radius=rad,
            thickness=0.5,color=color.cyan)
grid[30,30] = 1                  # Particle in center
sphere(pos=vector(4*30/3-40,-4*30/3+40,0),
         radius=.8,color=color.green)
ball=sphere(radius=0.8)                   # Moving ball

while True:                     # Generates new ball
    hit = 0                               # No hit
    angle = 2. *pi * random.random()
    x =rad * cos(angle)
    y =rad * sin(angle)
    dist = abs(gauss_ran() )          # Length of walk
    # print(dist)      # Uncomment to see start point
    # sphere(pos=(x,y),color=color.magenta)
    trav = 0
    ballcolor=(color.yellow)
    while( hit==0 and x<40 and x>-40 and y<40
```

```
              and y>-40 and trav < abs(dist)):
          if(random.random()  <0.5):  step=1
          else: step =- 1;
          xg = int(0.7*x+30)        # transform coord for indexes
          yg = int(-0.7*y+30+0.5)  # xg=m*x*b, 30=0*m+b, 58=m*40+b
          if ((grid[xg+1,yg] + grid[xg-1,yg]
             + grid[xg,yg+1] + grid[xg,yg-1]) >= 1):
             hit = 1                        # Ball hits fixed ball
             grid[xg,yg] = 1                # Position now occupied
             sphere(pos=vector(x,y,0),radius=0.8,color=color.yellow)
          else:
             if (random.random() < 0.5): x += step  # Move right
             else: y += step                # Prob = 1/2 to move up
             xp = 80*x/56.0-40
             yp = -80*y/56.+40
             ball.pos = vector(xp,yp,0)
             rate(10000)                    # Change ball speed
      trav = trav + 1      #increments distance, < dist
```

Listing 7.6. **DLA.py** simulates diffusion limited aggregation, a model for dendritic growth.

```
# Column.py:  Correlated ballistic deposition

from vpython import *
import random, numpy as np

graph1 = canvas(width = 500, height = 500,
        title = 'Correlated Ballistic Deposition', range = 250)
pts = points(color=color.green, radius=1)
hit = np.zeros((200))
maxi = 80000;  npoints = 200;  dist = 0;  oldx = 100; oldy = 0

for i in range(1, maxi):
    r = int(npoints*random.random() )
    x = r - oldx
    y = hit[r] - oldy
    dist = x*x  + y*y
    prob = 9.0/dist
    pp = random.random()
    if (pp < prob):
      if(r >0 and r< (npoints-1)):
          if((hit[r] >=  hit[r-1]) and (hit[r] >=  hit[r+1])):
              hit[r] = hit[r] + 1
          else:
              if (hit[r-1] > hit[r+1]): hit[r] = hit[r-1]
              else: hit[r] = hit[r + 1]
      oldx = r
      oldy = hit[r]
      olxc = oldx*2 - 200 # Linear transform
      olyc = oldy*4 - 200
      pts.append(pos=vector(olxc,olyc,0))
```

Listing 7.7. **Column.py** simulates correlated ballistic deposition of minerals onto substrates on which dendrites form.

```
# Permute.py accessible states versus energy

import matplotlib.pyplot as plt, numpy as np
from math import *

k = 0;   n = 25;   B = 1.;   mu = 1.;   i = 0;      Eo = -mu*B*n
Energy = [0.]*(13); Combo = [0.]*(13)
for k in range(0,26):
    c = factorial(n)/(factorial(n-k)*factorial(k))
    E =  -(n-k)*mu*B + k*mu*B
    print(k, E-Eo,c)
    if k < 13:                    # Only plot 1st 1/2 via symmetry
        Energy[i] = E - Eo
        Combo[i] = c
        i += 1
plt.subplot(121)                  # L: accessible states vs E-Eo
plt.plot(Energy,Combo)
plt.title('Number vs E-Eo')
plt.subplot(122)
plt.loglog(Energy, Combo)
plt.title('log(Number) vs log(E-Eo)')
plt.show()
```

Listing 7.8. **Permute.py** Permutations as number of accessible states.

```
# IsingVP.py: Ising model, with VPython

"""Dirichlet boundary conditions surrounding four walls
  Domain dimensions: WxH, with 2 triangles per square
  Based on FEM2DL_Box Matlab program in Polycarpou, Intro Finite
  Element Method in Electromagnetics, Morgan & Claypool (2006) """

from vpython import *; import random, numpy as np

scene = canvas(x=0,y=0,width=700,height=150, ↵
    range=10,title='Spins')
engraph = graph(y=200,width=700,height=150,
  title='E of Spin System', xtitle='iteration',
  ytitle='E',xmax=500, xmin=0, ymax=5, ymin=-5)
enplot = gcurve(color=color.blue, graph=engraph)
N = 30; B = 1.;   mu = .33;   J = .20;   k = 1.;   T = 100.
state = np.zeros((N));    S = np. zeros((N) ,float)
test = state
random.seed()

def energy (S) :
    FirstTerm = 0.
    SecondTerm = 0.
    for i in range(0,N-2):   FirstTerm += S[i]*S[i + 1]
    FirstTerm *= -J
    for i in range(0,N-1):   SecondTerm += S[i]
    SecondTerm *= -B*mu;
    return (FirstTerm + SecondTerm);

ES = energy(state)

def spstate(state):                                # Plots spins
```

```
    for obj in scene.objects: obj.visible = 0 # Erase arrows
    j = 0
    for i in range(-N,N,2):
        if state[j]==-1:   ypos = 5                  # Spin down
        else:              ypos = 0
            if  5*state[j]<0: arrowcol = vector(1,1,1)  # White = down
            else:                       arrowcol = vector(0.7,0.8,0)
            arrow(pos=vector(i,ypos,0),axis=vector(0,5*state[j],0),
                  color=arrowcol)
            j += 1

for  i  in range(0 ,N):  state[i] = -1  # Initial spins all down

for obj in scene.objects:   obj.visible = 0
spstate(state)
ES = energy(state)

for   j  in range (1,500):
    rate(3)
    test = state
    #print('ko')
    r = int(N*random.random());    # Flip spin randomly
    test[r] *= -1
    ET = energy(test)
    p = np.exp((ES-ET)/(k*T))    # Boltzmann test
    enplot.plot(pos=(j,ES))             # Adds segment to curve
    if p >= random.random():
            state = test
            spstate(state)
            ES = ET
```

Listing 7.9. **IsingVP.py** Applies the Metropolis algorithm to the Ising model and visualizes the results.

```
# MD1.py            Molecular dynamics in 1D

from vpython import *
import random, numpy as np

scene = canvas(x=0,y=0,width=400,height=150,
   title='Molecular Dynamics', range=4)        # Spheres
sceneK = graph(width=400,height=150,title='Average KE',ymin=0,
   ymax=0.3,xmin=0,xmax=100,xtitle='time',ytitle='KE avg')
Kavegraph=gcurve(color= color.red)  # plot KE
scenePE = graph(width=400,height=150,title=
   'Pot Energy', ymin=-0.6,ymax=0.0,xmin=0,xmax=100,
   xtitle='time',ytitle='PE')
PEcurve = gcurve(color=color.cyan)
Natom = 8;   Nmax =  8;   Tinit = 10.0;   t1 = 0; L = Natom
x =np.zeros( (Nmax), float);   vx = np.zeros( (Nmax), float)
fx = np.zeros( (Nmax, 2), float)
atoms = []

def twelveran():                 # Gaussian as average 12 randoms
    s = 0.0
    for i in range (1,13):  s += random.random()
    return s/12.-0.5
```

```
def initialposvel():               # Initial positions, velocities
    i = -1
    for ix in range(0, L):
            i = i + 1
            x[i] = ix
            vx[i] = twelveran()
            vx[i] = vx[i]*sqrt(Tinit)
    for j in range(0,Natom):
    xc = 2*x[j] - 7    # Linear TF places spheres
    atoms.append(sphere(pos=vector(xc,0,0),radius=0.5,color=color.red))

def sign(a, b):
    if (b >= 0.0): return abs(a)
    else: return - abs(a)

def Forces(t, PE):                                    # Forces
    r2cut = 9.                                        # Cutoff
    PE = 0.
    for i in range(0, Natom):  fx[i][t] = 0.0
    for i in range( 0, Natom-1 ):
        for j in range(i + 1, Natom):
            dx = x[i] - x[j]
            if (abs(dx) > L/2): dx = dx-sign(L, dx) # Image
            r2 = dx*dx
            if (r2 < r2cut):
                if (r2==0.): r2 = 0.0001    # Avoid 0 denom
                invr2 = 1./r2
                wij =  48.*(invr2**3 - 0.5) *invr2**3
                fijx = wij*invr2*dx
                fx[i][t] = fx[i][t]  +  fijx
                fx[j][t] = fx[j][t]  -  fijx
                PE = PE + 4.*(invr2**3)*((invr2**3)-1.)
    return PE

def timevolution():
    t1 = 0;   t2 = 1; KE = 0; PE = 0
    h = 0.038                         # Unstable if larger
    hover2 = h/2

    initialposvel()
    PE = Forces(t1,PE)
    for i in range(0,Natom):  KE = KE + (vx[i]*vx[i])/2
    t = 0
    while t<100:                                 # Time loop
        rate(1)
        for i in range(0,  Natom):
            PE = Forces(t1,PE)
            x[i] = x[i] + h*(vx[i] + hover2*fx[i][t1])
            if x[i] <= 0.: x[i] = x[i] + L    # Periodic BC
            if x[i] >= L : x[i] = x[i] - L
            xc = 2*x[i] - 8                   # Linear transform
            atoms[i].pos=vector(xc,0,0)
        PE = 0.0
        PE = Forces(t2,   PE)
        KE = 0.
        for  i  in range(0 , Natom):
            vx[i] = vx[i] + hover2*(fx[i][t1] + fx[i][t2])
            KE = KE + (vx[i]*vx[i] )/2
        T = 2*KE/(3*Natom)
        Itemp = t1
        t1 = t2
        t2 = Itemp
        Kavegraph.plot(pos=(t,KE))                   # Plot KE
        PEcurve.plot(pos=(t,PE),display=scenePE)  # Plot PE
```

```
        t += 1
timevolution()
```

Listing 7.10. **MD1D.py** perform a molecular dynamics simulation in one dimension.

```
# MD2D.py:                  Molecular dynamics in 2D

from vpython import *; import numpy as np; import random

scene = canvas(x=0,y=0,width=350,height=350,
   title='Molecular Dynamics', range=10)
sceneK = graph(x=0,y=350,width=600,height=150,title='Average KE',
   ymin=0.0,ymax=5.0,xmin=0,xmax=500,xtitle='time',ytitle='KE avg')
Kavegraph=gcurve(color= color.red)
sceneT = graph(x=0,y=500,width=600,height=150,title='Average PE',
   ymin=-60,ymax=0.,xmin=0,xmax=500,xtitle='time',ytitle='PE avg')
Tcurve = gcurve(color=color.cyan)
Natom = 25; Nmax = 25; Tinit = 2.0; dens = 1   # dens =1.2 for fcc
t1 = 0
x  = np.zeros( (Nmax), float); y  = np.zeros( (Nmax), float)
vx = np.zeros( (Nmax), float); vy = np.zeros( (Nmax), float)
fx = np.zeros( (Nmax, 2), float); fy = np.zeros( (Nmax, 2), float)
L = int(1.*Natom**0.5)                        # Side of lattice
atoms=[]

def twelveran():                          # Average 12 rands for Gaussian
    s=0.0
    for i in range (1,13): s += random.random()
    return s/12.0-0.5

def initialposvel():                                    # Initialize
    i = -1
    for ix in range(0, L):               # x-> 0  1  2  3  4
        for iy in range(0, L):           # y=0  0  5  10 15 20
            i = i + 1                     # y=1  1  6  11 16 21
            x[i]  = ix                    # y=2  2  7  12 17 22
            y[i]  = iy                    # y=3  3  8  13 18 23
            vx[i] = twelveran()           # y=4  4  9  14 19 24
            vy[i] = twelveran()           # numbering of 25 atoms
            vx[i] = vx[i]*sqrt(Tinit)
            vy[i] = vy[i]*sqrt(Tinit)
    for j in range(0,Natom):
        xc = 2*x[j] - 4
        yc = 2*y[j] - 4
        atoms.append(sphere(pos=vec(xc,yc,0), radius=0.5,
            color=color.red))

def sign(a, b):
    if (b >= 0.0): return abs(a)
    else: return - abs(a)

def Forces(t, w, PE, PEorW):                            # Forces
    # invr2 = 0.
    r2cut = 9.                           # Switch: PEorW = 1 for PE
    PE = 0.
    for i in range(0, Natom): fx[i][t] = fy[i][t]  = 0.0
    for i in range( 0, Natom-1 ):
        for j in range(i + 1, Natom):
```

```
                        dx = x[i] - x[j]
                        dy = y[i] - y[j]
                        if (abs(dx) > L/2): dx = dx - sign(L, dx) # Image
                        if (abs(dy) > L/2): dy = dy - sign(L, dy)
                        r2 = dx*dx + dy*dy
                        if (r2 < r2cut):
                             if (r2==0.): r2 = 0.0001 # Aavoid 0 denom
                             invr2 = 1./r2
                             wij = 48.*(invr2**3 - 0.5) *invr2**3
                             fijx = wij*invr2*dx
                             fijy = wij*invr2*dy
                             fx[i][t] = fx[i][t] + fijx
                             fy[i][t] = fy[i][t] + fijy
                             fx[j][t] = fx[j][t] - fijx
                             fy[j][t] = fy[j][t] - fijy
                             PE = PE + 4.*(invr2**3)*((invr2**3) - 1.)
                             w = w + wij
          if (PEorW == 1): return PE
          else: return w

def timevolution():
     avT = 0.; avP = 0.;    Pavg = 0.;    avKE = 0.;  avPE = 0.
     t1 = 0;  PE = 0.0; h = 0.031
     hover2 = h/2
     KE = 0.
     w = 0.
     initialposvel()
     for i in range(0,Natom): KE = KE+(vx[i]*vx[i]+vy[i]*vy[i])/2
     PE = Forces(t1,w,PE,1)
     time =1
     while 1:
          rate(100)
          for i in range(0, Natom):
               PE = Forces(t1,w,PE,1)
               x[i] = x[i] + h*(vx[i] + hover2*fx[i][t1])
               y[i] = y[i] + h*(vy[i] + hover2*fy[i][t1]);
               if x[i] <= 0.: x[i] = x[i] + L          # Periodic BC
               if x[i] >= L : x[i] = x[i] - L
               if y[i] <= 0.: y[i] = y[i] + L
               if y[i] >= L: y[i] = y[i] - L
               xc = 2*x[i] - 4
               yc = 2*y[i] - 4
               atoms[i].pos=vec(xc,yc,0)
          PE = 0.
          t2=1
          PE = Forces(t2, w, PE, 1)
          KE = 0.
          w = 0.
          for i in range(0 , Natom):
               vx[i] = vx[i] + hover2*(fx[i][t1] + fx[i][t2])
               vy[i] = vy[i] + hover2*(fy[i][t1] + fy[i][t2])
               KE = KE + (vx[i]*vx[i] + vy[i]*vy[i])/2
          w = Forces(t2, w, PE, 2)
          P = dens*(KE+w)
          T = KE/(Natom)   # increment averages
          avT = avT + T;   avP = avP + P
          avKE = avKE + KE;    avPE = avPE + PE
          time += 1
          t = time
          if (t==0): t=1
          Pavg = avP/t; eKavg = avKE/t; ePavg = avPE/t; Tavg = avT/t
          pre = (int)(Pavg*1000)
          Pavg = pre/1000.0
          kener = (int)(eKavg*1000)
```

```
        eKavg = kener/1000.0
        Kavegraph.plot(pos=(t,eKavg))
        pener = (int)(ePavg*1000)
        ePavg = pener/1000.0
        tempe = (int)(Tavg*1000000)
        Tavg = tempe/1000000.0
        Tcurve.plot(pos=(t,ePavg),display=sceneT)

timevolution()
```

Listing 7.11. **MD2D.py** perform a molecular dynamics simulation in two dimensions.

```
# MDpBC.py:  VPython MD with Periodic BC

from vpython import *
import random, numpy as np
L=1; Natom=16;   Nrhs=0;  dt=1e-6
dN = np.zeros((16))
scene = canvas(width=400,height=400,range=(1.3)   )
inside  = label(pos=vector(0.4,1.1,0),text='Particles here=',box=0)
inside2 = label(pos=vector(0.8,1.1,0),box=0)
border = curve(pos=[(-L,-L,0),(L,-L,0),(L,L,0),(-L,L,0),(-L,-L,0)])
half = curve(pos=[(0,-L,0),(0,L,0)],color=color.yellow) # middle
ndist = graph(ymax = 200, width=400, height=300,
    xtitle='Particles in right half', ytitle='N')
bars = gvbars(delta=0.8,color=color.red)
positions = []                            # position of atoms
vel = []                                       # vel of atoms
Atom = []                             # will contain spheres
fr  = [0]*(Natom)                          # atoms (spheres)
fr2 = [0]*(Natom)                          # second  force
Ratom = 0.03                              # radius of atom
pref = 5                             # a reference velocity
h = 0.01
factor = 1e-9                            # for lennRatomd jones

for i in range (0,Natom):                 # initial x's and v's
    col = vec(1.3*random.random(),1.3*random.random(),
         1.3*random.random())
    x = 2.*(L-Ratom)*random.random()-L+Ratom      # positions
    y = 2.*(L-Ratom)*random.random()-L+Ratom # border forbidden
    Atom = Atom+[sphere(pos=vec(x,y,0),radius=Ratom,color=col)]
    theta = 2*pi*random.random()               # select angle
    vx = pref*cos(theta)                  # x component velocity
    vy = pref*sin(theta)
    positions.append((x,y,0))             # add positions to list
    vel.append((vx,vy,0))                  # add momentum to list
    posi = np.array(positions)          # Ratomray with positions
    ddp = posi[i]
    if ddp[0]>=0 and ddp[0]<=L: Nrhs+=1  # count  atoms R half
    v = np.array(vel)                    # Ratomray of velocities

def sign(a, b):                          # sign function
    if (b >=  0.0):  return abs(a)
    else: return  - abs(a)
def forces(fr):
    fr=[0]*(Natom)
    for i in range( 0, Natom-1 ):
```

```
        for j in range(i + 1, Natom):
            dr=posi[i]-posi[j]
            if (abs(dr[0]) > L): dr[0] = dr[0]-sign(2*L,dr[0])
            if (abs(dr[1]) > L): dr[1] = dr[1]-sign(2*L, dr[1])
            r2=dr[0]**2+dr[1]**2+dr[2]**2
            if (abs(r2) < Ratom):   r2 = Ratom   # avoid 0 denom
            invr2 = 1./r2
            fij = invr2*factor*48.*(invr2**3-0.5)*invr2**3
            fr[i] = fij*dr+ fr[i]
            fr[j]= -fij*dr +fr[j]
    return fr

for t in range (0,1000):  Nrhs = 0      # begin at zero in ea time
    for i in range(0,Natom):
        rate(100)
        fr = forces(fr)
        dpos = posi[i]        # periodic BC
        if dpos[0] <= -L:  posi[i] = (dpos[0]+2*L,dpos[1],0)
        if dpos[0] >= L:   posi[i] = (dpos[0]-2*L,dpos[1],0)
        if dpos[1] <= -L:  posi[i] = (dpos[0],dpos[1]+2*L,0)
        if dpos[1] >= L:   posi[i] = (dpos[0],dpos[1]-2*L,0)
        dpos = posi[i]
        if dpos[0]>0 and dpos[0]<L: Nrhs+=1  # count particle right
        fr2 = forces(fr)
        fr2 = fr
        v[i] = v[i]+0.5*h*h*(fr[i]+fr2[i])        # velocity Verlet
        posi[i] = posi[i]+h*v[i]+0.5*h*h*fr[i]
        aa = posi[i]
        Atom[i].pos = vector(aa[0],aa[1],aa[2])   # plot positions
        posi[i]=aa
    dN[Nrhs] += 1
    inside2.text='%4s'%Nrhs                          # Atoms right side
    for j in arange(0,16): bars.plot(pos=(j,dN[j]))
```

Listing 7.12. MDpBC.py perform a molecular dynamics simulation with periodic boundary conditions.

8
Biological Models: Population Dynamics & Plant Growth

8.1 Chapter Overview

This chapter starts with the logistic map, a model for the variations in insect popu-lations with time. This map and others are simple yet provide surprisingly complex behaviors, and are also appropriate as problems in Classical Dynamics. In §8.3 we ex-tend the idea behind the logistic map to include predator and prey populations, and use differential equations instead of discrete mappings. The chapter ends with three models for growth of plants and proteins. Also of biological interest is molecular dynamics, covered in Chapter 7.

8.2 The Logistic Map

Biological populations often display complex dynamics that can be modeled rather simply. Here we explore several models, starting with the most elementary and adding refinements, with even the simplest model capable of producing surprising complex and chaotic dynamics. Although we focus on biological systems, the behaviors are rather general and applicable to a wide range of phenomena.

Recall the radioactive decay simulation in §6.3 where the discrete decay law, $\Delta N_i \Delta t = -\lambda N_i$, led to exponential-like decay. Clearly, then, reversing the sign of λ,

$$\frac{\Delta N_i}{\Delta t} = \lambda \, N_i, \tag{8.1}$$

leads to exponential-like *growth*. Yet realistically, for a given environment there tends to be a maximum population N_*, the *carrying capacity*, which implies that the growth must slow down as the population approaches N_*. Accordingly, we modify the ex-ponential growth model to incorporate an effective growth rate that decreases as the

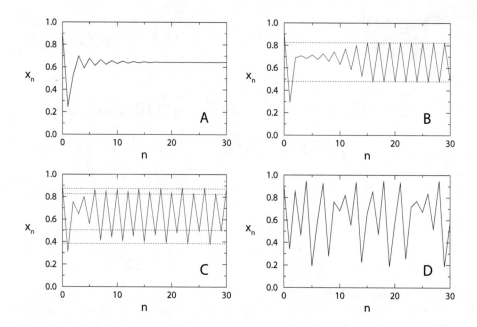

Figure 8.1. The insect population x_n versus the generation number n for the two growth rates: (A) $\mu = 2.8$, a single attractor; (B) $\mu = 3.3$, a double attractor; (C) $\mu = 3.5$, a quadruple attractor; (D) $\mu = 3.8$, nearly chaotic behavior.

population approaches N_*:

$$\frac{\Delta N_i}{\Delta t} = \lambda'(N_* - N_i)N_i. \tag{8.2}$$

This new law should produce exponential-like growth if N_i is small compared to N_*, slower growth as N_i approaches N_*, and actually decay if $N_i > N_*$.

1. Prove that with a change to dimensionless variables, (8.2) can be written in the standard form for the *logistic map*:

$$x_{i+1} = \mu x_i(1 - x_i), \tag{8.3}$$

   where the growth rate μ is a constant.

2. Why is the logistic map called nonlinear and one-dimensional?

3. For large N_*, how is x_i related to the population number N_i?

4. What is the range of values expected for x_i?

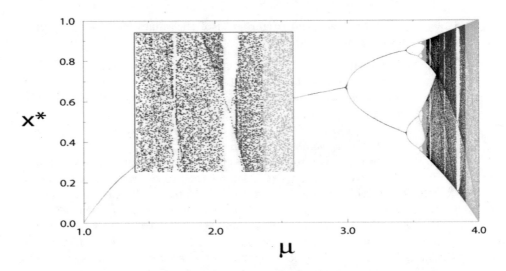

Figure 8.2. A bifurcation plot of attractor population x^* versus growth rate μ for the logistic map. The inset shows some details of a three-cycle window.

5. Calculate and plot x_{i+1} as a function of the generation number i for a range of μ values, $0 \leq \mu \leq 4$.

   a. The initial population x_0 is called the *seed*, with the interesting dynamics not sensitive to its particular value.
   b. Show that small μ values lead to extinction, while large values lead to instabilities.
   c. Confirm that negative values for μ lead to decaying populations.
   d. Look for the possibility of long-term stable states for $0 \leq \mu \leq 2$ in which the population remains unchanged after the transients die off.
   e. Verify that differing values of the seed x_0 do affect the transients, though *not* the values of the stable populations.

6. Use the analytic form of the logistic map to find a one-cycle fixed point for which

$$x_{i+1} = x_i = x_*. \tag{8.4}$$

7. Use the analytic form of the logistic map to find a two-cycle fixed point for which

$$\mu x_*(1 - x_*) = x_*. \tag{8.5}$$

   a. Verify numerically that your predictions for one- and two-cycle points are correct and independent of the seed used.
   b. Note the number of iterations needed for the transients to die out.

8. Create a *bifurcation diagram* (Figure 8.2) of the values found numerically for fixed points x_* versus growth rate μ.

   a. Proceed through the range $1 \leq \mu \leq 4$ into 1000 steps. These are the "bins" into which we will place the x_* values.
   b. Go through at least 200 iterations at each μ value to ensure that the transients have died out.
   c. In order not to miss any structures in your bifurcation diagram, loop through a range of initial x_0 values for each μ.
   d. Wait for the transients to die out so that the population is at an attractor x_*. Write the values (μ, x_*) to a file for each of several hundred iterations. Print out these x_* values to no more than three or four decimal places (the `int` command may be useful for that), and remove duplicate entries.
   e. Plot your file of x_* versus μ using small symbols for the points and with the points not connected. Our sample program `Bugs.py` is given in Listing 8.1.
   f. Enlarge (zoom in on) sections of your plot where the points are dense so that it displays bifurcations as well. This is called *self-similarity*.

9. At what value of μ does a very large number of populations suddenly change to a three-cycle population? This is an example of a *window*.

10. Look carefully at the logistic map's bifurcations occurring at

$$\mu_k \simeq 3,\ 3.449,\ 3.544,\ 3.5644,\ 3.5688,\ 3.569692,\ 3.56989,\ \ldots . \qquad (8.6)$$

You should find that the series ends in a region of chaos. Feigenbaum discovered that the μ values in this sequence converge geometrically when expressed in terms of the distance between bifurcations δ [Feigenbaum(79)]:

$$\mu_k \ \to \ \mu_\infty - \frac{c}{\delta^k}, \qquad \delta = \lim_{k \to \infty} \frac{\mu_k - \mu_{k-1}}{\mu_{k+1} - \mu_k}. \qquad (8.7)$$

Determine the constants in (8.7) and compare to those found by Feigenbaum:

$$\mu_\infty \simeq 3.56995, \quad c \simeq 2.637, \quad \delta \simeq 4.6692. \qquad (8.8)$$

Amazingly, the value of δ is universal for all second-order maps.

8.2.1 Other Discrete and Chaotic Maps

Here are several other maps, some of which are used in biological models. To study them, repeat the questions presented for the logistic map, only now for any of these other maps.

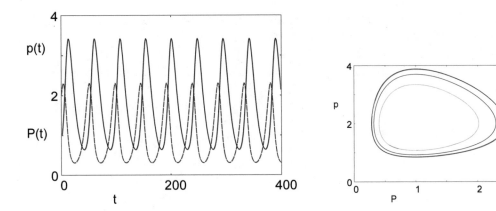

Figure 8.3. *Left:* The time dependencies of the populations of prey *p(t)* (solid curve) and of predator *P(t)* (dashed curve) from the Lotka-Volterra model. *Right:* A phase space plot of Prey population p as a function of predator population P. The different orbits correspond to different initial conditions.

The Ecology Map	$x_{n+1} = x_n e^{\mu(1-x_n)}$
The Gaussian Map	$x_{n+1} = e^{-bx_n^2} + \mu$
Quadratic Map	$x_{n+1} = 1 - 2x_n^2$
Ricker Model Map	$x_{n+1} = e^{[r(1-x_n/K)]} x_n$
Tent Map	$x_{n+1} = \begin{cases} 2x_n & \text{for } 0 \le x_n < \frac{1}{2} \\ 2(1-x_n) & \text{for } \frac{1}{2} \le x_n < 1 \end{cases}$
Baker's Map	$(x_{n+1}, y_{n+1}) = \begin{cases} (2x_n, y_n/2) & \text{for } 0 \le x_n < \frac{1}{2} \\ (2-2x_n, 1-y_n/2) & \text{for } \frac{1}{2} \le x_n < 1. \end{cases}$

8.3 Predator-Prey Dynamics

The logistic map is simple, though maybe too much so to be a realistic model of population dynamics. The Lotka-Volterra Model (LVM) extends the logistic map to predator and prey populations coexisting in the same geographical region:

$$p(t) = \text{prey density}, \qquad P(t) = \text{Predator density}. \tag{8.9}$$

If there were no interaction between the species, and the prey population p breeds at a per-capita rate of a, then there would be exponential growth:

$$\frac{dp}{dt} = ap \qquad \Rightarrow \qquad p(t) = p(0)e^{at}. \tag{8.10}$$

The rate at which predator eat prey requires both to be around, and so a model for the prey growth rate including both predation and breeding is:

$$\frac{\Delta p}{\Delta t} = a\,p - b\,p\,P, \qquad \frac{dp}{dt} = a\,p - b\,p\,P, \quad \text{(LVM-I for prey).} \qquad (8.11)$$

Nevertheless predators also tend to eat each other at a per-capita mortality rate m:

$$\left.\frac{dP}{dt}\right|_{mort} = -mP, \qquad \Rightarrow \quad P(t) = P(0)e^{-mt}. \qquad (8.12)$$

Likewise to the prey, the predator population grows at the rate

$$\frac{dP}{dt} = \epsilon\,b\,p\,P - m\,P \qquad \text{(LVM-I for predators),} \qquad (8.13)$$

where ϵ is a constant is a measure of the efficiency with which predators convert prey interactions into food.

Table 8.1. Typical Lotka-Volterra model parameter values

Model	a	b	ϵ	m	K	k
LVM-I	0.2	0.1	1	0.1	0	—
LVM-II	0.2	0.1	1	0.1	20	—
LVM-III	0.2	0.1	—	0.1	500	0.2

1. Express the two simultaneous ODEs (8.11) and (8.13) in the dynamical form for solution with the rk4 algorithm. A sample code to solve these equations is PredatorPrey.py in Listing 8.2 with some results shown in Figure 8.3.

2. Solve the ODEs for a range of initial conditions and parameter values. Some good starting values are given in Table 8.1.

   a. Create and examine several time series for prey and predator populations.
   b. Create and examine several phase space plots of predator versus prey populations.

3. Compute the equilibrium values for the prey and predator populations. Do you think that a model in which the cycle amplitude depends on the initial conditions can be realistic? Explain.

4. Show that the predator and prey populations oscillate out of phase with each other in time.

5. Construct a "phase space" plot of $P(t)$ versus $p(t)$.

6. What is the meaning of a closed orbit in this phase space plot?

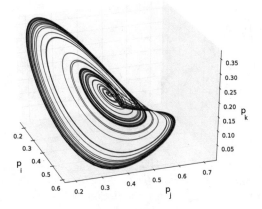

Figure 8.4. A chaotic attractor for a 4-D Lotka-Volterra model projected onto three axes.

7. If the prey are pests, what kind of structure would be required in phase space to be called "control" of the pests?

8. Because predators eat prey, one might expect the existence of a large number of predators to lead to the eating of a large number of prey. Explain why the maxima in predator population and the minima in prey population do not occur at the same time.

9. Why do the extreme values of the population repeat with no change in values?

10. Explain the meaning of the spirals in the predator-prey phase space diagram.

11. What different initial conditions lead to different phase-space orbits?

12. Discuss the symmetry or lack of symmetry in the phase-space orbits.

8.3.1 Predator-Prey Chaos

If three or more species are present, predator-prey models can become chaotic [Vano et al.(06), Cencini et al.(10)]. The basic ODEs for four competing populations p_i are

$$\frac{dp_i}{dt} = a_i p_i \left(1 - \sum_{j=1}^{4} b_{ij} p_j \right), \qquad i = 1, 4, \tag{8.14}$$

where a_i is a growth rate parameter for species i, and b_{ij} is a parameter related to the resources consumed by species j that otherwise would be used by i.

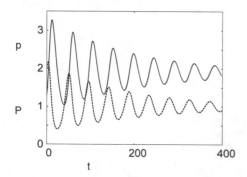
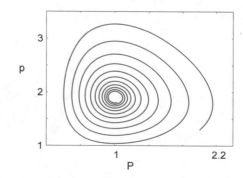

Figure 8.5. The Lotka-Volterra model II including prey limits. *Left:* Prey population $p(t)$ (solid curve) and predator population $P(t)$ (dashed curve) versus time. *Right:* Prey population p as a function of predator population P.

1. Because four species covers a very large parameter space, we suggest that you start your exploration using parameters close to those that [Vano et al.(06)] found produce chaos:

$$a_i = \begin{pmatrix} 1 \\ 0.72 \\ 1.53 \\ 1.27 \end{pmatrix}, \qquad b_{ij} = \begin{pmatrix} 1 & 1.09 & 1.52 & 0 \\ 0 & 1 & 0.44 & 1.36 \\ 2.33 & 0 & 1 & 0.47 \\ 1.21 & 0.51 & 0.35 & 1 \end{pmatrix}. \qquad (8.15)$$

Note: you may have to adjust the parameters or initial conditions slightly to obtain truly chaotic behavior. Some good starting values are given in Table 8.1.

2. Solve (8.14) with initial conditions that may lead to four species coexistence:

$$p_i(t = 0) = \begin{pmatrix} 0.3013, & 0.4586, & 0.1307, & 0.3557. \end{pmatrix}. \qquad (8.16)$$

3. Visualize the population dynamics by making plots of $p_1(t), p_2(t), p_3(t)$, and $p_4(t)$ versus time.

4. Create a data structure containing $[p_1(t_i), p_2(t_i), p_3(t_i), p_4(t_i)]$ for $i = 1, \ldots$. In order to avoid needlessly long files, you may want to skip a number of time steps before each file output.

5. Plot all possible 2-D phase space plots, that is, plots of p_i vs p_j, $i \neq j = 1$–3.

6. Plot all possible 3-D phase space plots, that is, plots of p_i vs p_j vs p_k. One such plot is shown in Figure 8.4.

8.3.2 Including Prey Limits

The initial assumption in the LVM that only predators limit the growth of prey needs to be extended to account for the restriction on growth arising from the depletion of the food supply as the prey population grows. Accordingly, as we did with the logistic map, we modify the constant growth parameter from a to $a(1-p/K)$, which accounts for growth decreasing as the population approaches the *carrying capacity* K:

$$\frac{dp}{dt} = a\,p\left(1 - \frac{p}{K}\right) - b\,p\,P, \qquad \frac{dP}{dt} = \epsilon\,b\,p\,P - m\,P, \qquad \text{(LVM-II).} \qquad (8.17)$$

1. Extend your solution of the two-population LVM to include prey limits. Some good starting values are given in Table 8.1. You should obtain solutions similar to that in Figure 8.5.

   a. Create and examine several time series for prey and predator populations.
   b. Create and examine several phase space plots of predator versus prey populations.
   c. Calculate numerical values for the equilibrium values of the prey and predator populations.
   d. Make a series of runs for different values of prey carrying capacity K. Can you deduce how the equilibrium populations vary with prey carrying capacity?
   e. Make a series of runs for different initial conditions for predator and prey populations. Do the cycle amplitudes depend on the initial conditions?

2. Show that both populations exhibit damped oscillations as they approach equilibrium.

3. Show that the equilibrium populations are independent of the initial conditions.

4. Explain why a spiral in the phase space plot would be considered "control" of the prey population.

8.3.3 Including Predation Efficiency

A yet more realistic predator-prey model decreases the rate bpP at which prey are eliminated in order to account for the time predators spend *handling* and digesting prey. If a predator spends time t_{search} searching for prey, then the *functional response* p_a is the probability of one predator finding one prey:

$$p_a = b\,t_{\text{search}}\,p \quad \Rightarrow \quad t_{\text{search}} = \frac{p_a}{bp}. \qquad (8.18)$$

If t_h is the time a predator spends handling a single prey, then the effective time a predator spends handling a prey is $p_a\,t_h$, and so the total time T that a predator spends finding and handling a single prey is

$$T = t_{\text{search}} + t_{\text{handling}} = \frac{p_a}{bp} + p_a t_h. \qquad (8.19)$$

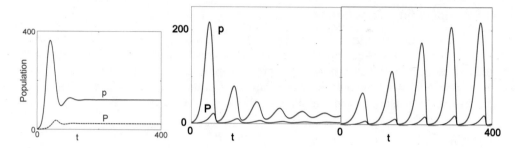

Figure 8.6. Lotka-Volterra model III with predation efficiency and prey limitations. The competing resource parameter b increases from left to right.

The effective *rate* of eating prey is then $p_a/T = bp/(1 + bpt_h)$. We see that as the number of prey $p \to \infty$, the efficiency in eating them $\to 1$. We include this efficiency in (8.17) by modifying the rate b at which a predator eliminates prey to $b/(1 + bpt_h)$:

$$\frac{dp}{dt} = ap\left(1 - \frac{p}{K}\right) - \frac{bpP}{1 + bpt_h}, \qquad \text{(LVM-III)}. \qquad (8.20)$$

And not to ignore our predator friends, we also limit the predator carrying capacity by making it proportional to the number of prey:

$$\frac{dP}{dt} = mP\left(1 - \frac{P}{kp}\right), \quad \text{(LVM-III)}. \qquad (8.21)$$

1. Extend your simulations to the extended model of (8.20) and (8.21).

   a. Create and examine several time series for prey and predator populations.
   b. Create and examine several phase space plots of predator versus prey populations.

2. As shown in Figure 8.6, adjust b so as to obtain three dynamic regimes.

3. Create phase space plots for each dynamical region.

4. Create a plot showing no overdamping.

5. Create a plot showing damped oscillations leading to stable equilibria.

6. Create a plot showing a limit cycle. The transition from equilibrium to a limit cycle is called a *phase transition*.

7. Calculate the critical value for b corresponding to a phase transition between the stable equilibrium and the limit cycle.

8. Which dynamical regime could be called satisfactory prey control?

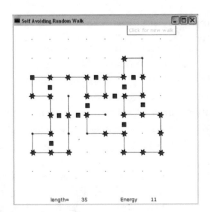 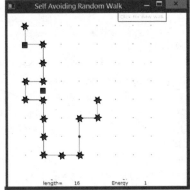

Figure 8.7. Two self-avoiding random walks that simulate protein chains with hydrophobic (H) monomers as stars, and polar (P) monomers in black. The squares indicates regions where two H monomers are not directly connected.

8.3.4 Two Predators, One Prey

An extension of the LVM contains two predator populations, P_1 and P_2, sharing the same prey population p:

$$\frac{dp}{dt} = ap\left(1 - \frac{p}{K}\right) - (b_1 P_1 + b_2 P_2)\, p, \qquad (8.22)$$

$$\frac{dP_1}{dt} = \epsilon_1 b_1 p P_1 - m_1 P_1, \qquad \frac{dP_2}{dt} = \epsilon_2 b_2 p P_2 - m_2 P_2. \qquad (8.23)$$

1. Extend your rk4 simulation to describe two predators and one prey.

   a. Use the following values for the model parameters and initial conditions:
      $a = 0.2$, $K = 1.7$, $b_1 = 0.1$, $b_2 = 0.2$, $m_1 = m_2 = 0.1$, $\epsilon_1 = 1.0$,
      $\epsilon_2 = 2.0$, $p(0) = P_2(0) = 1.7$, and $P_1(0) = 1.0$.
   b. Determine the time dependence for each population.
   c. Vary the characteristics of the second predator and calculate the equilibrium population for the three components.
   d. What does the model suggest as to the possibility of two predators coexisting while sharing the same prey?

2. Explore the parameter space to find examples of chaotic behavior.

8.4 Growth Models

8.4.1   Protein Folding as a Self-Avoiding Walk

A protein is a large biological molecule containing chains of nonpolar hydrophobic (H) monomers that are repelled by water and polar (P) monomers that are attracted by water. The structure of a protein results from a *folding process* in which random coils of chains rearrange themselves into a configuration of minimum energy. Your problem is to create a variation on the random walk problem that models the folding process and produces the lowest energy state of an H plus P sequence of various lengths [Yue et al.(04)]. This will be a *a self-avoiding walk* in which the walker does not visit the same point more than once.

We take the energy of a chain of monomers as $E = -\epsilon f$, where ϵ is a positive constant and f is the number of H–H neighbors *not* connected directly (P–P and H–P bonds do not count at lowering the energy). So if the neighbor next to an H is another H, it lowers the energy, though if it is a P, it does not. Accordingly, we expect the natural states of H–P sequences to be those with the largest possible number f of H–H contacts. We show a typical simulation result in Figure 8.7, where a square is placed halfway between two H neighbors (stars).

1. Program up the model for protein folding.

   a. Set up a random walk on a 2-D square lattice.
   b. At the end of each step, randomly choose an H or a P monomer, with the H more likely than the P.
   c. Place the monomer on the lattice site.
   d. Restrict the walk such that the positions available for each step are the three neighboring sites, with remaining on the present site not an option (this is the "self-avoiding" aspect).
   e. The walk stops at a corner, or if there are no empty sites available.
   f. Produce visualizations showing the positions occupied by the monomers, with the H and P monomer indicated by different color or shapes, as in Figure 8.7.
   g. When the walk ends, record the energy and length of the chain.

2. Run many simulations and save the outputs, cataloged by length and energy.

3. ⊙ Extend the folding to 3-D.

Our self-avoiding random walk code `ProteinFold.py` is given in Listing 8.3. A pseudocode of it is given below:

```
#  ProteinFold.py: Self avoiding random walk
Import packages
Assign parameters
Set up plotting

Function selectcol      # Select atoms' colors
    if random <= 0.7: red hydrophobic
    else white polar
```

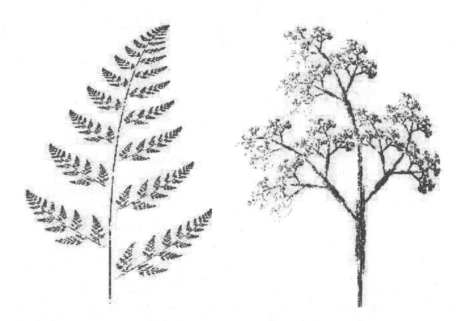

Figure 8.8. *Left:* A fractal fern generated by 30,000 iterations of the algorithm (8.27). Enlarging this fern shows each frond with a similar structure. *Right:* A fractal tree created with the algorithm (8.30).

```
Function findrest  # Check link energies
    ener = 1 if red unlinked neighbor

Function findenergy (energy)  # Fink energy of each link
    if white: pass
    if red:
        e = findrest for neighbor i-1,j
        energy = energy + e
        place yellow dot at center
        repeat for neighbor i+1,j
        repeat for neighbor i,j+1
        repeat for neighbor i,j-1

function grid  # Set up plot grid
```

8.4.2 Plant Growth Simulations

Although people are often attracted to the beauty of nature by its symmetries and regularities, there appears to be randomness at the very heart of natural growth, at least if you believe the algorithms to follow. These algorithms repeatedly map one set of points into another by a *self-affine connection* that includes scaling, translations, and rotations:

$$(x', y') = s(x, y) = (sx, sy) \quad \text{(scaling)},\tag{8.24}$$

$$(x', y') = (x, y) + (a_x, a_y) \quad \text{(translation)},\tag{8.25}$$

$$x' = x\cos\theta - y\sin\theta, \quad y' = x\sin\theta + y\cos\theta \quad \text{(rotation)}.\tag{8.26}$$

If successive points are related by different values of θ, there may be contractions and reflections, though there will still be an affine connection. In all cases, however, the objects created with these rules are self-similar; each step leads to new parts of the object bearing the same relation to the ancestor parts as the ancestors did to theirs, which leads to objects looking similar at all scales.

Below we present two Monte Carlo simulations that produce outputs resembling ferns and trees. For each simulation:

1. Determine the fractal dimension of various parts of the trees or ferns. (Fractal dimension determination is discussed in §2.9.)

2. Explore how variations in the models' parameters affect the produced structures.

3. Try your hand at varying the parameters or the structure of the rules, and see what you can come up with.

8.4.3 Barnsley's Fern

A Barnsley's fern is created by the mapping [Barnsley & Hurd(92)]:

$$(x, y)_{n+1} = \begin{cases} (0.5, 0.27y_n), & \text{with 2\% probability,} \\ (-0.139x_n + 0.263y_n + 0.57 \\ \quad 0.246x_n + 0.224y_n - 0.036), & \text{with 15\% probability,} \\ (0.17x_n - 0.215y_n + 0.408 \\ \quad 0.222x_n + 0.176y_n + 0.0893), & \text{with 13\% probability,} \\ (0.781x_n + 0.034y_n + 0.1075 \\ \quad -0.032x_n + 0.739y_n + 0.27), & \text{with 70\% probability.} \end{cases}\tag{8.27}$$

To select a transformation with probability $\mathcal{P}$, we select a uniform random number $0 \le r \le 1$ and perform the transformation if r is in a range proportional to $\mathcal{P}$:

$$\mathcal{P} = \begin{cases} 2\%, & r < 0.02, \\ 15\%, & 0.02 \le r \le 0.17, \\ 13\%, & 0.17 < r \le 0.3, \\ 70\%, & 0.3 < r < 1. \end{cases}\tag{8.28}$$

The rules (8.27) and (8.28) can be combined into one:

$$(x,y)_{n+1} = \begin{cases} (0.5, 0.27y_n), & r < 0.02, \\ (-0.139x_n + 0.263y_n + 0.57 \\ \quad 0.246x_n + 0.224y_n - 0.036), & 0.02 \leq r \leq 0.17, \\ (0.17x_n - 0.215y_n + 0.408 \\ \quad 0.222x_n + 0.176y_n + 0.0893), & 0.17 < r \leq 0.3, \\ (0.781x_n + 0.034y_n + 0.1075, \\ \quad -0.032x_n + 0.739y_n + 0.27), & 0.3 < r < 1. \end{cases} \quad (8.29)$$

Although (8.27) is clearer, (8.29) is easier to program, as we do in Listing 8.4. An interesting property of the fern in Figure 8.8 left is that it is not completely self-similar, as you can see by noting how different the stems and the fronds are, with each part having a different fractal dimension.

8.4.4 Self-Affine Trees

A tree-like structure, such as in Figure 8.8 right, is generated with the following self-affine transformation:

$$(x_{n+1}, y_{n+1}) = \begin{cases} (0.05x_n, 0.6y_n), & 10\% \text{ probability}, \\ (0.05x_n, -0.5y_n + 1.0), & 10\% \text{ probability}, \\ (0.46x_n - 0.15y_n, 0.39x_n + 0.38y_n + 0.6), & 20\% \text{ probability}, \\ (0.47x_n - 0.15y_n, 0.17x_n + 0.42y_n + 1.1), & 20\% \text{ probability}, \\ (0.43x_n + 0.28y_n, -0.25x_n + 0.45y_n + 1.0), & 20\% \text{ probability}, \\ (0.42x_n + 0.26y_n, -0.35x_n + 0.31y_n + 0.7), & 20\% \text{ probability}. \end{cases}$$
$$(8.30)$$

8.5 Code Listings

```
# Bugs.py The Logistic map

from vpython import *

m_min = 1.0;        m_max = 4.0;          step = 0.01
gcur=gcurve(color=color.red)
pts = gdots( color = color.green, size=0.00001)
lasty = int(1000 * 0.5)            # Eliminates some points
count = 0                          # Plot every 2 iterations
graph1 = graph(width=300, height=300, title='Logistic Map',
    xtitle='m', ytitle='x', xmax=4., xmin=1., ymax=1., ymin=0.,
    background=color.black)
```

```
for m in arange(m_min, m_max, step):
    y = 0.5
    for i in range(1,201,1):      y = m*y*(1-y)
    for i in range(201,402,1):    y = m*y*( 1 - y)
    for i in range(201, 402, 1):                    # Avoid transients
       oldy=int(1000*y)
       y = m*y*(1 - y)
       inty = int(1000 * y)                         # Avoid repeats
       if inty != lasty and count%2==0: pts.plot(pos=(m,y))
       lasty = inty
       count   += 1
```

Listing 8.1. Bugs.py produces the bifurcation diagram of the logistic map. A full program requires finer grids, a scan over initial values, and removal of duplicates.

```
# PredatorPrey.py:        Lotka-Volterra models

from vpython import *
import numpy as np

from rk4Algor import rk4Algor
Tmin = 0.0;   Tmax = 500.0;  Ntimes = 1000
y = np.zeros( (2), float)

y[0] = 2.0;   y[1] = 1.3
h = (Tmax - Tmin)/Ntimes
t = Tmin

def f( t, y):                # Modify function for your problem
       F=np.zeros((2),float)
       F[0] = 0.2*y[0]*(1 - (y[0]/(20.0) )) - 0.1*y[0]*y[1]
       F[1] = - 0.1*y[1] + 0.1*y[0]*y[1];
       return F

f(0,y)

graph1 = graph(width = 300, height = 300,
      title = 'Prey p & predator P vs time',xtitle = 't',
      ytitle = 'P, p',xmin=0,xmax=500,ymin=0,ymax=3.5)
funct1 = gcurve(color = color.magenta)
funct2 = gcurve(color = color.green)
graph2 = graph( width = 300, height = 300,
           title = 'Predator P vs prey p', xtitle = 'P',
           ytitle = 'p',xmin=0,xmax=2.5,ymin=0,ymax=3.5)
funct3 = gcurve(color = color.red)

for t in arange(Tmin, Tmax + 1, h):
    funct1.plot(pos = (t, y[0]) )
    funct2.plot(pos = (t, y[1]) )
    funct3.plot(pos = (y[0], y[1]) )
    rate(60)
    y = rk4Algor(t,h,2,y,f)
```

Listing 8.2. PredatorPrey.py computes the population dynamics for a group of predators and prey.

```
# ProteinFold.py: VPython, Self avoiding random walk
# Stops in corners or  occupied neighbors
# energy  = -f/eps, f=1 if neighbour = H, f=0 if p
# Yellow dot indicates unconnected neighbor

from vpython import *;  import random
import numpy as np

Maxx = 400;  Maxy = 400;  ran = 20; L = 100;  m = 100
size  = 8;  size2 = size*2;  nex = 0;  n = 100
M = [];  DD = []                    # Arrays for polymer & grid
clicked=True
graph1 = canvas(width=Maxx, height=Maxy, title='Protein Folding
  - To begin click in black screen', range=ran)
positions = points(color=color.cyan, radius = 2)

def selectcol():                       # Select atom's colors
    hp = random.random()                 # Select H or P
    if hp <= 0.7:
        col = vec(1,0,0)                 # Hydrophobic color red
        r = 2
    else:
        col = vec(1,1,1)                 # Polar color white
        r = 1
    return col,r

def findrest(m,length,fin,fjn):     # Check links energies
    ener = 0
    for t in range(m,length+1):    # Next link not considered
        if DD[t][0]==fin and DD[t][1]==fjn and DD[t][2]==2:
            ener = 1                # Red unlinked neighbor
    return ener

def findenergy(length,DD):          # Finds energy of each link
    energy = 0
    for n in range (0,length+1):
        i = DD[n][0]
        j = DD[n][1]
        cl = DD[n][2]
        if cl==1: pass                        # if white
        else:                                 # red
            if n < length+1:
                imin = int(i-1)           # Check neighbor i-1,j
                js = int(j)
                if imin >= 0:
                    e = findrest(n+2,length,imin,js)
                    energy = energy + e
                    if e==1:           # Yellow dot at neighbor
                        xol = 4*(i-0.5)-size2
                        yol = -4*j+size2
                        points(pos=vec(xol,yol,0),color=color.yellow,
                               radius=4)
                ima = i+1
                js = j
                if ima<=size-1:                # Check neighbor i+1,j
                    e = findrest(n+2,length,ima,js)
                    energy = energy+e
                    if e == 1:      # Yellow dot at neighbor
                        xol = 4*(i+0.5)-size2
```

```
                              yol = -4*j+size2
                              points(pos=vec(xol,yol,0),color=color.yellow,
                                         radius=4)
                  iss = i
                  jma = j+1
                  if jma <= size-1:          # Check neighbor i,j+1
                      e = findrest(n+2,length,iss,jma)
                      energy = energy+e
                      if e == 1:              # Yellow dot at neighbor
                          xol = 4*i-size2            # Start at middle
                          yol = -4*(j+0.5)+size2
                          points(pos=vec(xol,yol,0),color=color.yellow,
                                     radius=4)
                  iss = i
                  jmi = j-1
                  if jmi >= 0:               # Check neighbor i, j-1
                      e = findrest(n+2,length,iss,jmi)
                      energy = energy +e
                      if e==1:                # Yellow dot at neighbor
                          xol = 4*i-size2      # Start at middle
                          yol = -4*(j-0.5)+size2
                          points(pos=vec(xol,yol,0),color=color.yellow,
                                     radius=4)
        return energy

def grid():                                            # Plot grid
    for j in range(0,size):
        yp = -4*j+size2                      # World to screen coord
        for i in range (0,size):                 # Horizontal row
            xp = 4*i-size2
            positions.append(pos =vector(xp,yp,0))
grid()
length = 0

def  erase():
    graph1.visible=False
    #graph2.visible=True
    for obj in graph1.objects:          # Start new walk
            obj.visible = False              # Clear curve
    clicked=False
    return clicked  # Clicked=True

def handleclick(ev):
    graph1.visible = True # graph2.visible=False
    clicked = True
    return clicked
if clicked == True:
  while 1:
    rate(1)
    pts2 = label(pos=vec(-5, -18,0), box=0)
    length = 0
    grid =np.zeros((size,size))
    D = np.zeros((L,m,n))
    DD = []
    i = size//2                             # Center of grid
    j = size//2
    xol = 4*i-size2
    yol = -4*j+size2
    col,c = selectcol()
    grid[i,j] = c                          # Particle in center
    M = M+[points(pos=vec(xol,yol,0),color=col, radius=4)]
    DD = DD+[[i,j,c]]
    erase()
    while (i>0 and i<size-1 and j>0 and j<size-1
```

```
          and (grid[i+1,j] == 0
            or grid[i-1,j] == 0 or grid[i,j+1] == 0
            or grid[i,j-1] == 0)):
    r = random.random()
    if r < 0.25 :                           # Probability 25%
        if grid[i+1,j]==0:  i += 1    # Step R if empty
    elif 0.25 < r and r < 0.5:              # Step L
        if grid[i-1,j] == 0: i -= 1
    elif 0.50 < r and r < 0.75:                      # Up
        if grid[i,j-1]==0:  j -= 1
    else :                                  # Down
        if grid[i,j+1]==0:  j+=1
    if grid[i,j] == 0:
        col,c = selectcol()
        grid[i,j] = 2                       # Occupy grid point
        length += 1            # Increase length as occupied
        DD = DD+[[i,j,c]]
        xp = 4*i-size2
        yp = -4*j+size2
        curve(pos=[(xol,yol,0),(xp,yp,0)])# Connect last to new
        M = M + [points(pos=vec(xp,yp,0), color=col,radius=4)]
        xol = xp                            # Start new line
        yol = yp
    while (j == (size-1) and i != 0 and i != (size-1)):
        r1 = random.random()
        if r1 < 0.2:                        # Prob 20% move left
            if grid[i-1,j] == 0: i -= 1
        elif r1 > 0.2 and r1 < 0.4:         # Prob 20% move right
            if grid[i+1,j] == 0: i += 1
        else:                               # Prob 60% move up
            if grid[i,j-1] == 0:  j-=1
        if grid[i,j] == 0:
            col,c = selectcol()             # Increase length
            grid[i,j] = 2                   # Grid point occupied
            length += 1
            DD = DD + [[i,j,c]]
            xp = 4*i - size2
            yp = -4*j + size2
            curve(pos=[(xol,yol,0),(xp,yp,0)])
            M = M+[points(pos=vec(xp,yp,0),color=col,radius=4)]
            xol = xp
            yol = yp     # Last row; Stop if corner or neighbors
        if (i==0 or i==(size-1)) or (grid[i-1,size-1]!=0
                and grid[i+1,size-1]!=0):
            break
    while (j == 0 and i != 0 and i != (size-1)): # First row
        r1 = random.random()
        if r1 <0.2:
            if grid[i-1,j] == 0:  i -= 1
        elif r1 >0.2 and r1 <0.4:
            if grid[i+1,j]==0:      i += 1
        else:
            if grid[i,j+1]==0:      j += 1
        if grid[i,j]==0:
            col,c = selectcol()
            grid[i,j] = 2
            length += 1
            DD = DD + [[i,j,c]]
            xp = 4*i - size2
            yp = -4*j + size2
            curve(pos=[(xol,yol,0),(xp,yp,0)])
            M = M+[points(pos=vec(xp,yp,0),color=col,radius=4)]
            xol = xp
            yol = yp
```

```
            if i==(size-1) or i==0 or (grid[i-1,0]!=0
                and grid[i+1,0]!=0):
                break
        while (i==0 and j !=0 and j !=(size-1)):  # First column
            r1 = random.random()
            if r1<0.2:
                if grid[i,j-1] == 0:   j -= 1
            elif r1 > 0.2 and r1 < 0.4:
                if grid[i,j+1] == 0:   j += 1
            else:
                if grid[i+1,j] == 0:   i += 1
            if grid[i,j] == 0:
                col,c = selectcol()
                grid[i,j] = c
                length += 1
                DD = DD+[[i,j,c]]
                xp = 4*i - size2
                yp = -4*j + size2
                curve(pos=[(xol,yol,0),(xp,yp,0)])
                M = M +[points(pos=vec(xp,yp,0), ↵
                    color=col,radius=4)]
                xol = xp
                yol = yp
            if j==(size-1) or j==0 or (grid[0,j+1]!=0
                and grid[0,j-1]!=0):
                break
        while (i==(size-1) and j !=0 and j !=(size-1)):  # Last col
            r1 = random.random()
            if r1 < 0.2:
                if grid[i,j-1] == 0: j -= 1
            elif r1 > 0.2 and r1 < 0.4:
                if grid[i,j+1] == 0: j += 1
            else:
                if grid[i-1,j] == 0: i -= 1
            if grid[i,j] == 0:
                col,c = selectcol()
                grid[i,j] = c
                length += 1
                col,c=selectcol()
                DD = DD + [[i,j,c]]
                xp = 4*i - size2
                yp = -4*j + size2
                curve(pos=[(xol,yol,0),(xp,yp,0)])
                M = M +[points(pos=vec(xp,yp,0), ↵
                    color=col,radius=4)]
                xol = xp
                yol = yp
            if j == (size-1) or (grid[size-1,j+1]!=0
                and grid[size-1,j-1]!=0):
                break
label(pos=vec(-10, -18,0), text='Length=', box=0)
clabel=label(pos=vec(10,18,0), text='Click for new walk',
    color=color.red, display=graph1)
pts2.text = '%4s' %length
label(pos=vec(5,-18,0), text='Energy',box=0)
evalue=label(pos=vec(10, -18,0), box=0) # Energy
evalue.text = '%4s' %findenergy(length,DD)
clicked = False
graph1.bind('click',handleclick)
if(handleclick==True):
    continue
```

Listing 8.3. **ProteinFold.py** uses a self-avoiding random walk to simulate protein folding.

```
# Fern3D.py   based  on  M.F.  Barnsley ,  "Fractals  Everywhere"
# Press  mouse's  right  button  to  drag

from vpython import *
import random

imax = 20000                              # points  to  draw
x = 0.5;    y = 0.0;    z = -0.2;    xn = 0.0;    yn = 0.0
graph1 = canvas(width=400, height=400, forward=vec(-3,0,-1),
     title='Fern3D fractal (to rotate: drag right mouse button)',
     range=10)      # Range: -10< x,y,z<10,
graph1.show_rendertime = True
# Using points:  cycle=27 ms,  render=6 ms
# Using spheres:  cycle=750 ms,  render=30 ms
pts = points(color=color.green, radius=0.5)
for i in range(1,imax):
    r = random.random();                   # random  number
    if ( r <= 0.1):                        #10% probability
        xn = 0.0
        yn = 0.18*y
        zn= 0.0
    elif (r > 0.1 and r <= 0.7):           #60% probability
        xn = 0.85 * x
        yn =  0.85 * y + 0.1 * z + 1.6
        zn=-0.1*y + 0.85*z
    elif (r > 0.7 and r <= 0.85):          #15 % probability
        xn =   0.2 * x - 0.2* y
        yn = 0.2 * x + 0.2 * y + 0.8
        zn= 0.3*z
    else:
        xn = -0.2 * x +0.2 *y              #15% probability
        yn = 0.2 * x +0.2 *y + 0.8
        zn = 0.3*z
    x = xn
    y = yn
    z = zn
    xc = 4.0*x               # linear  transformations  for  plotting
    yc = 2.0*y-7
    zc = z
    pts.append(pos=vec(xc,yc,zc))
```

Listing 8.4. **Fern3D.py** simulates the growth of ferns in 3-D.

9

Additional Entry-Level Problems

9.1 Chapter Overview

Although this entire book contains problems and demos with parts that are appropriate for first- and second-year college courses, in this chapter we present additional problems aimed directly at the lower division. In several sections we give specific overviews and background materials, but you may also want to look at related sections other places in the book. And who's to say that these same problems can't be used for upper division courses?

9.2 Specular Reflection and Numerical Precision

For a perfectly reflecting surface, the basic law of optics tells us that the angle of incidence equals the angle of reflection (Figure 9.1 left). If no light is absorbed during a reflection, a light ray continues to reflect endlessly (Figure 9.1 right). With an origin placed at the center of the circular mirror, we locate the ray by the angle θ. For an

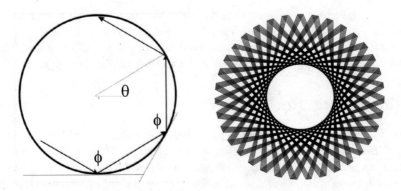

Figure 9.1. *Left:* Specular reflection within a circular mirror in which incident angle equals angle of reflection. *Right:* Infinite internal reflections between two circular mirrors.

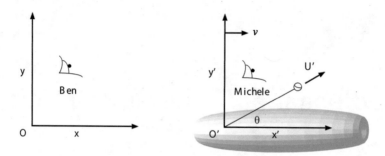

Figure 9.2. Observer Ben at rest on Earth (frame O) watches Michele on a rocket (frame O' moving to the right with velocity v) hit a golf ball.

initial angle $\phi < \pi$, the angle increases by 2ϕ after each reflection:

$$\theta_{\text{new}} = \theta_{\text{old}} + 2\phi. \tag{9.1}$$

Although this appears to indicate that θ increases endlessly, the addition or subtraction of 2π to θ does not change the location on the circle, and so if ϕ/π is a rational number (the ratio of two integers),

$$\frac{\phi}{\pi} = \frac{n}{m}, \tag{9.2}$$

the ray will fall upon itself and form a geometric figure (Figure 9.1 right).

1. Determine the path followed by a light ray for a perfectly reflecting mirror.

2. Plot the light trajectories for a range of values for the initial angle ϕ.

3. Repeat the previous calculation, though this time use just four places of precision. You can do this by using the Python command `round`, for instance, `round(1.234567,4) = 1.234`. You should find that a significant relative error accumulates. This concern increases as the number of calculational steps increases.

9.3 Relativistic Rocket Golf

Although this problem requires a level of relativity covered in elementary texts [Smith(65), Serway & Beichner(99)], some students may not feel comfortable if they have not seen the materials before. In that case, consider the problem as just an application, of equation 9.5. As a more advanced extension, the equivalent of which we do in §5.8.1, one may examine the golf ball's trajectory as viewed in different frames.

Michele, a golf fanatic obsessed with hitting further, hits her golf balls from a rocket moving at $c/2$, where $c = 299792458 m/s$ is the speed of light. As indicated in Figure 9.2, Michele drives a ball with a speed of $U' = 1/\sqrt{3}c$, at an angle $\theta = 30°$, as measured with respect to the moving rocket. She observes her drive to remain in the air for a hang time $T' = 2.6 \times 10^7$ seconds (which is almost a year!). Ben, an observer on the earth, sees her rocket moving to the right with a velocity $v = c/2$.

1. How would Ben, watching Michele's drive from the earth, describe the golf ball in terms of its speed, initial angle ϕ, and hang time T? (We present the theory below.)

2. How would the answer change if Michele hit her ball to the *left*, that is, at an angle $\pi - \theta$.

3. How would the answer change if Michele hit her ball to the left at an angle $60°$ below the horizon, namely, with $\theta = 240°$?

4. If Michele hit the ball with a speed $U = c$, how would the answers change?

5. Repeat the calculation of hitting to the left, though this time use just four places of precision. You can do this by using the Python command `round`, for example, `round(1.234567,4) = 1.234`. You should find a large relative error, always a concern with floating point numbers.

Being told that an object is traveling with a velocity near that of the speed of light implies that we need to apply Einstein's special theory of relativity [Smith(65),Serway & Beichner(99)]. As indicated in Figure 9.2, we start by considering Ben in a reference frame O with x-y axes attached to the earth, and Michele in frame O' with x'-y' axes attached to the rocket. Relative to Ben, frame O' (the rocket) is moving along the positive x axis with a uniform velocity v. Likewise, Michele feels herself at rest in the rocket and sees Ben moving to the left with velocity $-v$.

Ben and Michele may agree on the relative velocity of their frames, though they do not agree on measurements of times and distances. As an example, if Michele says that her golf ball was in the air for a period of time T', then Ben would say that the ball was in the air for a longer time

$$\Delta T = \gamma T', \qquad \gamma = \frac{1}{\sqrt{1 - v^2/c^2}} > 1. \qquad (9.3)$$

Furthermore, if Michele in the rocket O' sees her golf ball having velocity components

$$U' = (U'_x, U'_y), \qquad (9.4)$$

then Ben on earth O, will see her golf ball move with velocity components

$$U = (U_x, U_y), \qquad U_x = \frac{U'_x + v}{1 + vU'_x/c^2}, \qquad U_y = \frac{U'_y}{\gamma(1 + vU'_x/c^2)}. \qquad (9.5)$$

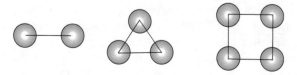

Figure 9.3. Static configurations for two, three, and four electric charges.

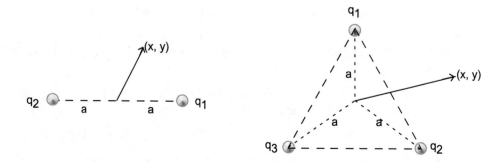

Figure 9.4. Coordinate systems for two- and three-charge configurations.

Notice that the numerator describes the usual addition of relative velocities, while the denominator is the relativistic correction.

9.4   Stable Points in Electric Fields

Consider the configurations of charges shown in Figure 9.3: a positive and a negative at coordinates $(1,0)$ and $(-1,0)$; three positive charges fixed to the corners of an equilateral triangle at coordinates $(0,1)$, $\sqrt{3}(1/2,-1/2)$, and $-\sqrt{3}(1/2,-1/2)$; and four positive charges fixed to the corners of a square at coordinates $(1,1)$, $(1,-1)$, $(-1,-1)$, and $(-1,1)$. As shown in Figure 9.4, the origin is taken as the center of each geometric figure.

1. The electric potential at a distance r from a single charge q is

$$V(r) = \frac{q}{r} = \frac{q}{\sqrt{x^2 + y^2}}. \tag{9.6}$$

    Determine and visualize the electric potential at the point (x, y) for each of these three configurations.

2. Decide if there is a point in space at which a free negative charge will remain at rest in stable equilibrium for each of the three configurations shown in Figure 9.3. (For the equivalent gravitational problem, these stable points are known

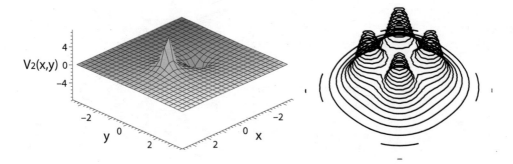

Figure 9.5. *Left:* Surface plot of the electric potential of a dipole. *Right:* Surface plot of the electric potential of four charges.

as Lagrange points and are the location of asteroids for the earth-sun system.) Surface plots of $V(x,y)$ versus x and y are recommended, for instance, Figure 9.5 left for the dipole.

3. Consider the quadrupole potential due to four positive charges placed at the corner of a square.

   a. Create a surface plot of the potential and decide if there is a stable point (right of Figure 9.5).
   b. Slice your surface plot through its center to verify if you have found a stable point.

9.5   Viewing Motion in Phase Space (Parametric Plots)

In Chapter 3 we give many examples of phase-space plots of the solutions of ordinary differential equations. No differential equations are used in this section.

The position x, velocity v, and acceleration a of a mass undergoing simple harmonic motion can all be viewed as functions of time:

$$x(t) = \sin \omega t, \quad v(t) = -\omega \cos \omega t, \quad a(t) = -\omega^2 \sin \omega t. \qquad (9.7)$$

This is what is obtained by solving the equations of motion.

1. Plot $x(t)$, $v(t)$, and $a(t)$ on the same graph. Comment on the phase relation among them.

2. Another way of looking at related functions such as these is to view them as functions of each other, for example, by plotting $x(t_i)$ along the abscissa and $v(t_i)$ along the ordinate, at successive times t_i, as done in Figure 9.6. This type

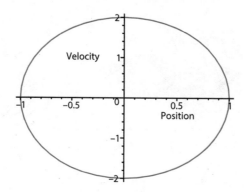

Figure 9.6. A phase space plot of velocity versus position for a harmonic oscillator.

of graph is called a *phase-space* or *parametric plot*, and is often illuminating since it replaces what could be some complicated time dependencies with a geometric figure that is easier to visualize.

a. Make a phase space plot of $v(t)$ versus $x(t)$.
b. Make a phase space plot of $a(t)$ versus $x(t)$.
c. Make a phase space plot of $a(t)$ versus $v(t)$.

3. Consider a spring with a nonlinear restoring force so that its potential energy is

$$V(x) = k\,x^6. \tag{9.8}$$

a. Use energy conservation to solve for the velocity as a function of position x.
b. Create a phase space plot of $v(t)$ versus $x(t)$ for this nonlinear oscillator.
c. Describe the differences and similarities in the phase space plots of the linear (harmonic) and nonlinear oscillators.

4. Make a phase space plot for a potential $V(x) = k\,x^3$.

5. Make a phase space plot for a potential $V(x) = k\,x^{20}$.

9.6   Other Useful Visualizations

1. Aside from perturbations, conic sections are the 2-D curves formed when a cone is cut (sectioned) by a plane. The orbits of planets and comets are known to follow conic sections, and are often expressed in polar form as

$$\frac{\alpha}{r} = 1 + \varepsilon\cos(\theta), \tag{9.9}$$

where ε is the eccentricity and $2\,\alpha$ is the *latus rectum* (inversely proportional to the planet's energy). An ellipse occurs when $0 < \varepsilon < 1$, a hyperbola for $\varepsilon > 1$, and a parabola for $\varepsilon = 1$.

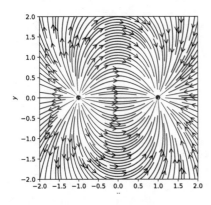 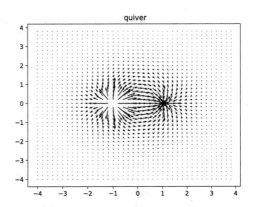

Figure 9.7. The electric field of a dipole as produced with *left:* Matplotlib's `streamplot`, as given by scipython.org, and *right:* `quiver`.

    a. Create polar plots of these three kinds of orbits using different values of ε.
    b. Create polar plots of these three kinds of orbits using different values of α.

2. Conic sections can also be expressed parametrically as:

    a. Hyperbola: $[x(s),\, y(s)] = [4\cosh(s),\, 1.4\sinh(s)]$
    b. Ellipse: $[x(s),\, y(s)] = [4\cos(s),\, 1.4\sin(s)]$
    c. Parabola: $[x(s),\, y(s)] = [s\cos(\theta) - s^2\sin(\theta),\, s^2\cos(\theta) + s\sin(\theta)]$,

where θ is an arbitrary parameter. Make plots of each of these conic sections as functions of s. Cover as much range in s as needed to obtain the full shapes.

3. Consider again the electric dipole in Figure 9.4. While before we examined the potential, a single number, here we want to examine the dipole's electric field, which is a vector quantity:

$$\mathbf{E} = \frac{(\mathbf{r} - \mathbf{r_1})}{|\mathbf{r} - \mathbf{r_1}|^2} - \frac{(\mathbf{r} - \mathbf{r_2})}{|\mathbf{r} - \mathbf{r_2}|^2}, \tag{9.10}$$

where $\mathbf{r}_1$ and $\mathbf{r}_2$ are the vector positions of the two charges.

    a. Determine the components of the electric field, for example,

$$E_x = \frac{x+1}{(x+1)^2 + y^2} - \frac{x-1}{(x-1)^2 + y^2}. \tag{9.11}$$

    b. Create a plot of the dipole's electric field. In Figure 9.7 we show such plots created simply with Matplotlib's `streamplot` and `quiver` functions:

```
import numpy as np, matplotlib.pyplot as plt
X,Y = np.meshgrid( np.arange(-4,4,.2), np.arange(-4,4,.2) )   # The fields
Ex = (X + 1)/((X+1)**2 + Y**2) - (X - 1)/((X-1)**2 + Y**2)
Ey = Y/((X+1)**2 + Y**2) - Y/((X-1)**2 + Y**2)
plt.figure()
plt.streamplot(X,Y,Ex,Ey)
plt.title('Streamplot')
plt.figure()
plt.quiver(X,Y,Ex,Ey,scale=50)
plt.title('Quiver')
```

4. A graphical approach to determining where two functions are equal to each other, is to draw both on a single plot and note the x values at which the two curves intersect (same y value). Alternatively, you can search for x value at which $f(x) - g(x) = 0$. Determine graphically the approximate solutions of the following equations:

   a. $\sin(x) = x^2$,
   b. $x^2 + 6x + 1 = 0$,
   c. $h^3 - 9h^2 + 4 = 0$.
   d. $h^4 - 9h^3 + 4 = 0$
   e. $n\sin(2x) = \sin(x)\cos(x)$,   find n.
   f. $n\cos(x)^2 = 1 + \cos(2x)$,   find n.
   g. $x^3 + x^2 + x = 1$.   (Solve too for complex roots.)

5. Demonstrate graphically the following mathematical facts. (*Hint:* To avoid overflow, you may want to use semilog plots.)

   a. The exponential e^x grows faster than any power x^n.
   b. The logarithm $\ln(x)$ grows slower than any power x^n.

6. A standing wave on a string is described by the equation

   $$y(x, t) = \sin(10x)\cos(12t),\qquad\qquad(9.12)$$

   where x is the distance along the string, y is the height of the disturbance, and t is the time.

   a. Create an animation (movie) of this function.
   b. Create a surface plot of this function, and see if you agree with us that it is not as revealing as the animation.

7. A traveling wave on a string is described by the equation

   $$y(x, t) = \sin(10x - 12t),\qquad\qquad(9.13)$$

   where x is the distance along the string, y is the height of the disturbance, and t is the time.

   a. Create an animation of this function.

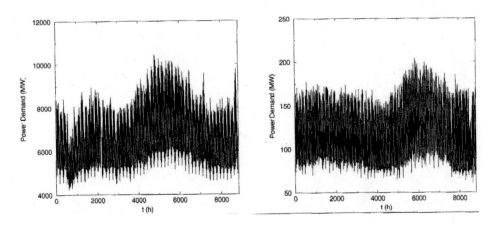

Figure 9.8. Electrical power use for Australia (*left*) and Mar del Plata (*right*) on an hourly basis over a one-year period.

b. Create a surface plot of this function, and see if you agree with us that it is not as revealing as the animation.

9.7 Integrating Power into Energy

Figure 9.8 left shows the hourly power consumption during a one-year period for Australia and for Mar del Plata [Degaudenzi & Arizmendi(20)]. The graphs are characterized by fluctuations due to the day-night cycle, the change of seasons, weather patterns, and possibly other causes. Your problem is to determine the total energy used over a three-year period. Since we are not given a table of numbers to integrate, we'll propose and try out three models:

$$P(t) = 4 + [2 + \sin(2\pi t)]\sin(\pi t/91), \tag{9.14}$$

$$P(t) = \left[4 + \frac{t}{365} + \frac{\sin(\pi t/91)}{2}\right]\left(2 + \frac{\sin(2\pi t)^2}{2}\right), \tag{9.15}$$

$$P(t) = \left[4 + \frac{t}{365} + \frac{\sin(\pi t/91)}{2}\right]\left(2 + e^{-\sin(2\pi t)}\right), \tag{9.16}$$

where the power P is in GW (10^9 watts) and the time t is in days. These formulas try to incorporate the various time dependencies of power usage, with $\sin(2\pi t)$ representing a daily variation of power, $\sin(\pi t/91)$ representing a twice-a-year variation, the linear terms accounting for a regular increase, and the exponential accounting for a daily variation.

1. Plot each of these models and comment on their similarity to the data.

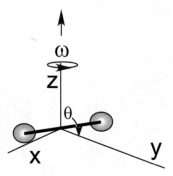

Figure 9.9. A barbell rotating about an axis pointing in the z direction.

2. Power is defined as the rate at which work is done, or energy used, and hence is the integral of power over a time period:

$$P = \frac{dW}{dt} \quad \Rightarrow \quad E(T) = \int_0^T P(t)\, dt. \qquad (9.17)$$

Consequently, if energy is measured in joules, then power is measured in joules/second or watts (W). (Since power companies charge for kilowatt-hours, they should be called *energy* companies.)

3. Express a kilowatt-hour in terms of joules (by calculation, not by look up).

4. Use a symbolic manipulation program or package to evaluate the total energy used over a three-year period for each of the three models. The highly oscillatory nature of the integrand is a challenge for automated integration routines, and a causal application of Maple and Mathematica gave us different answers.

5. Evaluate the total energy used over a three-year period for each of the three models numerically using Simpson's rule. Make sure to place multiple integration points within each oscillation, which means many points.

6. Evaluate the total energy used over a three-year period for each of the three models numerically using Gaussian quadrature. Make sure to place a good number of integration points within each oscillation, which means many points.

7. You should have found that these models show a rather linear increase in integrated energy over time. Modify the first and last models so that the total energy-use growth rate is approximately 10% per year.

8. Devise a model in which the total energy used becomes a constant in time.

9.8 Rigid-Body Rotations with Matrices

When a rigid body is rotated about a fixed point O, all parts have the same *angular velocity* $\boldsymbol{\omega}$. The linear velocity of a part i of the body is related to ω via:

$$\mathbf{v}_i = \boldsymbol{\omega} \times \mathbf{r}_i, \tag{9.18}$$

where $\mathbf{r}_i$ is a vector from O to part i, and $\times$ represents the vector cross product. Angular momentum is the rotational analog of linear momentum, and for the highly symmetric cases is parallel to the angular velocity:

$$\mathbf{L} = I\omega. \tag{9.19}$$

The proportionality constant I is called the *moment of inertia*. For general rotations, the angular momentum of a rigid body is calculated as the sum of the angular momentum of each particle in the body:

$$\mathbf{L} = \sum_i \mathbf{r}_i \times \mathbf{p}_i = \sum_i m_i \mathbf{r}_i \times (\omega \times \mathbf{r}_i). \tag{9.20}$$

Here $\mathbf{p}_i = m_i \mathbf{v}_i$ is the linear momentum of particle i, and the right-most vector cross product must be performed first. This sum over particles can be written succinctly as the multiplication of the angular velocity vector ω by the moment-of-inertia matrix (also called tensor):

$$\mathbf{L} = [I]\boldsymbol{\omega}. \tag{9.21}$$

If the vectors are represented by columns (the usual thing), (9.21) is explicitly

$$\begin{bmatrix} L_x \\ L_y \\ L_z \end{bmatrix} = \begin{bmatrix} I_{xx} & I_{xy} & I_{xz} \\ I_{yx} & I_{yy} & I_{yz} \\ I_{zx} & I_{zy} & I_{zz} \end{bmatrix} \begin{bmatrix} \omega_x \\ \omega_y \\ \omega_z \end{bmatrix}. \tag{9.22}$$

Performing the matrix multiplications results in three simultaneous linear equations:

$$\begin{aligned} L_x &= I_{xx}\omega_x + I_{xy}\omega_y + I_{xz}\omega_z, & (9.23) \\ L_y &= I_{yx}\omega_x + I_{yy}\omega_y + I_{yz}\omega_z, & (9.24) \\ L_z &= I_{zx}\omega_x + I_{zy}\omega_y + I_{zz}\omega_z. & (9.25) \end{aligned}$$

The diagonal elements I_{xx}, I_{yy}, and I_{zz} are called the *moments of inertia* about the x, y, and z axes, respectively, while the off-diagonal elements are *products of inertia*. It's the products of inertia that keep ω and $\mathbf{L}$ from being parallel.

1. Consider Figure 9.9 showing a barbell composed of a mass $m_1 = 2$ attached to mass $m_2 = 2$ by a massless rigid rod of length $2a = 8$. The barbell is fixed at an angle $\theta = \pi/4$ with respect to the z axis, and is being rotated with an angular velocity vector $\omega = 6$ radians/sec along the z axis (right hand rule).

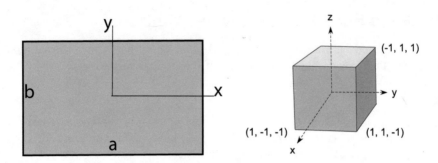

Figure 9.10. *Left:* A plate in the x-y plane with a 2-D coordinate system at its center. *Right:* A cube of side 1 with a 3-D coordinate system at its center.

    a. Use a symbolic manipulation program or Python to determine the vector angular momentum **L** of this barbell by adding up the angular momenta of the two masses.
    b. Evaluate the inertia matrix $[\mathbf{I}]$ for this barbell.
    c. Evaluate the matrix product $[\mathbf{I}]\boldsymbol{\omega}$ of the inertia tensor and angular velocity for the barbell. Check that you get the same value for **L** as above.
    d. Make 3-D plots of **L** and $\boldsymbol{\omega}$ for three different times, and thereby show that **L** rotates about a fixed $\boldsymbol{\omega}$. A changing **L** implies that an external torque is being applied to the barbell to keep it rotating.

2. Consider the two-dimensional rectangular metal plate shown on the left of Figure 9.10. It has sides $a = 2$ and $b = 1$, mass $m = 12$, and inertia tensor for axes through the center:

$$[I] = \begin{bmatrix} mb^2/12 & 0 \\ 0 & ma^2/12 \end{bmatrix} = \begin{bmatrix} 1 & 0 \\ 0 & 4 \end{bmatrix}. \tag{9.26}$$

The plate is rotated so that its angular velocity vector ω always remains in the xy plane. Examine the three angular velocities:

$$\omega = \begin{bmatrix} 1 \\ 0 \end{bmatrix}, \quad \omega = \begin{bmatrix} 0 \\ 1 \end{bmatrix}, \quad \omega = \begin{bmatrix} 1 \\ 1 \end{bmatrix}. \tag{9.27}$$

    a. Write a program that computes the angular momentum vector **L** via the matrix multiplication $\mathbf{L} = [I]\boldsymbol{\omega}$.
    b. Plot $\boldsymbol{\omega}$ and **L** for each case

3. Consider the rotation of the cube on the right of Figure 9.10. The cube has side $b = 1$, mass $m = 1$, and an inertia tensor for axes at the center:

$$[I] = \begin{bmatrix} +2/3 & -1/4 & -1/4 \\ -1/4 & +2/3 & -1/4 \\ -1/4 & -1/4 & +2/3 \end{bmatrix}. \tag{9.28}$$

The cube is rotated with the three angular velocities:

$$\boldsymbol{\omega} = \begin{bmatrix} 1 \\ 0 \\ 0 \end{bmatrix}, \qquad \boldsymbol{\omega} = \begin{bmatrix} 0 \\ 1 \\ 0 \end{bmatrix}, \qquad \boldsymbol{\omega} = \begin{bmatrix} 1 \\ 1 \\ 1 \end{bmatrix}. \tag{9.29}$$

a. Write a program that computes the angular momentum vector $\mathbf{L}$ via the requisite matrix multiplication.

b. Plot $\boldsymbol{\omega}$ and $\mathbf{L}$ for each case.

4. Consider the 2-D square plate of side 2 on the left of Figure 9.9 with a mass $m = 12$ at each corner. It is rotated with three values for the angular velocity:

$$\boldsymbol{\omega} = \begin{bmatrix} 1 \\ 0 \end{bmatrix}, \qquad \begin{bmatrix} 0 \\ 1 \end{bmatrix}, \qquad \begin{bmatrix} 1 \\ 1 \end{bmatrix}.$$

a. Determine the angular momentum vector $\mathbf{L}$ of the plate by adding up the angular momenta of the individual masses.

b. Evaluate the inertia matrix $[\mathbf{I}]$ for this plate. *Hint:* if this were a solid plate, its inertia tensor would be

$$I = \begin{bmatrix} 1 & 0 \\ 0 & 4 \end{bmatrix}. \tag{9.30}$$

c. Form the matrix product $[\mathbf{I}]\boldsymbol{\omega}$ of the inertia matrix $[\mathbf{I}]$ and angular velocity $\boldsymbol{\omega}$ for the plate, and check that you get the same value for $\mathbf{L}$ as before.

d. Make three 3-D plots of $\mathbf{L}$ and $\boldsymbol{\omega}$ for different times, and thereby show that $\mathbf{L}$ rotates about a fixed $\boldsymbol{\omega}$. From this we deduce that there must be an external torque applied to the plate.

e. *(Optional)* Determine the torque acting on the plate.

9.9 Searching for Calibration of a Spherical Tank

1. As shown in Figure 9.11, a spherical tank of radius $R = 3$ meters is used to hold fuel. You are given a stick to dip into the tank, and your problem is to place marks on the stick that indicate the volume for a given height.

a. Prove that the volume of the fluid in the tank as a function of height is

$$V(h) = \pi \left[Rh^2 - \frac{h^3}{3} \right]. \tag{9.31}$$

b. Your task requires you to invert this equation, that is, to determine $H(V)$. Show that for $R = 3$ and $V = 4/3\,\pi$, the equation to solve is

$$h^3 - 9h^2 + 4 = 0. \tag{9.32}$$

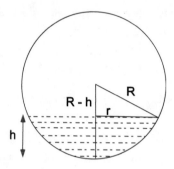

Figure 9.11. A spherical tank of radius R filled with a liquid to height h above its bottom.

   c. Show that the desired solution must lie in the range

$$0 \leq h \leq 2R = 6. \qquad (9.33)$$

   d. Find the solution to (9.32) using the closed-form expressions for the solutions of cubic equation.

   e. Use the bisection algorithm to search for a solution of (9.32) (a procedure you would have to follow if you did not have a closed-form solution).

   f. If you find more than one solution, which one is the physical solution?

   g. Use the bisection algorithm to calibrate the dipstick by determining the h's for which the tank is 1/4 full, 1/2 full, 3/4 full, and 99/100 full.

9.10 AC Circuits via Complex Numbers

9.10.1 Using Complex Numbers

1. Test your ability to program with complex numbers by checking that the following identities hold for a variety of complex numbers:

$$\begin{aligned} z + z &= 2z, & z + z* &= 2\,\mathrm{Re}\,z, \\ z - z &= 0, & z - z* &= 2\,\mathrm{Im}\,z, \\ zz* &= |z|^2, & zz* &= r^2. \end{aligned} \qquad (9.34)$$

Compare your output to some cases of pure real, pure imaginary, and simple complex numbers that you are able to evaluate by hand.

2. Complex mathematics is much easier if there is a symbol for the imaginary number $i = \sqrt{-1}$. Define a complex variable i in your program equal to i.

3. Now that you have a specific variable for i, create functions that compute the following functions of complex numbers:

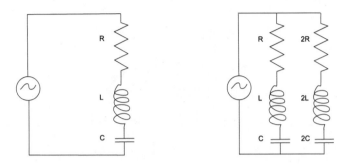

Figure 9.12. *Left:* An RLC circuit connected to an alternating voltage source. *Right:* Two RLC circuits connected in parallel to an alternating voltage. Observe that one of the parallel circuits has double the values of R, L, and C as does the other.

    a. $\exp(z) = e^{x+iy} = e^x e^{iy} = e^x(\cos y + i \sin y)$,
    b. $\sin(z) = (e^{iz} - e^{-iz})/2i$.

4. Verify that your functions work by trying some simple cases, for instance, $z_1, z_2 = 0, 1, i, 2, 2i, \ldots$, where you can figure out the answers.

5. Let $z = 3 + 3\sqrt{3}i$.

    a. Determine $|z|$ and check that you get 6.
    b. Determine the phase θ of z and check that you get $\theta = \pi/3$.

6. Find $\mathrm{Re}\,(2 - 3i)^2/(2 + 3i)$, $\mathrm{Im}\,(1/z^2)$, $|(1 + z)/(1 - z)|$, and $(2 - 3i)/(2 + 3i)$.

7. Consider the complex number $z = x + iy$. Make a surface plot of $\mathrm{Im}\,\cos(z)$ and $\mathrm{Re}\,\cos(z)$.

8. Consider the complex function $f(z) = z^3/(1 + z^4)$.

    a. Make a plot of $f(x)$ versus $x = \mathrm{Re}\,z$.
    b. Make surface plots of $\mathrm{Re}\,f(z)$ and $\mathrm{Im}\,f(z)$, restricting the range of x and y values to lie close to where there is the most variation.

9.10.2 RLC Circuit

The circuit on the left of Figure 9.12 contains a resistor R, an inductor L, and a capacitor C. All three elements are connected in series to an alternating voltage source $V(t) = V_0 \cos \omega t$. Kirchhoff's laws applied to this circuit lead to the differential equation

$$\frac{dV}{dt} = R\frac{dI}{dt} + L\frac{d^2 I}{dt^2} + \frac{I}{C},$$
(9.35)

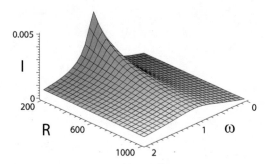

Figure 9.13. A surface plot of the current through an RLC circuit as a function of the resistance R and the frequency ω of the voltage source.

where $I(t)$ is the current in the circuit. We imagine the circuit being driven by a complex voltage source $V(t) = V_0 e^{-i\omega t}$ whose real part is the physical voltage. The resulting current, $I(t) = I_0 e^{-i\omega t}$, will also be complex, with its real part equal to the physical current.

1. Show that substitution of the $V(t)$ and $I(t)$ leads to the solution

$$V_0 e^{-i\omega t} = Z I_0 e^{-i\omega t}, \qquad Z = R + i\left(\frac{1}{\omega C} - \omega L\right). \tag{9.36}$$

   Here Z is called the *impedance* and $V = IZ$ is the AC generalization of Ohm's law, $V = IR$.

2. Show that in polar form the impedance is

$$Z = |Z|e^{i\theta}, \tag{9.37}$$

$$|Z| = \sqrt{R^2 + (1/\omega C - \omega L)^2}, \quad \theta = \tan^{-1}\left(\frac{1/\omega C - \omega L}{R}\right). \tag{9.38}$$

3. Show that substitution leads to the complex and physical currents

$$I(t) = \frac{V_0}{|Z|}e^{-i(\omega t + \theta)}, \quad I_{\text{phys}}(t) = \text{Re } I(t) = \frac{V_0}{|Z|}\cos(\omega t + \theta). \tag{9.39}$$

   We see that the phase of the current, relative to that of the voltage, is θ, and that its magnitude equals the magnitude of the voltage divided by the magnitude of the impedance.

4. Use complex arithmetic to determine the magnitude and time dependence of the current in this circuit as a function of the frequency ω of the external voltage. (We use Python's complex numbers in `DFTcomplex.py` in Listing 2.1.)

a. Assume a single value for the inductance and capacitance, three values for resistance, and a range of frequencies:

$$L = 1000 \text{ H}, \qquad C = \frac{1}{1000} \text{ F}, \qquad R = \frac{1000}{1.5}, \frac{1000}{2.1}, \frac{1000}{5.2} \ \Omega,$$

$$0 < \omega < 2/\sqrt{LC} = 2/s. \tag{9.40}$$

b. Compute and plot the magnitude and phase of the current as a function of the frequency $0 \le \omega \le 2$. Figure 9.13 shows the type of surface plot you should obtain, with the peak indicating a resonance.

c. Verify that the resonance peak is at the same frequency as the zero in the phase.

d. Verify that the smaller the resistance R, the sharper is the resonance peak.

5. Repeat all of the previous steps in the calculation but now for two RLC circuits in parallel, as shown on the right of Figure 9.12. The complex analysis is the same as that done with ordinary resistors:

$$\frac{1}{Z} = \frac{1}{Z_1} + \frac{1}{Z_2}, \qquad \text{(Parallel connection)}. \tag{9.41}$$

9.11 Beats and Satellites

1. If two tones close in frequency are played together, your ear hears them as a single tone with a varying amplitude. Make plots as a function of time of the results of adding the two sines $\sin(100\,t) + \sin(b\,t)$ functions:

a. Make a series of plots for b in the range $90 < b < 100$.

b. Plot for a long enough time to see at least three beats.

c. Make the frequency of the second sine wave progressively closer to that of the first and note the effect (Figure 9.14).

d. For the sum of two waves, deduce empirically a relation between the difference in the sine wave frequencies and the frequency of the oscillation of the amplitude.

2. Add together two periodic functions, one representing the position of the moon as it revolves around a planet, and the other describing the position of the planet as it revolves about the sun. Take the planet's orbits to have radius $R = 4$ and an angular frequency $\omega_p = 1$ radian/sec. Take the moon's orbit around the earth to have a radius $r = 1$ and an angular velocity $\omega_s = 14$ radians/sec. Show that if the planet position relative to the sun is

$$x_p = R \cos(\omega_p\, t), \qquad y_p = R \sin(\omega_p\, t), \tag{9.42}$$

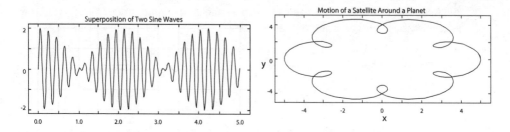

Figure 9.14. *Left:* Superposition of two waves with similar wave frequencies. *Right:* The trajectory of a satellite around a planet as seen from the sun.

then the position of the satellite relative to the sun is the sum of its position relative to the planet, plus the position of the planet relative to the sun:

$$x_s \;=\; x_p + r\,\cos(\omega_s\,t) = R\,\cos(\omega_p\,t) + r\,\cos(\omega_s\,t),$$
$$y_s \;=\; y_p + r\,\sin(\omega_s\,t) = R\,\sin(\omega_p\,t) + r\,\sin(\omega_s\,t).$$

a. Make plots of $x_s(t)$ and $y_s(t)$ versus the time t.
b. Make a parametric plot of $x(t)$ versus $y(t)$ and verify that you obtain an orbit similar to that in Figure 9.14 right.

A

General Relativity

A.1   Chapter Overview

These general relativity (GR) problems have been added after requests from readers, and extend the problems in special relativity in Chapter 5. These problems deal with the visualization of wormholes, the deflection of starlight by the sun, the gravitational lensing by highly massive stars, and the motion of a particle in a Newtonian potential with a GR correction, and the computation of some GR quantities.

A.2   Visualizing Wormholes

During Christopher Nolan's direction of the science fiction movie *Interstellar*, Kip Thorne (now a Noble laureate) helped develop the visualizations of rocket flight based on solutions of the equations of Einstein's theory of general relativity. The key element of the movie was that interstellar travel was possible in a single human lifetime if a spaceship passed through a wormhole (an Einstein–Rosen bridge), a tunnel-like structure in spacetime that connects one location in spacetime to another, or possibly to another universe [James et al. (15)]. Fig.(A.1) is the visualization of such a wormhole.

Although wormholes have never been observed, they may occur as quantum fluctuations on the Planck scale, $\sqrt{G\hbar/c^3} \sim 10^{-35}$ m. Furthermore, it just might be possible to have some type of exotic matter with negative energy density at the throat of the wormhole that would enlarge the wormhole to a macroscopic size that might permit a rocket ship pass through it. However, if our 4-D universe resided in a higher-dimensional space (bulk), such as the 5-D one imagined in *Interstellar*, then there might not be the need for exotic matter to hold the wormhole open. In either case, interstellar travel can be imagined to be possible. While all of this is unlikely (it is called science fiction after all), it is not strictly forbidden.

Morris and Thorn [Morris, M. S., K. S. Thorne, *Wormholes in spacetime and their use for interstellar travel: A tool for teaching General Relativity*, Am. J. Phys., **56**, 395-412, (1988).] discuss the fundamentals of space travel using wormholes as an exercise in general relativity. Your **problem** is to reproduce some stills of the wormhole visualizations that were created

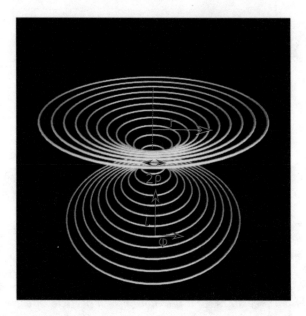

Figure A.1. The Ellis wormhole connecting an upper and lower (flatter) spaces. Note that this visualization has the wormhole's 4-D bulk embedded within a 3-D space. The throat diameter is 2ρ and the proper distance traveled in a radial direction is ℓ.

for the movie. As an alternative, you can reproduce some of the (different) visualizations found in [T. A. Roman, *The inflated wormhole: A MATHEMATICA animation*, Comp. in Phys., 480-487, (1994).]. You will not be asked to actually solve Einstein's equations, although we would encourage you to do so. Another extension would be the creation of videos that visualize what travel through a wormhole would look like if recorded by a camera on the space ship, or from outside the wormhole. Some such visualizations from the *Interstellar* movie can be found on *Youtube* [https://www.youtube.com/watch?v=f3ptQ0CPMmU.].

The equations that Thorne used to create the visualizations were expressed in geometrized units in which $G = 1$, $c = 1$, time is measured in length $1s = c*1 = 2.998 \times 10^8$ m, mass is also measured in length units, 1 kg= $G/c^2 \times$ 1 kg, so that 1 kg $= 0.742 \times 10^{-27}$ m, in which case the Sun's mass = 1.476 km. The wormhole consists of a 4-D cylinder with length $2a$ whose cross sections are spheres of radius ρ. In order to visualize the 4-D wormhole, it is embedded in a 3-D space so that the cross section are circles of radius ρ. The ends of the cylinder connect to flat 3-D spaces.

Thorne uses the Ellis extension of a spherical polar coordinates metric:

$$ds^2 = -dt^2 + d\ell^2 + r^2(d\theta^2 + \sin^2\theta d\phi^2). \tag{A.1}$$

Here the radius coordinate r is a function of ℓ, the physical distance (proper distance) traveled in a radial direction:

$$r(\ell) = \sqrt{\rho^2 + \ell^2},\tag{A.2}$$

where ρ is the radius of the throat in a cylindrical-shaped wormhole. Note that the time coordinate t enters the metric (A.1) with a negative sign. This means that for fixed ℓ, θ, and ϕ, t increases in the timelike direction. Accordingly, t is the proper time as measured by a person at rest in the spatial (ℓ, θ, ϕ) coordinate system.

Because $r^2(d\theta^2 + \sin^2\theta d\phi^2)$ is the familiar metric describing the surface of a sphere of radius r, the wormhole is spherically symmetric. This means that when $l \to +\infty$, as well as when $l \to -\infty$, the radius of the sphere within the wormhole approaches proper distance ℓ. This also means that as $l \to \pm\infty$ we would have two separate flat spaces connected by the wormhole. The transition between the two flat spaces via the wormhole's throat is made to resemble the transition to an external space in which a nonspinning black hole resides. This is described by the Schwarzschild or hole metric [James et al. (15)]:

$$ds^2 = -(1 - 2\mathcal{M}/r)dr^2 + \frac{dr^2}{1 - 2\mathcal{M}/r} + r^2\,(d\theta^2 + \sin^2\theta\,d\phi^2),\tag{A.3}$$

where $\mathcal{M}$ is the black hole's mass. With this metric, the radius r becomes the outward coordinate rather than the proper distance ℓ. The visualizations in the movie required a solution for $r(\ell)$, that is, a solution or an expression for the outward coordinate as a function of proper distance. To reduce the effort involved, the visualizations used an analytic expression for $r(\ell)$ outside the wormhole's cylindrical interior that is similar to the Schwarzschild $r(\ell)$:

$$r(|\ell| > a) = \rho + \frac{2}{\pi} \int_0^{|\ell|-a} \arctan\left(\frac{2\xi}{\pi\mathcal{M}}\right) d\xi\tag{A.4}$$

$$= \rho + \mathcal{M}\left[x\arctan x - \frac{1}{2}\ln(1 + x^2)\right], \qquad x = \frac{2|\ell| - a}{\pi\mathcal{M}}.\tag{A.5}$$

For cylindrical coordinates, the z coordinate is the embedding space height above the wormhole's midplane, and so the embedding space metric becomes

$$ds^2 = dz^2 + dr^2 + r^2\,d\phi^2.\tag{A.6}$$

In this case, the spatial metric of the wormhole's two-dimensional equatorial surface becomes:

$$ds^2 = d\ell^2 + r^2(\ell)\,d\phi^2.\tag{A.7}$$

Combining these equations lets us solve for $z(\ell)$:

$$dz^2 + dr^2 = d\ell^2,\tag{A.8}$$

$$z(\ell) = \int_0^{\ell} \sqrt{1 - (dr/dL)^2}\, dL.\tag{A.9}$$

You obtain the equations needed to program up the visualization of a wormhole by substituting (A.4) and (A.5) into (A.9).

1. In order to apply (A.9) we need to evaluate the derivative $dr/d\ell$. Use Python's symbolic algebra package `sympy` to show that

$$\frac{dr}{d\ell} = \frac{2}{\pi} \tan^{-1}\left(\frac{2\ell - a}{\pi \mathcal{M}}\right). \qquad (A.10)$$

   Our program `WormHole.py` in Listing A.1 evaluates this derivative.

2. Insert this $dr/d\ell$ into (A.9) and evaluate the $z(l)$ integral numerically for

$$\rho = 1, \quad a = 1, \quad \mathcal{M} = 0.5. \qquad (A.11)$$

3. The contour lines or rings shown in Fig. A.1 correspond to different values of ℓ. They were obtained by using Vpython in a Jupyter notebook with the program `VisualWorm.ipynb` given in Listing A.2[1]:

4. Make your own plot of the wormhole for $\ell = 1, \cdots, 11$.

5. Create a cylindrical wormhole of length $2L$ with a spherical cross sections of radius ρ. Visualize the wormhole with a 3-D embedding diagram in which the missing dimension results in the cross sections appearing as circles rather than spheres. Follow the same steps as used for the Ellis wormhole, (A.1), but now with

$$r(\ell) = \begin{cases} \rho, & |\ell| \leq L \quad \text{(Wormhole interior)}, \\ |\ell| - L + \rho, & |\ell| \geq L \quad \text{(Wormhole exterior)} \end{cases} \qquad (A.12)$$

A.3   Gravitational Deflection of Light

A *geodesic* is the shortest path between two points on a curved surface. General relativity assumes that light travels on geodesics, which are curved paths in a 4-D spacetime. To determine a geodesic, one starts with the infinitesimal 4-D path length (interval)

$$ds^2 = c^2 dt^2 - dx^2 - dy^2 - dz^2. \qquad (A.13)$$

Since light travel a distance ct in time t, the interval vanishes for light, and its path is therefore called a *null* geodesic. Since material particles move slower than light, their interval is positive (time-like). The path that light takes in spacetime is the solution of the geodesic equation

$$\frac{d^2 x^\beta}{d\lambda^2} + \Gamma^\beta_{\mu\nu} \frac{dx^\mu}{d\lambda} \frac{dx^\nu}{d\lambda} = 0. \qquad (A.14)$$

---

[1]While the previous version of Vpython explicitly called the Visual package and was run via an editor, we have been able to run the latest Vpython only within a Jupyter notebook.

Here $\Gamma^{\beta}_{\mu\nu}$ is the Christoffel symbol and would be obtained by solving the curvature equation $R_{\mu\nu} = 0$ for a given metric tensor $g_{\mu\nu}$. The curvature equation turns out to be a rather formidable set of ten nonlinear PDE's, which we are happy to leave for another time.

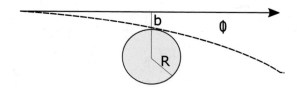

Figure A.2. A light ray being bent by an angle ϕ due to the gravitational effect of the sun.

One of the early tests of general relativity was its prediction of the angle of deflection ϕ for light starting at an impact parameter $b = R$ and just grazing the sun (Figure A.2). At first Newtonian mechanics solved this problem by calculating the orbit of a massive particle around the sun, and then taking the $m \to 0$ limit for the particle to obtain

$$\phi = 2\frac{GM}{Rc^2}. \tag{A.15}$$

Here G is the gravitational constant, M is the mass of the sun, and R is the radius of the sun. Later, Einsteinian mechanics was used to solve (A.14) approximately and obtained twice as large a value,

$$\phi = 4\frac{GM}{Rc^2}, \tag{A.16}$$

which agreed with the measurements.

Now let's try to calculate some numerical values for the deflection. In 1916 Schwarzschild found an exact solution of the Einsteinian equations using Schwarzschild metric [Moore(13)],

$$ds^2 = \left(1 - \frac{2GM}{c^2r}\right)c^2dt^2 - \left(1 - \frac{2GM}{c^2r}\right)^{-1}dr^2 - r^2(d\theta^2 + \sin^2\theta d\phi^2). \tag{A.17}$$

For this metric and for light just grazing the sun ($b = R$), the orbit equation takes the simple form

$$\left(\frac{1}{r}\frac{dr}{d\phi}\right)^2 = \left(1 - \frac{2M}{R}\right)\frac{1}{R^2} - \left(1 - \frac{2M}{r}\right)\frac{1}{r^2}. \tag{A.18}$$

A change of variable to $u = R/r$ produces an easier equation to solve:

$$\left(\frac{du}{d\phi}\right)^2 = 1 - u^2 - \frac{2M}{R}(1 - u^3). \tag{A.19}$$

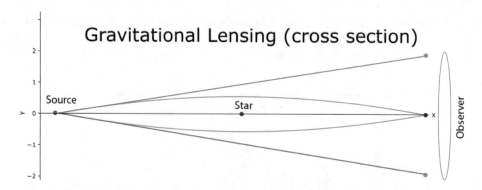

Figure A.3. Three trajectories of light showing the bending arising from the sun's mass. Note that there are three images formed on the right. Actually, as indicated by the ellipse, an observer would see a circle (an Einstein ring) obtained by rotating this figure along the x axis.

1. Verify that an approximate solution to (A.19) is

$$\phi \simeq 4\frac{GM}{Rc^2}. \tag{A.20}$$

2. Evaluate this expression to determine a numerical value for the angle of deflection for light grazing the sun's surface (*hint:* It's small). Use parameters $M = 2\times10^{23}$ grams, $R = 7 \times 10^{10}$ cm, and $G/c^2 = 7.4 \times 10^{-29}$ cm/gram.

3. Although the ODE (A.19) is nonlinear, that is not an obstacle for a numerical solution. Solve (A.19) numerically and compare your result with the value from the approximate analytic expression.

A.4   Gravitational Lensing

In a different approach to the deflection of a light due to a very massive star, [Moore(13)] assumes a Schwarzschild spacetime to describe the curved space outside of a spherically symmetric gravitational source (star). In terms of the inverse variable $u = 1/r$, the geodesic equation is now

$$\frac{d^2u}{d\phi^2} = 3GM\,u^2 - u. \tag{A.21}$$

1. Modify your ODE solver appropriate to this equation. Employ units such that mass is measured in meters, $GM=1477.1$ m, and $M = 28M_\odot$ ($M_\odot$ is a solar mass).

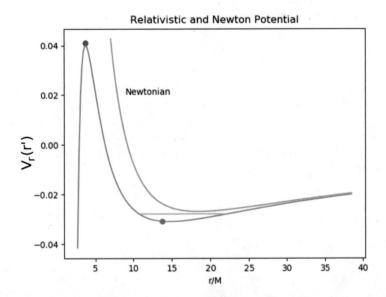

Figure A.4. Relativistic and Newtonian potential for $\ell/M = 4.3$. Different energies would correspond to differing values of the ordinate. One of the dots corresponds to the energy for a circular orbit.

2. Equation (A.21) is quite sensitive to the initial conditions. Assume that initially the light is very distant: $r = 10^6$, and $u(\phi = 0) = du(\phi)/d\phi = 10^{-6}$.

3. Convert your solution for $r(\phi)$ into one for (x, y), and plot up the photon paths for $0 \le \phi \le \pi$. Our plot is given in Figure A.3.

4. Employ the symmetry of this problem to rotate your solution about the x axis and thus obtain a circle. This is what an observer sees when viewing a distance light source lying behind a massive star that focuses the point source into a ring.

Our program `LensGravity.py` is given in Listing A.3.

A.5   Particle Orbits in GR Gravity

The classical solution of Newton's laws for the gravitational potential is just fine for most everything. However there are small corrections arising from relativity, and while small, these corrections are actually critical to the accuracy of modern gps devices. The usual approach is to determine an ODE with a GR correction to the familiar $1/r$ gravitational potential, and then solve the ODE approximately or numerically. We follow [Hartle(03)] and [Moore(13)] who derive an effective potential appropriate to the empty space external to a spherically symmetric star. For the Schwarzschild metric (A.3), they give the effective radial potential as

$$V_r(r) = -\frac{GM}{r} + \frac{\ell^2}{2r^2} - \frac{GM\ell^2}{r^3}, \tag{A.22}$$

where G is the gravitational constant, ℓ is the angular momentum per unit rest mass, M is the mass of the star, and the middle term is the usual angular momentum barrier. We see that (A.22) differs from the Newtonian potential by a $-GM\ell^2/r^3$ term that provides an strong attraction at very short distances, in addition to the usual $-GM/r$ attraction. We obtain a dimensionless, and simpler-to-compute, form of the potential by change of variables:

$$V_r(r') = -\frac{G}{r'} + \frac{\ell'^2}{2r'^2} - \frac{G\ell'^2}{r'^3}, \tag{A.23}$$

$$r' = \frac{r}{M}, \qquad \ell' = \frac{\ell}{M}. \tag{A.24}$$

1. Plot $V_r(r')$ versus r' for $\ell' = 4.3$ (FigureA.4).

2. Describe in words how the orbits within this potential change with energy. (*Hint:* one of the dots in Figure A.4 corresponds to the energy for a circular orbit.)

3. At what values of r' does the effective potential have a maximum and a minimum?

4. At what value of r' does a circular orbit exist?

5. Determine the range of r' values that occur for $\ell' = 4.3$.

6. Indicate the above range on your plot by a horizontal line, and describe the orbits.

7. Describe the orbit for energies corresponding to the maximum in the potential.

A.5.1   Orbit Computation

A fairly simple way to determine the orbits of massive particles in the effective potential (A.23) is based on energy conservation. It starts with the energy per unit mass expressed as the sum of kinetic and potential terms: [Moore(13)]:

$$E = \frac{1}{2}\left(\frac{dr}{d\phi}\right)^2 \frac{\ell^2}{r^4} - \frac{GM}{r} + \frac{\ell^2}{2r^2} - \frac{GM\ell^2}{r^3}, \tag{A.25}$$

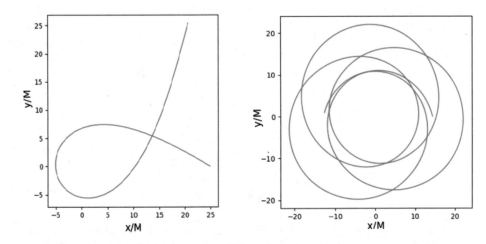

Figure A.5. *Left:* An orbit corresponding to an energy at the maximum of the potential. *Right:* A rapidly precessing orbit.

where ϕ is the polar angle. We obtain an ODE for the orbit by differentiating both sides of the equation with respect to ϕ:

$$\frac{d^2r}{d\phi^2} = -\frac{GM}{r^2} + \frac{\ell^2}{r^3} - \frac{3GM\ell^2}{r^4}, \tag{A.26}$$

where a common $dr/d\phi$ factor cancels out. The ODE is simplified by a change of variables:

$$\frac{d^2u'}{d\phi^2} = -u' + \frac{G}{\ell'^2} + 3Gu'^2, \tag{A.27}$$

$$u' = \frac{M}{r}, \quad \ell' = \frac{\ell}{M}. \tag{A.28}$$

As with Newtonian orbits, the energy of the system determines the orbit characteristics. For a numerical solution we use the energy integral to determine the initial conditions for the ODE. Specifically, the energy integral (A.25) can be solved for $du'/d\phi$:

$$\frac{du'}{d\phi} = \sqrt{\frac{2E}{\ell'^2} + 2\frac{Gu'}{\ell'^2} - u'^2 + 2Gu'^3}. \tag{A.29}$$

As you (should) have deduced qualitatively, the potential (A.23) produces qualitatively differing orbits depending upon the system's energy and angular momentum. The problems of this section ask you to use your ODE solver to explore numerically and graphically various orbits corresponding to differing initial conditions and energies. Our program `RelOrbits.py` is in Listing A.4 and runs in Spyder. Note that when you produce your graphs you should introduce some signal into your figures so that you can deduce in which direction the orbiting particle moves, something we have not done it in Figure A.5. Alternatively, you can produce animations or a time series of frames, in which case the direction of motion will be evident.

1. Set up your ODE solver appropriate for (A.29) using $G = 1$. *Hint:*

$$y[1] = \sqrt{\frac{2E}{\ell'^2} + 2\frac{Gu'}{\ell'^2} - u'^2 + 2Gu'^3}, \tag{A.30}$$

$$y[0] = \sqrt{\frac{2(-0.028)}{4.3^2} + 2\frac{y[0]}{4.3^2 - y[0]^2} + 2y[0]^3}. \tag{A.31}$$

2. Choose an energy corresponding to the maximum of the effective potential compute your version of Figure A.4, and an initial r value at which the potential is a maximum. As you may have deduced, this should lead to an unstable orbit such as on the left of Figure A.5.

3. See if you can find initial conditions that lead to a circular orbit. Is it stable?

4. Investigate the effect of gradually decreasing the angular momentum.

5. Choose an energy that corresponds to the minimum in the effective potential and plot the orbits. Examine the sensitivity of these orbits to the choice of initial conditions.

6. Determine an energy and initial conditions that produce a precessing perihelion, such as seen on the right of Figure A.5. In this case the massive particle moves between two turning points, as shown by the horizontal line in the potential well in Figure A.4.

7. Examine the orbits that occur if a particle is bound by the inner strong attraction. Can such a particle start at infinity and be captured?

A.6   Riemann and Ricci Tensors

Figure A.6 shows two free particles moving along the infinitesimally close geodesics $x^a(\tau)$ and $x^b(\tau)$. We consider x^a as the reference particle with $u^\mu \equiv dx^\mu/d\tau$ its 4-velocity. The two trajectories start off parallel at time $\tau = 0$ and are connected by the vector $n(\tau)$:

$$x^a = x^b + n^\alpha(\tau). \tag{A.32}$$

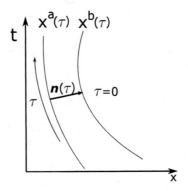

Figure A.6. Two free particles move along the infinitesimally close geodesics $x^a(\tau)$ and $x^b(\tau)$. The particles start off parallel at time $\tau = 0$ and are connected by the vector $n(\tau)$.

If the relative acceleration of the particles is zero, then the geodesics remain parallel and so:

$$\frac{d^2 n}{d\tau^2} = 0.$$

This derivative acts on the basics vectors, which in turn requires knowledge of the Christoffel symbols:

$$\left(\frac{d^2 n}{d\tau^2}\right)^\alpha = \left(\partial_\sigma \Gamma^\alpha_{\mu\nu} - \partial_\nu \Gamma^\alpha_{\mu\sigma} + \Gamma^\alpha_{\sigma\gamma}\Gamma^\gamma_{\mu\nu} - \Gamma^\alpha_{\nu\gamma}\Gamma^\gamma_{\mu\sigma}\right) u^\sigma u^\mu u^\nu.$$

The quantity in parenthesis is called the *Riemann tensor*:

$$R^\alpha_{\mu\nu\sigma} = \partial_\sigma \Gamma^\alpha_{\mu\nu} - \partial_\nu \Gamma^\alpha_{\mu\sigma} + \Gamma^\alpha_{\sigma\gamma}\Gamma^\gamma_{\mu\nu} - \Gamma^\alpha_{\nu\gamma}\Gamma^\gamma_{\mu\sigma}. \tag{A.33}$$

A.6.1 Problems

The following three problems can all be solved with variations of the same code.

1. Use **sympy** to evaluate the Riemann tensor for the Schwarzschild metric.

2. Use **sympy** to extract the Ricci curvature tensor, defined as the contraction

$$R_{\lambda\mu} \equiv R^\alpha_{\lambda\alpha\gamma} \tag{A.34}$$

   (*note* the implicit sum over α).

3. The Ricci scalar gives a single numerical measure of the curvature at each point in spacetime. It is defined as the contraction:

$$R \equiv R^\lambda_\lambda = g^{\lambda\gamma} R_{\lambda\gamma}. \tag{A.35}$$

   If a spacetime is flat, then $R = 0$ and the initially parallel geodesics remain so in time. If a spacetime is curved, then $R \neq 0$. Use **sympy** to extract the Ricci scalar from Ricci curvature tensor.

A.6.2 Help with Solution

1. Use the previously–developed code to create four matrices containing the Christoffel symbols, $\Gamma^0_{\mu\nu}$, $\Gamma^r_{\mu\nu}$, $\Gamma^\theta_{\mu\nu}$, and $\Gamma^\phi_{\mu\nu}$.

2. Define a 4-D array for the Riemann tensor $R^\alpha_{\lambda\alpha\gamma}$ with the indices corresponding to α, μ, ν, and σ. (There are four indices with each index having a range of 4.)

3. To deduce the Ricci curvature tensor $R_{\lambda\mu}$, define a 2-D array with each index having a range of 4.

4. Extract the Ricci scalar. Our version of said program is called `Ricci.py` and can be found in Listing A.5.

A.7 General Relativity Code Listings

```
# Wormhole.py: Symbolic evaluation of wormhole derivative

from sympy import *
L, x, M, rho, a, r,   lp= symbols('L x M rho a r lp')
x = (2*L-a)/(pi*M)
r = rho+M*(x*atan(x) -log(1+x*x)/2)
drdL = diff(r,L)
print ('drdL(raw) = ', drdL)
drdL = simplify(drdL)
print   (' And finally! dr/dL (simplified)=', drdL)
```

Listing **A.1.** **WormHole.py** evaluates symbolically a derivative needed in description of wormhole.

```
# VisualWorm.ipynb    Visualize wormhole with Vpython in notebook

from vpython import *
import numpy as np
import math

escene = canvas(width=400,height=400, range= 15)
a = 1 #2a is height inner cylinder
ring(pos=vector(0,0,0),radius=1,axis=vector(0,1,0),color=color.yellow)

def f(x):        # function to be integrated
    M = 0.5        # black hole mass
    a = 1          # 2a: cylinders height
    y = np.sqrt(1- (2*np.arctan(2*(x - a)/(np.pi*M))/np.pi)**2)
    return y

def trapezoid(Func,A,B,N):
    h = (B - A)/(N )             # step, A:initial, B:end
```

```
        sum = (Func(A)+Func(B))/2   # initialize, (first + last)/2
        for i in range(1, N):       # inside
            sum += Func(A+i*h)      #
        return h*sum                # sum times h

def radiuss(L):      # radius as function of L
    ro = 1           # radius of cylinder (a/ro=1)
    a = 1            # 2a: height of inner cylinder
    M = 0.5          # black hole (mass M/ro)=1
    xx = (2*(L-a))/(np.pi*M)
    p = M*(xx*np.arctan(xx))
    q = -0.5*M*math.log(1+xx**2)
    r = ro+ p+q
    return r

for i in range(1,12): # to plot 2 rings (at z ant -z)
    A = 0                   #limits of integration
    B = i
    N = 300         # trapezoid rule points
    if i>6: N = 600    # more points
    z = trapezoid(f,A,B,N) # returns z
    L = i+1
    rr = radiuss(L)            # radius
    ring(pos=vector(0,z,0),radius=rr,axis=vector(0,1,0),
        color=color.yellow)
    ring(pos=vector(0,-z,0),radius=rr,axis=vector(0,1,0),
        color=color.yellow)
```

Listing A.2. VisualWorm.ipynb A Vpython visualization of a wormhole from within Jupyter notebook.

```
# LensGravity.py      Deflection of light by sun wi Matplotlib

import numpy as np
import matplotlib.pyplot as plt
y = np.zeros((2),float);        ph = np.zeros((181),float)
yy = np.zeros((181),float);   xx = np.zeros((181),float)
rx = np.zeros((181),float);   ry = np.zeros((181),float)
Gsun = 4477.1                  # Sum mass x G (m)
GM = 28.*Gsun                  # Sun mass
y[0] = 1.e-6; y[1] = 1e-6      # Initial condition for u=1/r

def f(t,y):                              # RHS, can modify
    rhs = np.zeros((2),float)
    rhs[0] = y[1]
    rhs[1] = 3*GM*(y[0]**2)-y[0]
    return rhs

def rk4Algor(t, h, N, y, f):    # Do not modify
    k1 = np.zeros(N); k2=np.zeros(N); k3=np.zeros(N); ↵
        k4=np.zeros(N);
    k1 = h*f(t,y)
    k2 = h*f(t+h/2.,y+k1/2.)
    k3 = h*f(t+h/2.,y+k2/2.)
    k4 = h*f(t+h,y+k3)
    y = y+(k1+2*(k2+k3)+k4)/6.
    return y
```

```
f(0,y)                            # Initial conditions
dphi = np.pi/180.                 # 180 values of angle phi
i = 0
for phi in np.arange(0,np.pi+dphi,dphi):
    ph[i] = phi
    y = rk4Algor(phi,dphi,2,y,f)          # Call rk4
    xx[i] = np.cos(phi)/y[0]/1000000      # Scale for graph
    yy[i] = np.sin(phi)/y[0]/1000000
    i = i+1
m = (yy[180] - yy[165])/(xx[180]-xx[165])  # Slope of straight ↩
    line
b = yy[180] - m*xx[180]                    # Intercept of line
j = 0

for phi in np.arange(0,np.pi+dphi,dphi):
    ry[j] = m*xx[j] + b                    # Eqn straight line
    j=j+1
plt.figure(figsize=(12,6))
plt.plot(xx,yy)                   # Straight line light tajectroy
plt.plot(xx,-yy)                  # Symmetric for negative y
plt.plot(0,0,'ro')                # Mass at origin
plt.text(0.02,0.02, 'Sun')
plt.plot(0.98,0,'bo')             # Source
plt.plot(0.98,1.91,'go')          # Position source seen by O
plt.plot(0.98,-1.91,'go')
plt.text(1,0,'S observer')
plt.text(-1.00, 0.20,'Source')
plt.text(1.02, 1.91,"S' observer")
plt.text(1.02,-2,"S'' observer")
plt.plot([0],[3.])                # Invisible point
plt.plot([0],[-3.])               # Invisible point at -y
plt.plot(xx,ry)                   # Upper straight line
plt.plot(xx,-ry)                  # Lower straight line
plt.xlabel('x')
plt.ylabel('y')
plt.show()
```

Listing A.3. **LensGravity.py** solves for orbits of light around very massive star.

```
# RelOrbits.py: Reltiv orbits in gravitational pot (needs rk4)

import matplotlib.pyplot as plt
import numpy as np

dh = 0.04;  dt = dh;  ell = 4.3;  G = 1.0;  N = 2
E = 0.040139
phi = np.zeros((944),float)
rr = np.zeros((944),float)
y = np.zeros((2),float)
y[0] = 0.052
y[1] = np.sqrt(2*E/ell**2 + 2*G*y[0]/ell**2-G*y[0]**2+2*G*y[0]**3)

def f(t,y):
    rhs = np.zeros(2)
    rhs[0] = y[1]
    rhs[1] = -y[0]+G/ell**2 +3*G*y[0]**2
    return rhs
```

```
f(0,y)
i = 0
for fi in np.arange(0,5.8*np.pi,dt):
    y = rk4(fi,dt,N,y,f)
    rr[i] = (1/y[0])*np.sin(fi)  # Notice 1/r (=u)
    phi[i] = (1/y[0])*np.cos(fi)
    i = i+1
f1 = plt.figure()
plt.axes().set_aspect('equal')  # Aspect ratio equal
plt.plot(phi[:455],rr[:455])
plt.xlabel("r/M")
plt.show()
```

Listing A.4. RelOrbits.py solves for orbits of a massive particle in a gravitational potential with a GR correction.

```
# Ricci.py:   Riemann & Ricci tensors, Ricci scalar via sympy

from sympy import *
import numpy as np

t,r,th, fi, rg = symbols('t r th fi rg')  # Schwarzchild metric
print("contravariant")                    # Upper indices

# Inverse matrix
gT = Matrix([[1/(-1 + rg/r), 0, 0, 0], [0, 1 - rg/r, 0, 0],
    [0, 0, r**(-2), 0], [0, 0, 0, 1/(r**2*sin(th)**2)]])

# 4-Dim array for alpha, beta, mu, nu
Ri = [[[[[] for n in range(4)] for a in range(4)] for b in range(4)]
    for c in range(4)]
RT = [[[] for m in range(4)] for p in range(4)]          # Ricci tensor

# Christoffel symbols upper index t,r,theta and phi
Cht = Matrix([[0, 0.5*rg/(r*(r-rg)), 0, 0],
    [0.5*rg/(r*(r-rg)),0,0,0], [0, 0, 0, 0], [0, 0, 0, 0]])
Chr = Matrix([[0.5*rg*(r-rg)/r**3,0,0,0], ↵
    [0,-0.5*rg/(r*(r-rg)),0,0],
    [0,0, -1.0*r + 1.0*rg, 0], [0,0,0, (-1.0*r + rg)*sin(th)**2]])
Chth = Matrix([[0, 0, 0, 0], [0, 0, 1.0/r, 0], [0, 1.0/r, 0, 0],
    [0, 0, 0, -0.5*sin(2*th)]])
Chfi = Matrix([[0,0,0,0], [0, 0, 0, 1.0/r], [0,0,0,1./tan(th)],
    [0, 1./r, 1.0/tan(th), 0]])
for alpha in range(0,4): # Upper index
    if alpha == 0:       Chalp = Cht
    elif alpha == 1:     Chalp = Chr
    elif alpha == 2:     Chalp = Chth
    else:                Chalp = Chfi
    for be in range(0,4):      # Beta
     for mu in range(0,4):
         if mu == 0:     der2 = t
         elif mu == 1:   der2 = r
         elif mu == 2:   der2 = th
         elif mu == 3:   der2 = fi
         for nu in range(0,4):
             if nu == 0:    der1 = t  # Derivative
             elif nu == 1:  der1 = r
```

```
          elif nu == 2: der1 = th
          elif nu == 3: der1 = fi
          a1 = diff(Chalp[be,nu],der2)   # Christoffel symbol
          a2 = diff(Chalp[be,mu],der1)   # Symbol and derivative
          sump = 0 #   Einstein convention
          sumn = 0 #   Einstein convention
          for gam in [t,r,th,fi]:
              if gam == t:
                  Chgam = Cht
                  gama = 0
              elif gam == r:
                  Chgam = Chr
                  gama = 1
              elif gam == th:
                  Chgam = Chth
                  gama = 2
              elif gam == fi:
                  Chgam = Chfi
                  gama = 3
              sump = sump + Chalp[mu,gama]*Chgam[be,nu]
              sumn = sumn + Chalp[nu,gama]*Chgam[be,mu]
          R = simplify(a1-a2+sump-sumn) # Riemann tensor
          if R == 0:  Ri[alpha][be][mu][nu] = 0
          else :
              Ri[alpha][be][mu][nu] = R
              print("Ri[",alpha,"][",be,"][",mu,"][",nu,"]=",  ↩
                    Ri[alpha][be][mu][nu])
print("\n")
print("Ricci Tensor\n")
for ro in range(0,4):       # Find Ricci tensor
    for de in range (0,4):
        sum = 0
        for alp in range (0,4):    sum = sum+Ri[alp][ro][alp][de]
        RT[ro][de] = simplify(sum)
        print("RT[",ro,"][",de,"]",RT[ro][de]) # Ricci's tensor
sumR = 0  # Ricci Scalar
for be in range(0,4):
    for nu in range(0,4):      sumR = sumR+gT[be,nu]*RT[be][nu]
        print(sumR)
RS = (sumR)
print("RS",RS)      # Ricci Scalar R
```

Listing A.5. **Ricci.py** uses sympy to evaluate the Riemann tensor, the Ricci curvature tensor, and the Ricci scalar for the Schwarzschild metric. .

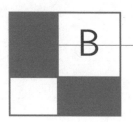

Python Code Directory

Name	Listing	Description	Name	Listing	Description
AdvecLax.py	4.5	Advection Eq	Beam.py	4.9	Navier-Stokes
Bisection.py	2.6	Bisection Algor	BoundCall.py	6.5	P Space Bound
Bugs.py	8.1	Logistic Map	CatFriction.py	4.2	Catenary Waves
CavityFlow.py	4.11	Cavity Flow	CentralValue.py	1.11	Central Value
CircPolarztn.py	5.11	Circular Polarized	CoulWF.py	6.16	Coulomb WF
CWT.py	2.4	Wavelet TF	DecaySound.py	7.3	Spontan Decay
DFTcomplex.py	2.1	DTF	DielectVis.py	??	FDTD Dielectric
DielectVis.py	??	FDTD Animate	DLA.py	7.6	Aggregation
DWT.py	2.5	Discrete Wavelets	EasyMatPlot.py	1.3	MatPlotLib e.g.
EasyVisualVP.py	1.1	VPthonl Plot	Entangle.py	6.21	Entangled QM
EqHeat.py	7.1	Heat Eq	EqHeatMov.py	7.2	Heat Eq Animate
EqStringMovMat.py	4.1	String Animated	FDTD.py	5.9	FDTD
Fern3D.py	8.4	Fern Fractal	FFTmod.py	2.3	FFT
Fit.py	2.8	Least Sq Fit	ForcedOscillate.py	3.1	ODE eg
GaussPoints.py	1.17	Gauss Points	GlauberState.py	6.19	Glauber States
GradesMatplot.py	1.4	MatPlotLib eg	Hdensity.py	6.8	H Density
HOmovSlow.py	6.9	HO Packet Slow	HOmov.py	6.10	HO Animate
HOanal.py	6.2	HO Analytic	HOchargeMat.py	6.13	Charged HO
HOchargeMat.py	6.13	Charged HO	HOnumeric.py	6.1	HO WF Numer
HOpacket.py	6.11	HO Packet Viz	HOpacketMat.py	6.12	HO Packet Mat
Hyperfine.py	6.20	H Hyperfine	ImagePlaneMat.py	5.5	Images Plane
ImagePlaneVP.py	5.6	Images Plane	ImageSphereVP.py	5.7	Images Sphere
IntegGaussCall.py	1.16	Gauss Integrate	IsingVP.py	7.9	Ising Model
LaplaceCyl.py	5.4	Lapace's Eq	LaplaceDisk.py	5.3	Laplace Eq Disk
LaplaceLine.py	5.2	Laplace's Eq Line	LaplaceTri.py	5.1	Laplace Triangle
LensGravity.py	A.3	Gravitational Lens	LorentzFieldVP.py	5.18	Lorentz TF Fields
MatPlot2figs.py	1.5	MatPlotLib eg	NewtonCall.py	2.7	Newton Search
MD1D.py	7.10	1-D MD	MD2D.py	7.11	2-D MD
MDpBC.py	7.12	Periodic BC MD	Permute.py	7.8	State Counting
Plm.py	6.7	Legendre Polys	PondMatPlot.py	1.7	Scatter Plot
PredatorPrey.py	8.2	Predator Prey	ProjectileAir.py	3.4	Projectile ODE
ProteinFold.py	8.3	Protein Folding	QMC.py	6.23	Path Integration
QuantumEigenCall.py	6.3	QM Eigenvalue	QuantumNumerov.py	6.4	Numerov Mth
QuarterPlate.py	5.12	1/4 Wave Plate	RelOrbits.py	A.4	Relatvist Orbits
Ricci.py	A.5	Ricci Tensor	rk4Algor.py	1.14	rk4 Algorithm
rk4Call.py	1.12	Calls rk4	rk4Duffing.py	1.13	rk4 e.g.
SincFilter.py	2.2	Noise Filtering	ScattSqWell.py	6.14	Sq Well Scatt
Scatter3dPlot.py	1.8	Scatter Plot	Simple3Dplot.py	1.6	Surface Plot
SlidingBox.py	1.10	Animate	Soliton.py	4.6	KdeV Soliton
SolitonAnimate.py	4.7	Soliton Movie	SqBilliardCM.py	3.3	Square Billiards
SU3.py	6.22	SU(3) Quarks	–		

Name	Listing	Description	Name	Listing	Description
3GraphVP.py	1.2	VPython Plot	3QMdisks.py	6.17	3 QM Disks
TeleMat.py	5.13	Telegraph Eq	TeleVis.py	5.14	Telegraph Viz
ThinFilm.py	5.15	Thin Film	Torricelli.py	4.10	Torricelli Flow
TrapMethods.py	1.15	Trapezoid Integ	TwoCharges.py	5.16	Lorentz TF Q's
TwoFields.py	5.17	Lorentz Field	TwoDsol.py	4.8	2-D Soliton
TwoForces.py	1.9	Animate e.g.	TwoWells.py	6.18	T-Dep 2 Wells
UranusNeptune.py	3.5	Uranus Orbit	VisualWorm.ipynb	A.2	Worm Hole Viz
Walk.py	7.4	Random Walk	Walk3D.py	7.5	3D Random Walk
Waves2D.py	4.4	2-D Waves Num	Waves2Danal.py	4.3	2-D Waves Anal
WormHole.py	A.1	Worm Hole Derv			

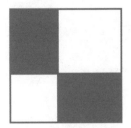

Bibliography

Bibliography

[Abarbanel et al.(93)] ABARBANEL, H. D. I., M. I. RABINOVICH AND M. M. SUSHCHIK (1993), *Introduction to Nonlinear Dynamics for Physicists*, World Scientific, Singapore.

[Abramowitz & Stegun(72)] ABRAMOWITZ, M. AND I. A. STEGUN (1972), *Handbook of Mathematical Functions*, 10th Ed., U.S. Govt. Printing Office, Washington.

[Alonso & Finn(67)] ALONSO, M. AND E. J. FINN (1967), *Fundamental University Physics*, Addison-Wesley, Reading.

[Anderson et al.(113)] ANDERSON, E., Z. BAI, C. BISCHOF, J. DEMMEL, J. DONGARRA, J. DU CROZ, A. GREENBAUM, S. HAMMARLING, A. MCKENNEY, S. OSTROUCHOV AND D. SORENSEN (2013), *LAPACK Users' Guide*, 3rd Ed., SIAM, Philadelphia; http://www.netlib.org.

[Argyris(91)] ARGYRIS, J., M. HAASE AND J. C. HEINRICH (1991), Computer Methods in Applied Mechanics and Engineering, **86**, 1.

[Askar & Cakmak(78)] ASKAR, A. AND A. S. CAKMAK (1978), *Explicit integration method for the time-dependent Schrödinger equation for collision problems*, J. Chem. Phys. **68**, 2794.

[Atkins & Elliot(10)] ATKINS, L. J. AND R. C. ELLIOT (2010), *Investigating thin film interference with a digital camera*, Am. J. Phys., **78**, 1248.

[Barnsley & Hurd(92)] BARNSLEY, M. F. AND L. P. HURD (1992), *Fractal Image Compression*, A. K. Peters, Wellesley.

[Bayley & Townsend(21)] BAYLEY, V.A. AND J. S. TOWNSEND (1921), *The motion of electrons in gases*, Philosophical Magazine, **6**, 42.

[Beazley(09)] BEAZLEY, D. M. (2009), *Python Essential Reference*, 4th Ed., Addison-Wesley, Reading.

[Becker(54)] BECKER, R. A. (1954), *Introduction to Theoretical Mechanics*, McGraw-Hill, New York.

[Becker(86)] BERGÉ, P., Y. POMEAU AND CH. VIDA (1986), *Order within Chaos*, Wiley, New York.

[Benenti et al.(04)] BENENTI, G., C. CASATI AND G. STRINI (2004), *Principles of Quantum Computation and Information*, World Scientific, Singapore.

[Beu(13)] BEU, T. A. (2013), *Introduction to Numerical Programming*, CRC Press, Taylor and Francis Group, Boca Raton.

[Bevington & Robinson(02)] BEVINGTON, P. R., AND D. K. ROBINSON (2002), *Data Reduction and Error Analysis for the Physical Sciences*, 3rd Ed., McGraw-Hill, New York.

[Blehel et al.(90)] BLEHER, S., C. GREBOGI AND E. OTT (1990), *Bifurcations in chaotic scattering*, Physica D, **46**, 87.

[Brans(91)] BRANSDEN, B. H. AND C. J. JOACHAIN (1991), *Quantum Mechanics*, 2nd Ed., Cambridge, Cambridge.

[Burgers(74)] BURGERS, J. M. (1974), *The Non-Linear Diffusion Equation; Asymptotic Solutions and Statistical Problems*, Reidel, Boston.

[Carruthers & Nieto(65)] CARRUTHERS P. AND M. M. NIETO (1965), *Coherent states and the forced quantum oscillator*, Am. J. Phys, **33**, 537-544.

[Cencini et al.(10)] CENCINI, M., F.CECONNI AND A. VULPIANI (2010), *Chaos from Simple Models to Complex Systems*, World Scientific, Singapore.

[Christiansen & Lomdahl(81)] CHRISTIANSEN, P. L. AND P. S. LOMDAHL (1981), *Numerical solutions of 2 + 1 dimensional Sine-Gordon solitons*, Physica D, **2**, 482.

[Christiansen & Olsen(78)] CHRISTIANSEN, P. L. AND O. H. OLSEN (1978), *Ring-shaped quasi-soliton solutions to the two- and three-dimensional Sine-Gordon equation*, Phys. Lett. **68A,** 185; (1979), *Ring-shaped quasi-soliton solutions to the two- and three-dimensional Sine-Gordon equation*, Physica Scripta **20**, 531.

[CiSE(07,11)] PEREZ, F., B. E. GRANGER AND J. D . HUNTER (2011), *Python: an ecosystem for scientific computing*, Computing in Science & Engineering, **13**, March-April 2011, 13-21;

OLIPHANT, T. E. (2007), *Python for scientific computing*, Computing in Science & Engineering, **9**, May-June 2007.

[COBE(16)] NASA (2016), *Cosmic background explorer*,
    https://science.nasa.gov/missions/cobe.

[Cohen(06)] COHEN-TANNOUDJI, C., B. DIU, F. LALOE (2006), *Quantum Mechanics*, Wiley-VCH, Weinheim.

[Courant et al.(28)] COURANT, R., K. FRIEDRICHS AND H. LEWY (1928), *On the partial difference equations of mathematical physics*, Math. Ann. **100**, 32.

[Cooley & Tukey(65)] COOLEY, J. W. AND J. W. TUKEY (1965), *Search Results An algorithm for the machine calculation of complex Fourier series*, Math. Comput., **19**, 297.

[Darwin(20)] DARWIN, C. G. (1920), *The dynamical motions of charged particles*, Phil. Mag. **39**, 537.

[Degaudenzi & Arizmendi(20)] DEGAUDENZI, M. E. AND C. M. ARIZMEND. (2000), *Wavelet-based fractal analysis of electrical power demand*, Fractals. **8**, 239-245.

[DeJong(92)] DEJONG, M. L. (1992), *Chaos and the simple pendulum*, The Phys. Teacher **30**, 115.

[Donnelly & Rust(05)] DONNELLY, D. AND B. RUST (2005), *The fast fourier transform for experimentalists*, Comp. in Science & Engr. **7**, 71.

[Enns(01)] ENNS, R. H. AND G. C. MCGUIRE, (2001), *Nonlinear Physics with Mathematica for Scientists and Engineers*, Birkhauser, Boston.

[Essen(96)] ESSEN, H. (1996), *Darwin magnetic interaction energy and its macroscopic consequences*, Phys. Rev. E **53**, 5228.

[Falkovich & Sreenivasan(06)] FALKOVICH, G. AND K. R. SREENIVASAN (2006), *Lesson from hydrodynamic turbulence*, Phys. Today, **59**, 43.

[Feigenbaum(79)] FEIGENBAUM, M. J. (1979), *The universal metric properties of nonlinear transformations*, J. Stat. Phys. **21**, 669.

[Fetter & Walecka(80)] FETTER, A. L. AND J. D. WALECKA (1980), *Theoretical Mechanics of Particles and Continua*, McGraw-Hill, New York.

[Feynman & Hibbs(65)] FEYNMAN, R. P. AND A. R. HIBBS (1965), *Quantum Mechanics and Path Integrals*, McGraw-Hill, New York.

[Feynman(65)] FEYNMAN, R., R. B. LEIGHTON, M. SANDS (1963), *The Feynman Lectures on Physics*, Vol II, Addison-Wesley, Reading.

[Fosdick et al.(96)] FOSDICK, L. D., E. R. JESSUP, C. J. C. SCHAUBLE AND G. DOMIK (1996), *An Introduction to High Performance Scientific Computing*, MIT Press, Cambridge.

[Fraunfelder & Henley(91)] FRAUENFELDER, H. AND E. M. HENLEY (1991), *Subatomic Physics*, Prentice Hall, Upper Saddle River, New Jersey.

[Gas(14)] GASPARD, P. (2014), *Quantum chaotic scattering*, Scholarpedia, **9(6)**, 9806, http://www.scholarpedia.org/article/Quantum_chaotic_scattering.

[Gottfried(66)] GOTTFRIED, K. (1966), *Quantum Mechanics*, Benjamin, New York.

[Gould et al.(06)] GOULD, H., J. TOBOCHNIK AND W. CHRISTIAN (2006), *An Introduction to Computer Simulations Methods*, 3rd Ed., Addison-Wesley, Reading.

[Hartle(03)] HARTLE, J. B. (2003), *Gravity, An Introduction to Einstein's General Relativity*, Addison Wesley, San Francisco.

[Hartley(82)] HARTLEY, J. G. AND R. JOHN (1982), *Coherent states for the time-dependent harmonic oscillator*, Phys. Rev **D 25**, 382-386.

[Hartmann(98)] HARTMANN, W. M. (1998), *Signals, Sound and Sensation*, AIP Press, Springer, New York.

[Hoffmann & Chiang(00)] HOFFMANN, K. A. AND S.T. CHIANG (2000), *Computational Fluid Dynamics*, Engineering Education Systems, Wichita.

[Hubble(29)] HUBBLE, E. (1929), *A relation between distance and radial velocity among extra-galactic nebulae*, Proc. Nat. Academy of Sciences of the United States of America, **15**, 168-173.

[Inan & Marshall(11)] INAN, U. S. AND R. A. MARSHALL (2011), *Numerical Electromagnetics, The FDTD Method*, Cambridge, Cambridge.

[Jackson(88)] JACKSON, J. D. (1988), *Classical Electrodynamics*, 3rd Ed., Wiley, New York.

[Jackson(91)] JACKSON, J. E. (1991), *A User's Guide to Principal Components*, Wiley, New York.

[Jolliffe(01)] JACKSON, J. E. (2001), *Principal Component Analysis*, 2nd Ed., Springer, New York.

[Kaye et al.(07)] KAYE, P., R. LAFLAMME AND M. MOSCA (2001), *An Introduction to Quantum Computing*, Oxford University Press, Oxford.

[Keller(59)] KELLER, J. B. (1959), *Large amplitude motion of a string*, Am. J. Phys. **27**, 584.

[Kittel(05)] KITTEL, C. (2005), *Introduction to Solid State Physics*, 8th Ed., Wiley, New York.

[Kov(11)] KOVACIC, I., M. J. BRENNAN (EDS.) (2011), *The Duffing Equation*, Wiley, New York.

[Korteweg & deVries(1895)] KORTEWEG, D. J. AND G. DEVRIES (1895), *On the change of form of long waves advancing in a rectangular canal, and on a new type of long stationary waves*, Phil. Mag., **39**, 4.

[Kreyszig(98)] KREYSZIG, E. (1998), *Advanced Engineering Mathematics*, 8th Ed., Wiley, New York.

[Landau(96)] LANDAU, R. H. (1996), *Quantum Mechanics II, A Second Course in Quantum Theory*, 2nd Ed., Wiley, New York.

[LPB(15)] LANDAU, R. H., M. J. PÁEZ AND C. C. BORDEIANU (2015), *Computational Physics, Problem Solving with Python*, 3nd Ed., Wiley-VCH, Weinheim.

[LPB(08)] LANDAU, R. H., M. J. PÁEZ AND C. C. BORDEIANU (2008), *A Survey of Computational Physics*, Princeton University Press, Princeton.

[Landau & Lifshitz(77)] LANDAU, L. D. AND E. M. LIFSHITZ (1976), *Mechanics*, 3rd Ed., Butterworth-Heinemann, Oxford.

[Lang & Forinash(98)] LANG, W. C., K. FORINASH (1998), *Time-frequency analysis with the continuous wavelet transform*, Amer. J. of Phys, **66**, 794.

[Langtangen(08)] LANGTANGEN, H. P. (2008), *Python Scripting for Computational Science*, Springer, Heidelberg.

[Langtangen(09)] LANGTANGEN, H. P. (2009), *A Primer on Scientific Programming with Python*, Springer, Heidelberg.

[Liu(14)] LIU, C. (2014) *The Three-Body Problem*, Macmillan, New York.

[Maestri et al.(00)] MAESTRI, J. J. V., R. H. LANDAU AND M. J. PÁEZ (2000), *Two-particle Schrödinger Equation animations of wave packet-wave packet scattering*, Am. J. Phys., **68**, 1113; movies at http://science.oregonstate.edu/~landaur/nacphy/ComPhys/PACKETS/.

[Mannheim(83)] MANNHEIM, P. D. (1983), *The physics behind path integrals in quantum mechanics*, Am. J. Phys., **51**, 328.

[Maor(74)] MAOR, E. (1974), *A discrete model for the vibrating string*, Int. J. Math. Ed. in Sci. & Tech., **6**, 345.

[Marion & Thornton(03)] MARION, J. B. AND S. T. THORNTON (2003), *Classical Dynamics of Particles and Systems*, 5th Ed., Harcourt Brace Jovanovich, Orlando.

[McMahon(08)] McMahon, D. (2008), *Quantum Computing Explained*, Wiley, New York.

[Metropolis et al.(53)] Metropolis, M., A. W. Rosenbluth, M. N. Rosenbluth, A. H. Teller and E. Teller (1953), *Equation of state calculations by fast computing machines*, J. Chem. Phys. **21**, 1087.

[Moore(13)] Moore, T. A. (2013), *A General Relativity Workbook*, University Science Books, Mill Valley.

[Motter & Campbell(13)] Motter, A. and D. Campbell (2013), *Chaos at fifty*, Phys. Today, 66(5), 27.

[Napolitano(18)] Napolitano, J. (2018), *A Mathematica Primer for Physicists*, CRC Press, Boca Raton.

[Page & Adams(45)] Page, L. and N. I. Adams Jr. (1945), *Action and reaction between moving charges*, Amer. J. Phys **13**, 141.

[Patrignani et al.l(16)] Patrignani, C. et al. (Particle Data Group) (2016), *Review of particle physics*, Chin. Phys. C **40**, 100001.

[Peitgen et al.(94)] Peitgen, H.-O., H. Jürgens and D. Saupe (1992), *Chaos and Fractals*, Springer, New York.

[Penrose(07)] Penrose, R. (2007), *The Road to Reality*, Vintage, New York.

[Pguide(14)] Python for Programmers (2014),
    https://wiki.python.org/moin/BeginnersGuide/Programmers.

[Plearn(14)] Interactive Python Tutorial (2014),
    http://www.learnpython.org/.

[Press et al.(94)] Press, W. H., B. P. Flannery, S. A. Teukolsky and W. T. Vetterling (1994), *Numerical Recipes*, Cambridge University Press, Cambridge.

[Ptut(14)] The Python Tutorial (2014), http://docs.python.org/2/tutorial/.

[Ram & Town(21)] Bailey, V. A. and J. S. Townsend (1921), *The motion of electrons in gases*, Phil. Mag., **42**, 873.

[Rapaport(97)] Rapaport, D.C. (1997), *The Art of Molecular Dynamics Simulation*, Cambridge University Press, Cambridge.

[Reif(67)] Reif, F. (1967), *Statistical Physics, Berkeley Physics Course* **5**, McGraw-Hill, New York.

[Resnick(68)] RESNICK, R. (1968), *Introduction to Special Relativity*, Wiley, New York.

[Row 2004] SPROTT, J. C., G. ROWLAND, AND D. HARRISON (2004), *Flash Animations for Physics*, https://faraday.physics.utoronto.ca/PVB/Harrison/Flash/.

[Sakurai(67)] SAKURAI, J. J. (1967), *Advanced Quantum Mechanics*, Addison Wesley, Reading.

[Satoh(11)] SATOH, A. (2011), *Introduction to Practice of Molecular Simulation*, Elsevier, Amsterdam.

[Schiff(68)] SCHIFF, L. I. (1968), *Quantum Mechanics*, 3rd Ed., McGraw-Hill, New York.

[Serway & Beichner(99)] SERWAY, R. A. (1999), *Physics for Scientists and Engineers*, Fifth Ed., Saunders, Orlando.

[Smith(65)] SMITH, J. H. (1965), *Introduction to Special Relativity*, Benjamin, New York.

[Smith(02)] SMITH, L. I. (2002), *A Tutorial on Principal Components Analysis*, http://snobear.colorado.edu/Markw/BioMath/Otis/PCA/principal_components.ps.

[Smith(99)] SMITH, S. W. (1999), *The Scientist and Engineer's Guide to Digital Signal Processing*, California Technical Publishing, San Diego.

[Stephen(87)] STEPHEN, H. AND R. SAKAT (1987), *Coherent states of a harmonic oscillator*, Am. J. Phys, **55**, 1109.

[Stetz et al.(73)] STETZ, A., J. CARROLL, N. CHIRAPATPIMOL, M. DIXIT, G. IGO, M. NASSER, D. ORTENDAHL AND V. PEREZ-MENDEZ (1973), *Determination of the axial vector form factor in the radiative decay of the pion*, LBL 1707.

[Squirtes(02)] SQUIRES, G. L. (2002), *Problems in Quantum Mechanics with Solutions*, Cambridge University Press, Cambridge.

[Sullivan(00)] SULLIVAN, D. (2000), *Electromagnetic Simulations Using the FDTD Methods*, IEEE Press, New York.

[Tabor(89)] TABOR, M. (1989), *Chaos and Integrability in Nonlinear Dynamics*, Wiley, New York.

[Taghipour et al.(14)] TAGHIPOUR, R., T. AKHLAGHI AND A. NIKKAR (2014), *Explicit solution of the large amplitude transverse vibrations of a flexible string under constant tension*, Lat. Am. J. Solids Struct **11**, http://dx.doi.org/10.1590/S1679-78252014000300010.

[Thijssen(99)] THIJSSEN, J. M. (1999), *Computational Physics*, Cambridge University Press, Cambridge.

[James et al. (15)] JAMES, O., E. VON TUNNZELMAN, P. FRANKLIN, K. S. THORNE (2015), *Visualizing Interestellar's wormhole*, Am. J. Phys., **83**, 486-499, (2015).

[Vano et al.(06)] VANO, J. A., J. C. WILDENBERG, M. B. ANDERSON, J. K. NOEL AND J. C. SPROTT (2006), *Chaos in low-dimensional lotka-volterra models of competition*, Nonlinearity **19**, 2391.

[Visscher(91)] VISSCHER, P. B. (1991), *A fast explicit algorithm for the time-dependent Schrödinger equation*, Comput. in Phys., **5**, 596.

[Ward et al.(05)] WARD, D. W. AND K. A. NELSON (2005), *Finite difference time domain, fdtd, simulations of electromagnetic wave propagation using a spreadsheet*, Computer Apps. in Engr. Education, **13**, 213.

[Wiki(14)] Wikipedia (2014), *Principal component analysis*, http://wikipedia.org/wiki/Principal_component_analysis.

[Witten & Sander(83)] WITTEN, T. A. AND L. M. SANDER (1981), *Diffusion-limited aggregation, a kinetic critical phenomenon*, Phys. Rev. Lett. **47**, 1400; (1983), *Diffusion-limited aggregation*, Phys. Rev. B, **27**, 5686.

[Yue et al.(04)] YUE, K., K. M. FIEBIG, P. D. THOMAS, H. S. CHAN, E. I. SHAKHNOVICH AND A. DILL (1995), *A test of lattice protein folding algorithms*, Proc. Natl. Acad. Sci. USA, **92**, 325.

Index